Sustainable Agriculture and Forestry

Sustainable Agriculture and Forestry

Editor: Thelma Bosso

R CALLISTO
REFERENCE
www.callistoreference.com

Callisto Reference,
118-35 Queens Blvd., Suite 400,
Forest Hills, NY 11375, USA

Visit us on the World Wide Web at:
www.callistoreference.com

ISBN: 978-1-63239-982-3 (Hardback)

Cataloging-in-Publication Data

Sustainable agriculture and forestry / edited by Thelma Bosso.
    p. cm.
Includes bibliographical references and index.
ISBN 978-1-63239-982-3
1. Sustainable agriculture. 2. Sustainable forestry. 3. Agriculture.
4. Forests and forestry. I. Bosso, Thelma.
S494.5.S86 S87 2018
630--dc23

TABLE OF CONTENTS

Preface..IX

Chapter 1 **Breeding Importance of the Hybrid Depression Problem and Possible Ways of its Overcoming**..1
Ruzanna Robert Sadoyan

Chapter 2 **Water Stress Tolerance of Six Rangeland Grasses in the Kenyan Semi-arid Rangelands**..5
Koech Oscar Kipchirchir, Kinuthia Robinson Ngugi, Mureithi Stephen Mwangi, Karuku George Njomo, Wanjogu Raphael

Chapter 3 **Evaluation of Some Performance Traits and Carcass Characteristics of *Archachatina marginata* Snails Fed Plant Wastes**...13
Olubukola Omolara Babalola

Chapter 4 ***Photorhabdus Luminescens*: Virulent Properties and Agricultural Applications**...................................18
Elizabeth Gerdes, Devang Upadhyay, Sivanadane Mandjiny, Rebecca Bullard-Dillard, Meredith Storms, Michael Menefee, Leonard D. Holmes

Chapter 5 **Ecology of Basal Stem Rot Disease of Oil Palm (*Elaeis guineensis* Jacq.) in Cameroon**...25
Afui Mathias Mih, Tonjock Rosemary Kinge

Chapter 6 **Effect of a Hormone Containing Nitrobenzene in Combination with Fertilizers on Early Flower Induction of *Ixora coccinea* Hybrids under Outdoor and Shaded Conditions**..33
Hapu Arachchige Ruwani Kalpana Jayawardana, Mohamed Cassim Mohamed Zakeel, Channa De Zoysa

Chapter 7 **Effect of GA$_3$, Girdling or Pruning on Yield and Quality of 'Parletta' Seedless Grape**..36
Ismail Ali Abu-Zinada

Chapter 8 **Comparative Effects of Foliar Application of Gibberellic Acid and Benzylaminopurine on Seed Potato Tuber Sprouting and Yield of Resultant Plants**...........................40
Martin Kagiki Njogu, Geofrey Kingori Gathungu, Peter Muchiri Daniel

Chapter 9 **Humic Acid to Decrease Fertilization Rate on Potato (*Solanum tuberosum* L.)**...................50
Ismail Ali Abu-Zinada, Kamal Soliman Sekh-Eleid

Chapter 10 **Utilization of Mangrove Forest Plant: Nipa Palm (*Nypa fruticans* Wurmb.)**...................55
Md. Farid Hossain, Md. Anwarul Islam

Chapter 11 **Soil Fertility Status of Rice Field in Paundi Watershed, Lamjung District, Nepal**...................60
Ram Kumar Shrestha

Chapter 12   **Supply Chain to Industrialization of Pig Excrete by Biotransformation to Increase Corn Performance**..64
José Antonio Valles Romero, Emilio Raymundo Morales Maldonado

Chapter 13   **Estimates of Tree Biomass, and its Uncertainties through Mean-of-Ratios, Ratio-of-Means, and Regression Estimators in Double Sampling: A Comparative Study of Mecrusse Woodlands**...69
Tarquinio Mateus Magalhães, Thomas Seifert

Chapter 14   **Comparative Study on the Effect of *Citrillus lanatus* and *Cucumis sativus* on the Growth Performance of *Archachatina marginata***..79
Ufele Angela Nwogor

Chapter 15   **Analysis of Genetic Diversity using Simple Sequence Repeat (SSR) Markers and Growth Regulator Response in Bio ield Treated Cotton (*Gossypium hirsutum* L.)**...............................84
Mahendra Kumar Trivedi, Alice Branton, Dahryn Trivedi, Gopal Nayak, Mayank Gangwar, Snehasis Jana

Chapter 16   **Herbaceous Species Diversity in Kanawa Forest Reserve (KFR) in Gombe State, Nigeria**...90
Abba Halima Mohammed, Sawa Fatima Binta Jahun, Gani Alhassan Mohammed, Abdul Suleiman Dangana

Chapter 17   **Effects of Lime-Aluminium-Phosphate Interactions on Maize Growth and Yields in Acid Soils of the Kenya Highlands**..101
Esther Mwende Muindi, Jerome Mrema, Ernest Semu, Peter Mtakwa, Charles Gachene

Chapter 18   **Chemical Composition, Bio-Diesel Potential and uses of *Jatropha curcas* L. (*Euphorbiaceae*)**..110
Temesgen Bedassa Gudeta

Chapter 19   **Effect of Nitrogen and Phosphorus Fertilizer Rates on Yield and Yield Components of Barley (*Hordeum Vugarae* L.) Varieties at Damot Gale District, Wolaita Zone, Ethiopia**...124
Mesfin Kassa, Zemach Sorsa

Chapter 20   **The Assessment of Implementing Conventional Cotton: A Regression Analysis of Meta-Data**...129
Julian Witjaksono, Dahya, Asmin

Chapter 21   **Resource Domestication: An Introduction to Biodiversity and Wildlife in Agriculture**..134
Benjamin E. Uchola

Chapter 22   **Economics of Rural Livelihoods: A Case Study of Bitter Kola Marketing in Akwa Ibom State, Nigeria**...141
Asa Ubong Andem, Daniel Enwongo Aniedi, Ebong Effiong Okon

Chapter 23   **Floristic Composition and Vegetation Structure of Woody Species in Lammo Natural Forest in Tembaro Woreda, Kambata-Tambaro Zone, Southern Ethiopia**...........................145
Melese Bekele Hemade, Wendawek Abebe

Chapter 24   **Inhibitory Effects of Oligochitosan on Pathogenic Fungi Isolated from**
Zanthoxylum bungeanum...152
Peiqin Li, Zhou Wu, Tao Liu, Yanan Wang

Chapter 25   **Modification, calibration and validation of APSIM to suit maize (Zeamays L.)**
production system: A case of Nkango Irrigation Scheme in Malawi........................160
John Mthandi, Fredrick C. Kahimba, Andrew K. P. R. Tarimo, Baandah. A. Salim,
Max W. Lowole

Chapter 26   **Technological Gaps in Adoption of Improved Soybean Production Technology**
by Soybean Growers in Dahod District, Gujarat...171
R. G. Machhar, S. K. Patel, H. L. Kacha, U. M. Patel, G. D. Patel, R. Radha Rani

Chapter 27   **Optimization of Minituber Size and Planting Distance for the Breeder Seed**
Production of Potato..175
Md. Altaf Hossain, Abdullah-Al-Mahmud, Md. Abdullah-Al-Mamun,
Md. Shamimuzzaman, Md. Mizanur Rahman

Chapter 28   **Bio-organic Fertilizer on Pechay Homegarden in Cotabato**...................................182
Mosib B. Tagotong, Onofre S. Corpuz

Chapter 29   **Utilization of Urea Super Granule in Raised Bed Versus Prilled Urea in**
Conventional Flat Method for Transplanted Aman Rice (Oryza Sativa)...................186
Md. Halim Mahmud Bhuyan, Most. Razina Ferdousi, Md. Toufiq Iqbal,
Ahamed Khairul Hasan

Chapter 30   **Morphological and Molecular Analysis using RAPD in Biofield Treated Sponge**
and Bitter Gourd..195
Mahendra Kumar Trivedi, Alice Branton, Dahryn Trivedi, Gopal Nayak,
Mayank Gangwar, Snehasis Jana

Chapter 31   **Traditional Rice Farming Ritual Practices of the Magindanawn in**
Southern Philippines...202
Saavedra M. Mantikayan, Esmael L. Abas

Permissions

List of Contributors

Index

Preface

The concept of sustainable agriculture embraces methods that will enhance the environmental quality. Sustainable forestry also aims at the optimum utilization of resources. Some of the practices of sustainable agriculture are soil amendment and crop rotation. New sustainable methods and techniques are constantly being developed to make soil more productive. This book includes some of the vital pieces of work being conducted across the world, on various topics related to sustainable agriculture and forestry. With state-of-the-art inputs by acclaimed experts of this field, this book targets students and professionals.

Significant researches are present in this book. Intensive efforts have been employed by authors to make this book an outstanding discourse. This book contains the enlightening chapters which have been written on the basis of significant researches done by the experts.

Finally, I would also like to thank all the members involved in this book for being a team and meeting all the deadlines for the submission of their respective works. I would also like to thank my friends and family for being supportive in my efforts.

Editor

Breeding Importance of the Hybrid Depression Problem and Possible Ways of its Overcoming

Ruzanna Robert Sadoyan

Scientific Center of Agriculture, Ministry of Agriculture, Echmiadzin, Republic of Armenia

Email address:

ruzannasad@mail.ru

Abstract: Widely spread natural negative mutations in the Triticum genus, the divergence of members of the complementary systems of depression, their concentration in different eco-geographical zones, depending on their biotype and variety leading to the appearance of depressive hybrid plants cause serious difficulties in breeding process. To overcome hybrid dwarfism, the impact of mineral fertilizers (N90P90K75) on the growth and development of Dwarf I and Dwarf II wheat hybrids was studied. Creation of suitable growing conditions revealed, that contrary the increase in vegetative mass of hybrid Dwarf I (plant height, number of leaves, weight per plant) and elements of productivity of hybrid Dwarf II (plant height, productive tillering, the length of the main ear, grain yield per plant) in none of the hybrids the overcoming of depression was registered.

Keywords: Wheat, Hybrid Depression, Breeding, Mineral Fertilizers, Vegetative Mass

1. Introduction

On purpose to enrich the hybrid population genetically and to obtain a wide range of variability, the breeders often use the methods of distant eco-geographical, interspecific and intergeneric hybridization. In breeding program the hybrid depression is presented as serious problem. In the research of genes lethality in the Triticum genus three main aspects of breeding and genetic importance problem were identified. The first aspect is the prevalence of genes lethality at a certain combination originated in the case of disturbance of regulatory mechanisms of vitality. The second aspect is the diversity of lethality types and multiple alleles of genes determining this phenomenon. The third aspect is the extensive polymorphism of complementary genes of hybrid depression of some species in different eco-geographical zones [1, 2, 3].

In the breeding aspect it is important also the linkage of depression genes not only among themselves but also with the genes controlling the immune system and economically important traits as weight and size of grain and rust resistance [4]. According to Austin [5], the adoption of wheat cultivars of shorter stature was one of the important factors for obtaining greater yields, due to resistance to lodging and earlier ripening. The introgression of dwarfing (Rht) genes greatly promoted the partitioning of wheat biomass to grains [6]. The scheme of linkage of depression genes and economically important traits is presented on the Figure 1.

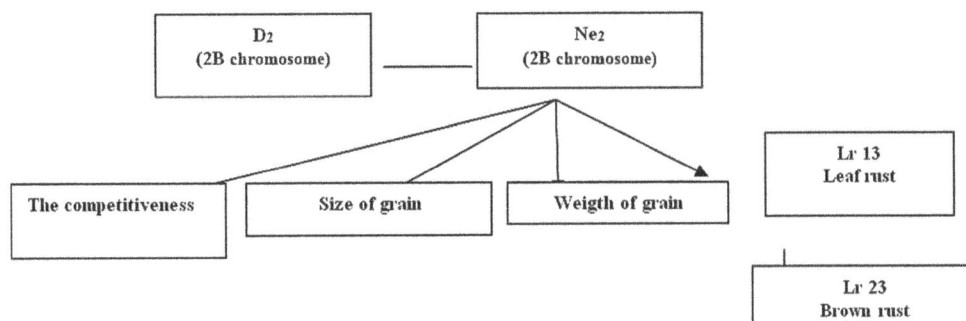

Fig. 1. The scheme of linkage of depression genes and economically important traits.

Significance of the hybrid depression problem is increasing due to the requirements of modern breeding of intensive forms of wheat. The simultaneous presence in one genotype of few dominant genes leads to the manifestation of different types of depression, which undoubtedly affects the efficiency of the selection process. Expression of the gene alleles individually determining each type of hybrid depression can cause different levels of depression of hybrid generation up to the plants lethality [7]. Naturally, when the genes of depression lead to the essential decrease of viability, the possibility of heterosis is minimized [8].

Widely spread natural negative mutations in the Triticum genus, their complementary nature, the divergence of members of the complementary systems of depression, their concentration in different eco-geographical zones, depending on their biotype and variety leading to the appearance of depressive hybrid plants cause serious difficulties in breeding process. One of the optimal methods to overcome depression and to get healthy hybrid generation is the application of natural heterogeneous variety on dominant genes of hybrids depression, that exhibit low level of depression in F1. Another way to overcome hybrid depression is the creation of favorable conditions for cultivation [8]. In particular, it concerns the creation of optimal conditions for the growth and development of hybrids.

Zhang et al. [9] discussed the physiological domains in crop drought resistance as a major factor in the stabilization of crop performance in drought prone environments. The tolerance to stress is controlled by many genes, and their simultaneous selection is difficult. At the same time, the semi-dwarf varieties, also tolerant to stress because of their resistance to flattering by wind, rain and effectiveness in converting fertilizer input into higher yield [9], have the genetic basis that can be traced to a small number of genes. The tendency to taller dwarf cultivars with better resource (light, water, minerals) capture efficiency than shorter dwarf cultivars was discussed by Hawkesford [10]. It was revealed, that the genes, responsible for selection dwarfism, are mostly involved in the biosynthesis and signaling pathways of gibberellins [11].

The objective of our research was the application of mineral fertilizers for overcoming the wheat hybrid depression.

2. Materials and Methods

Our research on overcoming the hybrid depression of Dwarf I and Dwarf II was conducted on the basis of the Armenian Scientific Center of Agriculture of Ministry of Agriculture, located in Echmiadzin at Ararat region. The region is characterized by dry and sharply continental climate and the cultivation of agricultural crops is conducted under irrigation.

Considering agrochemical parameters of irrigated meadow-brown soils of the experimental area, under winter seeding of hybrids were introduced by following fertilizers: ammonium nitrate - 34% N, simple powdered superphosphate - 17% P2O5, potassium salt - 40% K_2O_4. The norm of fertilizer per 1 ha is $N_{90}P_{90}K_{75}$, while on $1m^2$ accounted N-26,5g, P-53g, K-18,7g. The fertilizers were applied in the following terms: during basic tillage a full dose of potassium 18,7g and 70% phosphorus (37,1g), during pre tillage 8,8g (33%) N, simultaneously with seeding the rest of phosphorus 15,9g (30%) and in spring 17.7 g (67%) of nitrogen, as feeding after the snow, followed by harrowing the soil. The experiments were carried out in triplicate, the distance between rows was 20 cm, between plants 10 cm, the size of the accounting plots was $1m^2$.

The scheme of experiment was as followed:

1. Dwarf I (♀Frisco x Amby♂) F_1 lethal form, as control
Dwarf I (♀Frisco x Amby♂) F_1 lethal form, with fertilizer
2. Dwarf II (♀Amby x Delfi♂) F_1 semi lethal form, as control
Dwarf II (♀Amby x Delfi♂) F_1 semi lethal form, with fertilizer

Statistical data processing was performed by Student's t-test

3. Results and Discussion

The influence of mineral fertilizers on the hybrid Dwarf I was detected. Based on the obtained data of lethal form of Dwarf I, the mineral fertilizers increased vegetative mass about 1.5-2 times, but the type of depression was not displaced (*Table 1*).

Table 1. *The influence of mineral fertilizers ($N_{90}P_{90}K_{75}$) on the hybrid Dwarf I.*

Samples	Height of plants, cm	Quantity of leaves per/plant	Weight per plant, gr
Dwarf I (Frisco x Amby) F1 Control	12,70±1,40	8,90±1,50	9,30±0,66
Dwarf I (Frisco x Amby) F1, with fertilizer	23,70±1,0*	18,90±1,20*	14,30±0,8*

Average value ± standard deviation (n=3), * the differences are significant as compared to the control hybrid Dwarf I (p< 0.05).

According to the obtained data, the differences between control plants of hybrid Dwarf I and hybrid Dwarf I treated with fertilizer, for all morphological parameters of studied traits were revealed. For hybrid Dwarf I treated with fertilizer in comparison with the control plants significantly higher parameters of height of plants, quantity of leaves and weight per plant were detected.

Semi lethal form of dwarfness was undergone to certain,

but small changes. All parameters of investigated traits with fertilizer increased by 1.2 - 1.6 times. The hybrid depression type semi lethal form of Dwarf II was not changed (Table 2).

Table 2. *The influence of mineral fertilizers ($N_{90}P_{90}K_{75}$) on elements of productivity of hybrid Dwarf II.*

Samples	Height of plants, cm	Product. tillering per/plant	The length of the ear, cm	Grain yield per plant, gr
Dwarf II (Amby x Delfi) F1 Control	25,3±0,9	4,0±0,2	6,7±0,3	0,75±0,09
Dwarf II (Amby x Delfi) F1 with fertilizer	31,0±1,44*	5,0±0,17**	8,6±0,3*	1,1±0,10*

Average value ± standard deviation (n=3), * the differences are significant as compared to the control hybrid Dwarf II ($p< 0.05$),** the differences are insignificant as compared to the control hybrid Dwarf II ($p> 0.05$).

On the base of obtained results, the differences between control plants of hybrid Dwarf II and hybrid Dwarf II treated with fertilizer, for all morphological parameters of studied traits were revealed. Regarding to the parameters of height of plants, length of ear and yield per plant significantly higher level in comparison with control plants was detected. In contrast, for productive tillering, the detected difference was insignificant. The results obtained support the statement, that the yields of both genetic and environmental manipulations are important for advanced crop production to maximize growth rate and yield [12].

Laghari et al. [13] compared three wheat varieties (TD-1, T.J-83 and Mehran-89) treated with nine levels of NPK fertilizer (0-0-0, 60-60-00, 60-60-30, 120-60-00, 120-60-60, 180-60-00, 180-60-90, 240-60-00 and 240-60-120 kg/ha-1), that significantly enhanced growth, yield and nutrient uptake traits of wheat. Optimal was application of 120-60-60 NPK kg/ha-1 to TD-1 variety that induced maximum of investigated parameters.

According to the Jan et al. [14] nitrogen presents major element of the fertilizer for a good yield, being closely linked to the vegetative growth and hence determining the fate of reproductive cycle. It was demonstrated, that the each increment of nitrogen fertilizer responded better to growth and yield of cultivars [12]. Laghari et al. [13] state, that most favorable nitrogen efficiency in wheat can be reached by optimal fertilizer rate application and management techniques. Data for various parameters of the crops were collected and analyzed by them to determine the influence of varying nitrogen levels (zero, 70, 140 and 210 kg N ha-1) applied to wheat cultivars. By Loddo and Gooding [15] the modern shorter varieties of wheat justify, economically, a greater use of N fertilizers due to increase of land-use efficiency (yield/ha). The introduction of dwarfing genes by breeders allowed the production of varieties with high leaf N content and enhanced sink capacity [16].

Physiological and genotypic differences in K efficiency (the capacity of a genotype to grow and yield well in soils low in available K), and their application in breeding programs to enhance K efficiency were discussed in review by Rengel and Damon [17].

Not less important is the problem of enhancing the P efficiency for wheat growth and yield due to genotypic differences. By Ortiz-Monasterio et al.[18] "many soils have large reserves of total phosphorus, but low levels of "available" phosphorus". Authors cite the paper of Al-Abbas and Barber [19] that total soil P is often 100 times higher than the fraction of soil P available to crop plants.

4. Conclusion

We have studied the influence of mineral fertilizers (N90P90K75) on the growth and development of hybrids Dwarf I (lethal form) and Dwarf II (semi lethal form) wheat hybrids as possible way to overcome the hybrid depression. It has been shown that the application of mineral fertilizers (N90P90K75), in the case of hybrid Dwarf I (lethal form) lead to the increase in vegetative mass, but the type of depression was not changed. The semi lethal form of Dwarf II although has undergone some changes, but retained the hybrid type of dwarfism. We suppose that this is the result of strong allelic complementation of hybrid depression genes that lead to blocking of endogenous active substances and disturbance of several physiological processes.

References

[1] Tsunewaki K., Nakai Y. (1967). Distribution of necrosis genes in wheat. I. Common wheat from Central Asia. Canad. Jour. Genet. and Cytol., 9 (1): 69-74.

[2] Zeven A. C. (1970). Geographical distribution of genes causing hybrid dwarfness in hexaploid wheat of the old world. Euphytica, 19: 33-39.

[3] Pukhalskiy V.A., Martynov S.P., Dobrotvorskaya T.V. (2000). Analysis of geographical and breeding-related distribution of hybrid necrosis genes in bread wheat (Triticum aestivum L.). Euphytica, 114: 233–240.

[4] Zeven A. C. (1981). Eight supplementary list of wheat varieties classified according to their genotype for hybrid necrosis. Euphytica, 30: 521-539.

[5] Austin R.B. (1999). Yield of wheat in the United Kingdom: recent advances and prospects. Crop Science 39, 6:1604–1610.

[6] Miralles D. J. and Slafer G. A. (2007). Sink limitations to yield in wheat: how could it be reduced? The Journal of Agricultural Science, 145:139–149.

[7] Hermsen J. G. (1966). Hybrid necrosis and red hybrid chlorosis in wheat. Hereditas, suppl., 2: 439-452.

[8] Hermsen J. G. (1963). Hybrid necrosis as a problem for the wheat breeder. Euphytica, 12(1): 1-16.

[9] Zhang X., Chen X., Wu Z., Zhang X., Huang C., Cao M. (2005). A dwarf wheat mutant is associated with increased drought resistance and altered responses to gravity. African Journal of Biotechnology, 4 (10):1054-1057.

[10] Hawkesford M. J. (2014). Reducing the reliance on nitrogen fertilizer for wheat production. Journal of Cereal Science ,59:276-283.

[11] Claeys H., De Bodt S., Inzé D. (2013). Gibberellins and DELLAs: central nodes in growth regulatory networks. Trends in Plant Science, 1–9.

[12] Ali H., Ahmad S., Ali H., Hassan F. S. (2005). Impact of Nitrogen Application on Growth and Productivity of Wheat (Triticum aestivum L.) J. Agri. Soc. Sci, 1(3):216–218.

[13] Laghari G. M., Oad F. C., Tunio S., Gandahi A. W., Siddiqui M. H., Jagirani A. W., Oad S. M. (2010). Growth, yield and nutrient uptake of various wheat cultivars under different fertilizer regimes. Sarhad J. Agric, 26 (4):489-497.

[14] Jan T., Jan M. T., Arif M., Akbar H., Ali S. (2007). Response of wheat to source, type and time of nitrogen application. Sarhad J. Agric., 23(4):871-880.

[15] Loddo S. and Gooding M.J. (2012). Semi-dwarfing (Rht-B1b) improves nitrogen-use efficiency in wheat, but not at economically optimal levels of nitrogen availability Cereal Research Communications, 40(1):116-121.

[16] Makino A. (2011). Photosynthesis, Grain Yield, and Nitrogen Utilization in Rice and Wheat. Plant Physiology, 155(1): 125–129.

[17] Rengel Z. and Damon P. M. (2008). Crops and genotypes differ in efficiency of potassium uptake and use. Physiologia Plantarum, 133: 624–636.

[18] Ortiz-Monasterio J.I., B.Manske G.G., van Ginkel M. (2001). Nitrogen and Phosphorus Use Efficiency in "Application of Physiology in Wheat Breeding"Reynolds, M.P., J.I. Ortiz-Monasterio, and A. McNab (eds.). Mexico, D.F.: CIMMYT.

[19] Al-Abbas A.H. and Barber S.A. (1964). A soil test for phosphorus based upon fractionation of soil phosphorus. I. Correlation of soil phosphorus fractions with plant-available phosphorus. Soil Science Society of America Proceedings, 28:218-221.

Water Stress Tolerance of Six Rangeland Grasses in the Kenyan Semi-arid Rangelands

Koech Oscar Kipchirchir[1,*], Kinuthia Robinson Ngugi[1], Mureithi Stephen Mwangi[1], Karuku George Njomo[1], Wanjogu Raphael[2]

[1]Department of Land Resource Management and Agricultural Technology, University of Nairobi, Nairobi, Kenya
[2]National Irrigation Board, Mwea Irrigation Agricultural Development (MIAD) Centre, WANGURU, Kenya

Email address:

okkoech@uonbi.ac.ke (K. O. Kipchirchir), okkoech@yahoo.com (K. O. Kipchirchir)

Abstract: This study evaluated six grass species in terms of water stress responses by visual quality and living ground cover attributes and the recovery responses post water stress grown at 80, 50, 30% field capacity soil moisture contents. The grass species evaluated were *Chloris roxburghiana*, *Eragrostis superba*, *Enteropogon macrostachyus*, *Cenchrus ciliaris*, *Chloris gayana*, and *Sorghum sudanense*. The grasses demonstrated varied levels of water stress tolerance as evaluated by quality ratings based on colour (greenness) and uniformity of colour, leaf firing, living matter and wilting signs. All species declined in visual quality rating with prolonged water stress treatment with exception of *Sorghum sudanense* and *Cenchrus ciliaris* that had better quality ratings of six after 42 days water stress period. *Sorghum sudanense*, *Chloris gayana* and *Cenchrus ciliaris* had accelerated recovery in quality, attaining a visual rating of eight at 21 days of water stress period. The three soil moisture content treatments had higher quality ratings than rainfed conditions which represented water deficit. *Sorghum sudanense* and *Chloris gayana* had higher quality ratings and water use efficiency under rainfed compared to the other species. All the grasses showed higher living ground cover greater than 40% at recovery period of 28 days, when irrigation was resumed at the prescribed level, and attained living cover of over 60% by day 42. *Sorghum sudanense*, *Chloris gayana* and *Cenchrus ciliaris* were able to withstand water stress longer and had also a quick recovery among the six grasses. These three species are recommended for pasture establishment in semi-arid lands where water supply uncertainties exist, owing to their high tolerance to water stress.

Keywords: Drought Tolerance, Water Stress Tolerance, Range Grasses, Pasture Irrigation, Kenya

1. Introduction

Plants survival in the arid and semi arid lands (ASALs) is determined by their adaptive capacities to the prevailing unpredictable and highly variable climatic conditions (Kimani & Pickard, 1998; Doss *et al.*, 2008). Range grasses have evolved in these uncertainties and developed their inherent resistance or tolerance levels to the frequent dry seasons and droughts (Kabubo-Mariara, 2008). Understanding water stress tolerance of grasses is crucial in pasture management when it comes to water supply and choice of adapted species depending on climatic conditions. Different grass species have inherent genetic composition that enhances their varied adaptation to water stress during droughts. Some of the adaptation mechanisms are related to; rooting depth, pattern and distribution; seed germination rates, leaf characteristics, and stem:leaf ratios among others (Rünk *et al.* 2014). During the dry seasons and droughts, pastures are exposed to water stress and they respond by among others; reducing transpiration rates to minimize losses, leaf rolling, growing leaf hairs etc. However, this may reduce pasture yields, but enhance survival which is more critical in the arid environments. This process is different among grass species depending on plant root and leaf characteristics (Hanson, 1988; Gibbens & Lenz, 2001; Vicente-Serrano *et al.*, 2010). For example, *Cynodon dactylon (Bermuda* grass) and *Medicago sativa* (Alfafa) have deep roots that enhance utilization of water in the lower soil profile (Schenk & Jackson, 2002a; Schenk & Jackson, 2002b). Water stress tolerance of grasses is one of the considerations in selecting drought tolerant species suitable for drylands in the face of climate variability and change. This gives opportunities for dryland pasture establishment and reseeding with the most adapted species as a result of the increasing unreliable rainfall in the recent past. Tworkoski & Glenn (2001) also observed variability in grass species tolerance to droughts which influence their individual survival rates and tolerance to water stress. Droughts have negative impacts on plant's performance and productivity, but the intensity depends on their adaptation mechanisms and responses (Passioura, 2007; Farooq *et al.*, 2009).

Grass species have a wide range of overlapping adaptive strategies to water stress (Ludlow, 1980). This includes; escape mechanism where annual grasses set seeds early to avoid dry seasons, storage organs like rhizomes, dormant buds and resurrection leaves that can become active with availability of precipitation. Other strategies are adjustments by plants to reduce leaf area through leaf firing, dropping of leaves and leaf rolling, adjusting the water uptake from roots e.g. root elongation and branching, maintaining turgor pressure, reduced growth during drought (dormancy) and rapid growth when water stress is reduced. Some species maintain maximum number of plants tillers and/or leaves during dry season as documented by Ludlow, (1980). Grass plants have also been observed to have physiological responses to droughts that have sustained them in the natural environment under conditions of uncertainties (Larcher, 2003).

Pasture productivity is of great concern for livestock producers in the ASAL environments and there is need to promote proper pasture management and choice of species that are adapted to frequent water deficits that ensures reliable supply of good quantity and quality forage in the face of climate change and variability. This study evaluated the water stress tolerance and determined the living basal ground cover during recovery of the grass species. The species evaluated were *Chloris roxburghiana, Eragrostis superba, Enteropogon macrostachyus, Cenchrus ciliaris, Chloris gayana, Sorghum sudanense*. The evaluation aimed at determining the species response to water stress to aid in species selection for pasture establishment in drylands of Kenya as well as help in planning the management of the same species with regards to water supply.

2. Materials and Methods

2.1. Study Area

The study was carried out is in Tana River County (figure 1), within coordinates 1°30′S, 40°0′E, 1.5°S 40°E. The climate of the area is hot and dry with daily temperatures ranging between 20 and 38°C. Rainfall is bimodal in distribution with long rains occurring in April-June and short rains in November-December. Long-term average rainfall ranges from 220 to 500 mm and is erratic in distribution. Temperatures are highest between February and April and September to October. The County is divided into three

livelihood zones; namely, pastoral, agro-pastoral (mixed farming) and marginal mixed farming.

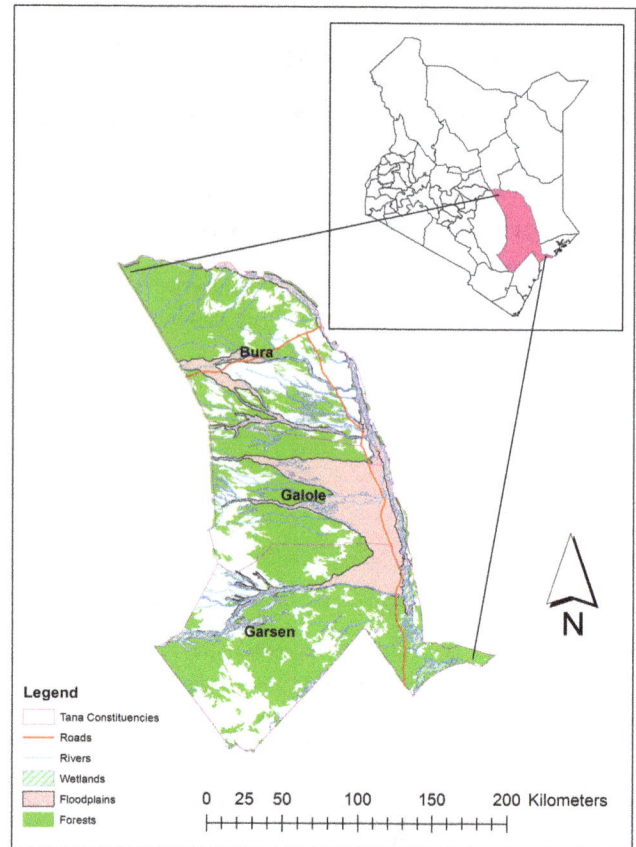

Figure 1. Study Country - Kenya (top right) in relation to study area- Tana River County.

The soil types are vertisols and vertic fluvisols associated with swelling and forming ponds during wet seasons with low infiltration rates from the sealing by high clay content. During dry seasons, the soil dry out and develop cracks. At the hinterlands are shallow and have undergone seasons of trampling by livestock, thus are easily eroded during rainy seasons. Pastoralism and agropastoralism are the main economic activities in the study area, with two established National irrigation schemes, Hola and Bura, with the latter being the experimental study site.

2.2. Experimental Layout and Design

One-acre parcel of land that had not been cultivated during the last season was identified within Bura irrigation scheme, National Irrigation Board (NIB) research site. The land was cleared of all bushes, ploughed and harrowed to a fine tilth. The area was then divided into 4 main plots of 39 m x 11 m size each. The plots within the one acre were demarcated to be 5 metres apart to minimize lateral seepage among the main plots. Each main plot was then sub-divided into 30 sub-plots measuring 3 m x 3 m with 1 m boundary.

The experimental design was factorial experiment in a completely randomised design comprising two factors, grass species and soil moisture content at 6 and 4 levels, respectively. Main plots demarcated were each randomly assigned a watering schedule as first treatment where treatment one (T1) was 80% FC, treatment two (T2) was 50% FC, treatment three (T3) was 30% FC and treatment four (T4) was the control (rain fed). The second treatment level was grass species randomly assigned to the 30 sub-plots within each of the 4 main plots. The grass species treatments were; *Chloris roxburghiana* - CR, *Eragrostis superba*- ES, *Enteropogon macrostachyus* -EM, *Cenchrus ciliaris* -CC, *Chloris gayana* -CG, *Sorghum sudanense* -SB. The species were randomly allocated to the sub-plots.

2.3. Experimental Materials, Sowing and Irrigation

Gypsum blocks (GBs) were used to determine different soil moisture content levels and monitoring soil moisture changes. The method was also used in determining soil moisture recharge times to maintain prescribed moisture contents of 80, 50 and 30% Field capacity (FC). GBs were installed at the centre of each sub plot, at two depths, 15 and 30 cm in separate holes which were dug using a 50 mm soil auger. Prior to installation they were soaked overnight as recommended. Before installation, moisture readings corresponding to 80%, 50%, and 30% FC soil moisture content was calibrated for all the GBs using moisture meter which aided in determining prescribed soil moisture content for the main blocks. After installation, wire ends originating from the installed blocks were carefully supported by vertical sticks for ease of taking readings and identification of installation points.

The source of grass seeds was Kenya Agricultural Research Institute (KARI), Kiboko Range Research Station. Before planting, the seeds were tested for germination percentage using the standard seed test by germination method as described by ISTA (1976) before planting. The germination rates obtained were used to determine the mixing and sowing rates of the species. Sowing was done manually in the finely prepared seedbeds. Phosphate fertilizer was applied to all the treatments at the recommended rate of 200 kg ha-1 to enhance establishment. Thereafter, no fertilizer application was done for the whole data collection period. All other routine pasture husbandry practices such as weeding were done for all the treatments. For each treatment, soil moisture was maintained at the prescribed level through irrigation at the prescribed soil moisture content by means of the Delmhorst Soil Moisture Meter Gypsum Blocks (GBs) installed within each sub-plots.

2.4. Data Collection on Grass Responses to Water Stress

At week 16, when the grasses were fully established, water stress condition was applied for 49 days. This was done by suspending irrigation for all the three soil moisture content until the grasses showed signs of water stress (wilting symptoms, colour change, colour uniformity and leaf firing). The grass species were rated by looking at the colour of leaves by visual quality in terms of greenness estimated on a 1-9 scale, with 1 being brown dead grass, six being

minimally acceptable and nine being optimal green colour and uniformity following the procedure described by Morris & Shearman, (2006) and Tarawali et al., (1995). This was done at 14th, 21st, 28th, 35th, 42nd and 49th days from the day of water deprivation. After the 49 days water stress treatment, a recovery period was evaluated by resumption of irrigation for the three respective levels of 80, 50, 30% FC and the ratings done at the same days from resumption. The rates of each species took to return to normal vegetative state (recovery of leaves and re-growth, uniformity and green colour) was recorded. The grass species living basal cover determination during water stress recovery phase was estimated by point frame method as described by Evans & Love (1957).

3. Results

3.1. Grass Responses to Water Stress

Table 1. Grass visual quality ratings (1-9) based on green colour, leaf firing and uniformity from water stress responses for 49 days at 80, 50 and 30% FC soil moisture content and rainfed.

%Soil moisture content	Days of drought tolerance test application					
80% FC	**14**	**21**	**28**	**35**	**42**	**49**
C R	9.0[a]	8.2[a]	6.1[ab]	4.1[b]	3.2[c]	2.0[c]
E S	9.0[a]	9.0[a]	7.0[ab]	7.0[ab]	4.0[b]	3.2[c]
EM	9.0[a]	9.0[a]	6.2[ab]	6.1[ab]	5.2[ab]	4.0[b]
CC	9.0[a]	9.0[a]	9.0[a]	8.2[a]	6.0[ab]	4.0[b]
CG	8.0[a]	8.2[a]	5.0[ab]	4.2[b]	3.1[c]	3.2[c]
SB	9.0[a]	9.0[a]	9.0[a]	8.0[a]	6.3[ab]	6.0[ab]
50% FC						
C R	8.0[a]	8.0[a]	6.4[ab]	4.3[b]	3.1[c]	3.5[c]
E S	9.0[a]	9.0[a]	7.0[ab]	7.2[ab]	4.2[b]	3.4[c]
EM	8.0[a]	9.0[a]	6.3[ab]	6.0[ab]	5.4[ab]	3.2[c]
CC	9.0[a]	9.0[a]	9.0[a]	8.0[a]	6.1[ab]	4.3[b]
CG	8.1[a]	6.0[a]	5.3[ab]	4.0[b]	3.0[c]	3.2[c]
SB	9.0[a]	8.2[a]	9.0[a]	8.1[a]	6.0[ab]	6.0[ab]
30 % FC						
C R	8.2[a]	8.2[a]	5.1[ab]	4.2[b]	3.0[c]	2.0[c]
E S	9.0[a]	9.0[a]	8.2[a]	6.0[ab]	4.2[b]	2.2[c]
EM	8.1[a]	9.0[a]	6.1[ab]	6.0[ab]	5.0[ab]	3.0[c]
CC	9.0[a]	9.0[a]	8.4[a]	6.2[ab]	6.2[ab]	3.0[c]
CG	8.2[a]	5.1[ab]	5.0[ab]	3.2[c]	3.0[c]	3.0[c]
SB	9.0[a]	8.2[a]	8.0[a]	7.0[b]	6.2[b]	5.0[ab]
Rainfed						
C R	6.2[ab]	5.0	4.0[b]	3.0[c]	3.1[c]	2.0[c]
E S	6.4[ab]	5.4[b]	4.0[b]	3.2[c]	3.3[c]	2.2[c]
EM	6.0[ab]	5.4[b]	4.1[b]	3.1[c]	3.0[c]	2.0[c]
CC	6.0[ab]	6.2[ab]	4.2[b]	3.2[c]	3.2[c]	2.1[c]
CG	5.3[ab]	4.0[b]	4.0[b]	3.2[c]	3.0[c]	2.0[c]
SB	6.0[ab]	6.1[b]	5.2[ab]	3.1[c]	3.0[c]	3.2[c]

Means within the same columns with different superscripts are significantly different at $p<0.05$.
Key: CR=Chloris roxburghiana, ES= Eragrostis superba, EM= Enteropogon macrostachyus, CC= Cenchrus ciliaris, CG= Chloris gayana, SB= Sorghum sudanense. Visual quality ratings 1 to 9 was used as visual rating scale with 1 being complete wilting, 100% leaf firing, complete dormancy or no plant recovery; and 9 being no wilting, no leaf firing, 100% Green

The visual quality ratings of grasses grown at 80, 50, and 30% FC soil moisture content and rainfed treatments are presented in Table 1. At 14th day of water stress the irrigated grasses treatments showed higher visual quality ratings 0f 8.0 - 9.0 compared to rainfed treatment with quality rating of 5.3-6.4. At 28th day, *C. ciliaris* and *S. Sudanense* had quality rating of over eight with all the species being below seven. At the same time, rainfed grass species showed lower quality ratings than the irrigated at ratings of 4.0 - 4.2 except *S. sudanense* which had a rating of 5.2. *C. ciliaris* and *S. sudanense* had visual quality rating of six after 42 days, with over half of the specie showing green colour. *C. roxburghiana* and *C. gayana* had the lowest rating of three at the same period. After 49 days of water stress, *S. sudanense* had the highest tolerance to water stress with a quality rating of six at 80 and 50% FC soil moisture content and rating of five at 30% FC soil moisture content. At the same period rainfed treatment showed lower quality ratings for all the grass species compared to the irrigated treatments, however, *S. sudanense* still had higher quality rating of three with the rest having two.

Table 2. Grass visual quality ratings (1-9) based on green colour, leaf firing and uniformity during recovery period following 49 days water stress treatment at varying soil moisture contents.

%Soil moisture content	Days of drought tolerance recovery					
80% FC	14	21	28	35	42	49
C R	5.2^b	6.4^{ab}	7.2^{ab}	8.2^d	8.2^d	9.0^d
E S	5.4^b	5.0^b	9.0^b	9.0^d	9.0^d	9.0^d
EM	4.5^b	5.0^b	6.0^{ab}	6.2^{ab}	8.0^d	9.0^d
CC	5.1^b	8.4^d	9.0^d	9.0^d	9.0^d	9.0^d
CG	4.3^b	5.3^{ab}	8.3^d	8.2^d	9.0^d	9.0^d
SB	6.2^{ab}	8.0^d	8.4^d	9.0^d	9.0^d	9.0^d
50% FC						
C R	5.1^{ab}	6.2^{ab}	7.1^{ab}	8.2^d	9.0^d	9.0^d
E S	6.3^{ab}	5.0^b	8.2^d	9.0^d	9.0^d	9.0^d
EM	4.2^b	5.2^b	6.4^{ab}	8.0^d	8.0^d	9.0^d
CC	5.3^{ab}	8.0^d	9.3^d	9.0^d	9.0^d	9.0^d
CG	5.1^{ab}	6.0^{ab}	8.1^d	9.0^d	9.0^d	9.0^d
SB	$6.3a^b$	8.4^d	8.2^d	9.0^d	9.0^d	9.0^d
30 % FC						
C R	4.4^b	6.0^{ab}	6.5^{ab}	6.2^{ab}	9.0^d	9.0^d
E S	4.3^b	5.0^b	8.4^d	8.1^d	9.0^d	9.0^d
EM	4.0^b	5.4^b	6.1^{ab}	8.3^d	8.2^d	9.0^d
CC	5.2^{ab}	8.4^d	9.0^d	8.4^d	9.0^d	9.0^d
CG	3.1^c	5.1^b	6.3^{ab}	8.2^d	8.1^d	9.0^d
SB	5.0^b	8.2^d	8.3^d	9.0^d	9.0^d	9.0^d
Rainfed						
C R	2.3^c	2.4^c	2.0^c	2.1^c	3.2^{bc}	4.3^b
E S	2.4^c	2.2^c	2.1^c	2.3^c	3.4^{bc}	4.1^b
EM	2.0^c	2.3^c	2.2^c	2.3^c	3.0^{bc}	4.3^b
CC	2.1^c	2.2^c	2.1^c	2.2^c	3.0^{bc}	4.2^b
CG	2.4^c	2.0^c	2.0^c	2.0^c	3.2^{bc}	5.1^b
SB	3.5^c	3.2^{bc}	2.3^c	2.3^c	4.1	5.0^b

Key: CR=Chloris roxburghiana, ES= Eragrostis superba, EM = Enteropogon macrostachyus, CC = Cenchrus ciliaris, CG = Chloris gayana, SB= Sorghum sudanense. Visual quality ratings 1 to 9 was used as visual rating scale with 1 being complete wilting, 100% leaf firing, complete dormancy or no plant recovery; and 9 being no wilting, no leaf firing, 100% Green-no dormancy, or 100% recovery.

The quality ratings for recovery of the grasses for the 49

days after water stress tolerance are presented in Table 2. At 14 days after irrigation resumption for the respective soil moisture contents of 80, 50 and 30% FC, all the grasses recovered having quality rating of between 4.0 and 5.5 except for *C. gayana* at 30% FC that had rating of 3.1. All the grass species under irrigation attained visual quality ratings greater than six after 28 days of recovery and a rating of nine at 49th day. *C. ciliaris* and *S. sudanense* had accelerated recovery by day 21, having quality rating of over eight at all the soil moisture treatments. Grass species under rainfed did not have any recovery up to day 35 due to lack of rainfall. However, at day 42 and 49 recovery period, the six grass species showed recovery after receiving rains amounting to 107mm allowing *S. sudanense* and *C. gayana* attain quality rating of five at 49 days while the other species rated four.

3.2. Grass Basal Ground Cover During Water Stress Recovery

The percentage living basal ground cover of the selected grass species during recovery period after water stress tolerance treatment, are presented in Table 3. There was an increase in the living basal cover with recovery periods at the irrigated treatments. Basal cover for 80 and 50% soil moisture content was >75% and significantly ($p \leq 0.05$) higher than 30% FC (<75%) and rainfed treatment at the end of 49 days recovery period. Rainfed treatment had significantly ($p \leq 0.05$) lower living basal cover compared to the irrigated treatments at the end of recovery period (<40%). There was no observed significant difference in basal cover among the individual grass species at specific soil moisture content treatments.

4. Discussions

4.1. Water Stress Tolerance

The findings of this study demonstrate that range grass species have varied adaptation capacities to water stress tolerance which mimic drought effects. The observed decline in grass species quality with prolonged water stress was from the increased evapotranspiration demands that are not met hence most grass physiological processes reduced. The grasses respond by adjusting their photosynthesis process to minimize excessive water loss through leaf firing and rolling as well as wilting which reduce vegetative growth which can be fatal if it is prolonged water stress to attain permanent wilting point (PWP) (Croser *et al.*, 2003). Dodd & Orr, (1995) assessed drought tolerance and recovery of 11 species of perennial legumes for 18 months and also observed *T. pratense* lines to be more susceptible to water stress than others (*T. semipilosum* lines, and *T. tumens*) which were highly tolerant and recovered well from simulated drought.

The higher tolerance of *S. sudanense* to water stress could be attributed to the higher tiller numbers and deep rooting observed in this study published (Koech *et al.*, 2014). The deeper rooting improves drought tolerance of perennial temperate C4 grasses (Kemp & Culvenor, 1994). These factors have been reported to have a contribution to water stress tolerance of grasses by having enough reservoir of water in tissues (Ludlow, 1980). Eneji *et al.* (2008) working on effects of silicon application on growth and water stress responses of *Festuca arundinacea, Phleum pretense, Chloris gayana,* and *Sorghum sudanense* reported *S. sudanense* to have been least affected by water stress compared to the three species which was attributed to its deep rooting system. Chen *et al.* (2008) also reported *S. sudanense* to have large root biomass that makes it competitive for water and nutrient absorption when grown in mixtures and hence increased drought tolerance. The findings of this study reveal that for established pastures in the semi-arid environments should consider the species' responses to water stress tolerance and droughts for improved productivity. This has also been reported to be important considerations for pasture breeding (Kemp & Culvenor, 1994).

Table 3. Percentage of living basal ground cover ratings during recovery period after 49 days water stress treatment at varying soil moisture contents

	Days of recovery					
80% FC	**14**	**21**	**28**	**35**	**42**	**49**
C R	24.3 ±3.3	31.3±7.8	44.8±7.5	55.5±14.5	77.5±9.5	77.5±9.5
E S	34.7±9.3	36.4±2.4	45.2±9.8	64.5±8.2	82.5±5.0	82.5±5.0
EM	20.0 ±4.1	28.0±6.2	33.5±4.7	58.0±12.1	77.0±23	98.0±23
CC	31.4±6.1	40.1±9.8	54.2±11.8	62.3±19.5	69.5±7.5	96.5±19.3
CG	15.0±6.2	33.3±8.9	48.4±7.9	66.0±12.4	71.0±22.0	92.4±24.1
SB	45.5±6.5	54.0±2.3	60.5±21.4	76±7.9	82.4±15.7	94±15.37
50% FC						
C R	20.4±6.1	31.2±9.4	45.4±11.2	57.5±6.4	68.2±19.1	81.5±19.5
E S	31.3±7.4	43.3±5.6	44.4±9.6	59±12.0	77±21.0	87.5±15.0
EM	17.7±6.2	28.4±3.1	38.9±7.9	53.0±5.2	68.7±15.1	97.0±31.3
CC	28.8±3.2	34.5±11.3	52.1±11.1	60.5±22.5	70.2±9.5	92.5±14.5
CG	11.0±3.2	24.3±14.2	26.2±4.8	45.5±11.1	97.5±5.0	90.1±32.2
SB	46.1±11.2	51.8±13.5	58.5±3.5	72.5±19.5	85.5±29.3	98.0±25.7
30% FC						
C R	14.2±4.6	17.3±6.3	22.0±6.2	37.3±9.4	55.5±11.1	67.5±9.5
E S	33.2±9.3	35.2±7.6	44.4±9.6	61.3±16.1	65.0±11.2	69.4±14.0
EM	12.8±2.9	19.4±7.0	38.9±7.9	49.0±15.0	56.7±9.8	69.8±5.0
CC	31.2±7.6	35.2±9.3	52.1±11.1	58.3±13.1	67.8±9.5	71.5±9.5
CG	8.6±1.5	16.5±8.2	26.2±4.8	40.5±9.1	67.9±9.8	74.5±5.0
SB	48.2±23.1	48.1±12.3	55.5±3.5	62.5±19.5	72.4±11.6	74.5±19.5
Rainfed						
C R	14.2±4.1	11.3±6.2	8.0±3.0	7.5±3.0	15.3±2.3	22.5±3.1
E S	13.2±2.4	11.5±4.2	10.4±9.6	7.5±2.4	16.5±4.2	28.0±3.2
EM	12.8±2.4	12.4±3.1	10.9±7.9	6.8±3.2	18.0±2.5	27.0±4.8
CC	11.2±4.1	10.5±3.3	10.1±11.1	7.5±3.5	19.5±3.2	38.8±3.5
CG	18.6±2.5	13.5±8.2	10.2±4.8	7.5±3.2	20.5±2.2	28.9±2.5
SB	28.2±3.3	24.0±2.4	10.5±3.1	8.5±3.0	18.0±2.4	29.5±3.5

Key: CR=Chloris roxburghiana, ES= Eragrostis superba, EM= Enteropogon macrostachyus, CC= Cenchrus ciliaris, CG= Chloris gayana, SB= Sorghum sudanense, ±Standard deviation

The observed higher water stress tolerance of *S. sudanense, E. macrostachyus* and *C. gayana* compared to *C. roxburghiana, C. ciliaris* and *E. superba* in this study further emphasizes the species inherent genetic constitution in adapting to water stress. This was also observed by Guenni *et al.* (2002) working with five Bracharia species (*B. brizantha* (CIAT 6780), *B. decumbens* (CIAT 606), *B. dictyoneura* (CIAT 6133), *B. humidicola* (CIAT 679) and *B. mutica*) under simulated drought reported wilting occurring after 14 days for *B. brizantha, B. decumbens* and *B. mutica* and after 28 days in *B. humidicola* and *B. dictyoneura*.

Despite *C. ciliaris* having lower water stress tolerance than *S. sudanense* and *E. macrostachyus* in this study, the species had better and quicker recovery than the rest. This finding therefore suggests that grass species may have low water stress tolerance but be adapted to accelerate recovery as a strategy; therefore, caution should be exercised in selecting species based on hardiness to water stress only. Nawazish *et al.* (2006) evaluated water stress tolerance of *C. ciliaris* from different ecotypes (drought hit habitat and irrigated soil) under three moisture regimes of 100% FC (control), 75% FC and 50% FC. They reported that ecotype from drought hit habitat was adapted to moderate and high moisture deficit. This species was also noted to depict adaptation against severe water deficits by having thick epidermal layer and cuticle that reduced evapotranspiration. The observed different adaptation of *C. ciliaris* to drought tolerance depending on ecotype is a consideration that was not factored in this study when evaluating water stress tolerance and should be considered in future studies.

Other study also reported drought tolerance of *C ciliaris* made it invasively the dominant grass species over *Heteropogon contortus* in Hawaii native grasslands (Daehler & Goergen 2005). This invasiveness was associated with its adaptability to droughts and grazing than most native species in the area. De la Barrera (2008) also reported *C. ciliaris* to be an invader in southern Sonoran Desert due to its drought tolerance. The results of this study similarly indicated that *C. ciliaris* was the most resilient species after droughts due to accelerated recovery, this could be the reason this studies found the species to be an invader as a result of quick recovery than the other plants. The drought tolerance of *C. ciliaris* has been documented by others (Lazarides *et al.,* 1997; Bhattarai *et al.,* 2008, Marshall *et al.,* 2012). The species has been identified as high value feed for livestock (Kumar *et al.,* 2004; Guevara *et al.,* 2009). *C. ciliaris* has also been reported as the most suitable for reseeding degraded arid saline soils (Lazarides *et al.,* 1997). However, Marshall *et al.* (2012) named it as one of the remarkable threats to biodiversity in drylands due to its invasive nature. Despite the varied views of *C. ciliaris* as a weed or invader in other parts of the world, it is ranked the best and most preferred livestock feed by pastoral communities in the Kenyan rangelands (Reed *et al.,* 2008; Ndathi *et al.,* 2011). For similar reasons, *C. ciliaris* is being promoted for reseeding denuded grazing lands in Kenya (Mnene, 2006; Kirwa *et al.,* 2010; Mganga *et al.,* 2010; Verdoodt *et al.,* 2010; Mureithi *et al.,* 2014; Koech, *et al.,* 2014).

There is need for long-term investigations on the drought responses of the grasses evaluated in this study since it only represented short-term dry seasons of 49 days and responses under prolonged dry seasons may be different. The responses and mechanism observed in this study can be used as reference point for irrigated pastures management, where short-term water shortages and the expected effects on grass species performance can be used to make management decisions and for making choice of species to cultivate. The high temporal variability of rainfall was also reported by Ifejika *et al.* (2008) in Makindu, a semi-arid agro-pastoral area in Kenya, with limited pasture production. Rainfall variability has been identified as one of the determinants of livestock productivity in the Kenyan rangelands (Davis *et al.,* 2006; Orindi *et al.,* 2007; Theisen, 2012). This calls for innovative ways of improving water resource utilization for fodder production sustainability, for instance, integration of water harvesting and pasture production with pastoralism.

4.2. Living Basal Ground Cover

The observed higher basal cover for irrigated treatments compared to rainfed is attributed the fact that the species under this condition were not adversely affected by water stress and showed quick recovery. These findings highlight the benefits of water supply in increasing ground cover therefore enhancing soil conservation. Ground cover in the semi-arid environments determines the soil hydrological properties, soil moisture and also influences present and future productivity. The observed variations in percent

ground cover among the species with *S. Sudanense* attaining higher cover at day 28 could be attributed to genetic variability. Hu *et al.* (2010) subjected two genotypes of Kentucky bluegrass (Poa pratensis L.), 'Midnight' (tolerant) and 'Brilliant' (sensitive) which differ in drought resistance to drought stress for 15 days. They then re-watered for 10 days and observed that single-leaf net photosynthetic rate, stomatal conductance and transpiration rate decrease during drought, with a less rapid decline in 'Midnight' than in 'Brilliant', which they attributed to genetic variations. These findings could explain the observed variability in cover for the different species in this study. Malinowski & Belesky, (2000) also reported drought tolerance in grasses to be influenced by both physiological and biochemical adaptations of species hence varying responses to water stress.

Study by Chai *et al.* (2010) looking at physiological traits of two C_3 perennial grass species, Poa pratensis and Lolium perenne, for drought survival after well watering before 20 days drought through withholding irrigation observed that seven days of re-watering, drought-damaged leaves were rehydrated and recovered fully in P. pratensis but could not fully recover in L. perenne. P. pratensis. The species also produced a greater number of new roots, while L. perenne had more rapid elongation of new roots after 16 days of re-watering. The observed low ground cover for rainfed at the end of recovery period indicate that natural grasslands productivity is limited by moisture supply and irrigation can be used to bridge the gap. All the species except *E. macrostachyus* attained living ground cover greater than 40% by day 28 and by 42 days the cover was over 70%, at the irrigated moisture content levels. This finding suggests ecological adaptability of the evaluated grasses to the highly variable environment which has perpetuated their survival under frequent droughts over the years. Mganga *et al.* (2013) also recommended *C. ciliaris*, *E. superba* and *E. macrostachyus* as suitable for range reseeding and rehabilitation due to their high drought tolerance. This study has showed *E. macrostachyus* to be slow in recovery among the six species, but interestingly, at the end of 49 days recovery phase, the species had almost similar percentage living ground cover to other species.

5. Conclusions

C. ciliaris and *S. sudanense* emerged to be best species adapted to water stress while *C. roxburghiana* and *C. gayana* had the lowest tolerance to water stress. Notably, still *S. sudanense, C. ciliaris* and *C. gayana* had higher recovery rates from water stress in that order which suggest better adaptation to droughts than *Chloris roxburghian*. Two species have shown greater candidates for drought tolerance in this study, namely; *C. ciliaris* and *S. sudanense*. These findings also indicate that relying on rainfall for pasture production is not reliable and yields are bound to be affected by rainfall variability and the unpredicted droughts which is common in the drylands. Therefore, irrigation can be considered as one way of improving pasture production for reliable fodder supply in the ASALs of Kenya.

Acknowledgement

This work was supported by the National Irrigation Board (NIB) Kenya, National Council of Science, Technology and Innovation (NCSTI), German Academic Exchange Service (DAAD) -Kenya and the Centre for Sustainable Dryland Ecosystems and Societies (CSDES)-University of Nairobi, Kenya.

References

[1] Bhattarai, S. P., Fox, J., and Gyasi-Agyei, Y. (2008). Enhancing buffel grass seed germination by acid treatment for rapid vegetation establishment on railway batters. *Journal of Arid Environments, 72(3), 255-262.*

[2] Chai, Q., Jin, F., Merewitz, E., and Huang, B. (2010). Growth and physiological traits associated with drought survival and post-drought recovery in perennial turf grass species. *Journal of the American Society for Horticultural Science, 135(2), 125-133.*

[3] Chen, M., Chen, B., and Marschner, P. (2008). Plant growth and soil microbial community structure of legumes and grasses grown in monoculture or mixture. *Journal of Environmental Sciences, 20(10), 1231-1237.*

[4] Croser, J. S., Clarke, H. J., Siddique, K. H. M., and Khan, T. N. (2003). Low-temperature stress: implications for chickpea (Cicer arietinum L.) improvement. *Critical Reviews in Plant Sciences, 22(2), 185-219.*

[5] Daehler, C. C. and Goergen, E. M. (2005). Experimental restoration of an indigenous Hawaiian grassland after invasion by buffel grass (Cenchrus ciliaris). *Restoration Ecology, 13(2), 380-389.*

[6] Davis, R., Gichere, S., Mogaka, H., and Hirji, R. (2006). Climate variability and water resource degradation in Kenya: improving water resources development and management. *Washington, DC: World Bank.*

[7] De la Barrera, E. (2008). Recent invasion of buffel grass (Cenchrus ciliaris) of a natural protected area from the southern Sonoran Desert. *Revista Mexicana de Biodiversidad, 79(2), 385-392.*

[8] Dodd, M. B., and Orr, S. J. (1995). Seasonal growth, phosphate response, and drought tolerance of 11 perennial legume species grown in a hill country soil. *New Zealand journal of agricultural research, 38(1), 7-20.*

[9] Doss, C., McPeak, J., and Barrett, C. B. (2008). Interpersonal, intertemporal and spatial variation in risk perceptions: Evidence from East Africa. *World Development, 36(8), 1453-1468.*

[10] Eneji, A. E., Inanaga, S., Muranaka, S., Li, J., Hattori, T., An, P., and Tsuji, W. (2008). Growth and nutrient use in four grasses under drought stress as mediated by silicon fertilizers. *Journal of Plant Nutrition, 31(2), 355-365.*

[11] Evans, R.A. and Love, R.M. (1957). The step-point method of sampling- A practical tool in range research. Journal of Range Management 10:208-212.

[12] Falkenmark, M. (2007). Shift in thinking to address the 21st Century hunger gap, moving focus from blue to green water management. *Water Resource Management, 21(1): 3–18.*

[13] Farooq, M., Wahid, A., Kobayashi, N., Fujita, D., and Basra, S. M. A. (2009). Plant drought stress: effects, mechanisms and management. *In Sustainable Agriculture (pp. 153-188). Springer Netherlands.*

[14] Gibbens, R. P., and Lenz, J. M. (2001). Root systems of some Chihuahuan Desert plants. *Journal of Arid Environments, 49(2), 221-263.*

[15] Guenni, O., Marín, D., and Baruch, Z. (2002). Responses to drought of five Brachiaria species. I. Biomass production, leaf growth, root distribution, water use and forage quality. *Plant and soil, 243(2), 229-241.*

[16] Guevara, J. C., Grünwaldt, E. G., Estevez, O. R., Bisigato, A. J., Blanco, L. J., Biurrun, F. N., and Passera, C. B. (2009). Range and livestock production in the Monte Desert, Argentina. *Journal of Arid Environments, 73(2), 228-237.*

[17] Hanson, R. L. (1988). Evapotranspiration and droughts. Paulson, RW, Chase, EB, Roberts, RS, and Moody, DW, Compilers, National Water Summary, 99-104.

[18] Hu, L., Wang, Z., and Huang, B. (2010). Diffusion limitations and metabolic factors associated with inhibition and recovery of photosynthesis from drought stress in a C3 perennial grass species. *Physiologia plantarum, 139(1), 93-106.*

[19] Ifejika S., C., Kiteme, B., and Wiesmann, U. (2008). Droughts and famines: the underlying factors and the causal links among agro-pastoral households in semi-arid Makueni District, Kenya. *Global Environmental Change, 18(1), 220-233.*

[20] Kabubo-Mariara, J. (2008). Climate change adaptation and livestock activity choices in Kenya: An economic analysis. In Natural Resources Forum, 32(2): 131-141. *Blackwell Publishing Ltd.*

[21] Kemp, D. R., and Culvenor, R. A. (1994). Improving the grazing and drought tolerance of temperate perennial grasses. *New Zealand Journal of Agricultural Research, 37(3), 365-378.*

[22] Kimani, K., and Pickard, J. (1998). Recent trends and implications of group ranch sub-division and fragmentation in Kajiado District, Kenya. Geographical Journal, 202-213.

[23] Kirwa, E. C., Mnene, W. N., Kubasu, D., Kimitei, R. K., Kidake, B., and Manyeki, J. K. (2010). Assessing the performance of established range grass species in southern Kenya rangelands. In Proceedings of the 12th KARI Biennial Scientific Conference. 8th–12th November, Nairobi, Kenya (pp. 871-876).

[24] Kumar, P., Kumar, S., Sharma, K. D., Choudhary, A., and Gehlot, K. (2004). Lignite mine spoil characterization and approaches for its rehabilitation. *Arid Land Research and Management, 19(1), 47-60.*

[25] Larcher, W. (2003). Physiological plant ecology: ecophysiology and stress physiology of functional groups. *Springer. http://tinyurl.com/qh3tdph. Accessed 6th May 2013.*

[26] Lazarides, M., Cowley, K., and Hohnen, P. (1997). CSIRO Handbook of Australian Weeds. CSIRO publishing.

[27] Ludlow, M. M. (1980). Stress physiology of tropical pasture plants. *Tropical grasslands, 14(3), 136-145.*

[28] Maingi, J. K., and Marsh, S. E. (2002). Quantifying hydrologic impacts following dam construction along the Tana River, Kenya. *Journal of Arid Environments, 50(1), 53-79.*

[29] Malinowski, D. P., Zuo, H., Kramp, B. A., Muir, J. P., & Pinchak, W. E. (2005). Obligatory summer-dormant cool-season perennial grasses for semiarid environments of the southern Great Plains. *Agronomy journal, 97(1), 147-154.*

[30] Marshall, V. M., Lewis, M. M., and Ostendorf, B. (2012). Buffel grass (Cenchrus ciliaris) as an invader and threat to biodiversity in arid environments: A review. *Journal of Arid Environments, 78, 1-12.*

[31] Mganga, K. Z., Musimba, N. K. R., Nyariki, D. M., Nyangito, M. M., and Mwang'ombe, A. W. (2013). The choice of grass species to combat desertification in semi arid Kenyan rangelands is greatly influenced by their forage value for livestock. *Grass and Forage Science.* http://onlinelibrary.wiley.com/doi/10.1111/gfs.12089/full. Accessed 1st Aug 2013.

[32] Mganga, K. Z., Musimba, N. K., Nyariki, D. M., Nyangito, M. M., Mwang'ombe, A. W., Ekaya, W. N., and Muiru, W. M. (2010). Dry matter yields and hydrological properties of three perennial grasses of a semi-arid environment in East Africa. *African Journal Plant Science 4(5), 138-144.*

[33] Mnene, W. N. (2006). Strategies to increase success rates in natural pasture development through reseeding degraded rangelands of Kenya (Doctoral dissertation, Ph. D. Thesis, University of Nairobi, Nairobi, Kenya).

[34] Morris, K.N., and Shearman, R.C. (2006). NTEP Evaluation Guidelines. National Turfgrass Evaluation Programme. Available at: *http://www.ntep.org/pdf/ratings.pdf. Accessed on 6th June 2012.*

[35] Mureithi, S. M., Verdoodt, A., Gachene, C. K. K., Njoka, J. T., Wasonga, V. O., De Neve, S., Meyerhoff, E., and Van Ranst, E. (2014). Impact of enclosure management on soil properties and microbial biomass in a restored semi-arid rangeland, Kenya. *Journal of Arid Land, doi: 10.1007/s40333-014-0065-x.*

[36] Nawazish, S.M., Hameed, M. and Naurin, S. (2006). Leaf anatomical adaptations of Cenchrus ciliaris L. from the Salt Range, Pakistan against drought stress. *Pak. J. Bot, 38(5), 1723-1730.*

[37] Ndathi, A. J., Nyangito, M. M., Musimba, N. K., and Mitaru, B. N. (2011). Farmers' preference and nutritive value of selected indigenous plant feed materials for cattle in drylands of south-eastern Kenya. *http://www.lrrd.cipav.org.co/lrrd24/2/ndat24028.htm. Accessed on 6th June 2012).*

[38] Orindi, V. A., Nyong, A., and Herrero, M. (2007). Pastoral livelihood adaptation to drought and institutional interventions in Kenya. *Human Development Report Office, Occasional Paper, 54.*

[39] Passioura, J. (2007). The drought environment: physical, biological and agricultural perspectives. *Journal of experimental Botany, 58(2), 113-117.*

[40] Reed, M. S., Dougill, A. J., and Baker, T. R. (2008). Participatory indicator development: what can ecologists and local communities learn from each other. *Ecological Applications, 18(5), 1253-1269.*

[41] Rünk, K., Pihkva, K., and Zobel, K. (2014). Desirable site conditions for introduction sites for a locally rare and threatened fern species< i> Asplenium septentrionale</i>(L.) *Hoffm. Journal for Nature Conservation, 22(3), 272-278.*

[42] Schenk, H. J., and Jackson, R. B. (2002a). Rooting depths, lateral root spreads and below - ground/above - ground allometries of plants in water limited ecosystems. *Journal of Ecology, 90(3), 480-494.*

[43] Schenk, H. J., and Jackson, R. B. (2002b). The global biogeography of roots. *Ecological monographs, 72(3), 311-328.*

[44] Tarawali S.A., Tarawali G., Larbi A. and Hanson J. (1995). Methods for the Evaluation of Legumes, Grasses and Fodder Trees for Use as Livestock Feed. ILRI Manual 1. *ILRI (International Livestock Research Institute), Nairobi, Kenya. pp. 51.*

[45] Theisen, O. M. (2012). Climate clashes? Weather variability, land pressure, and organized violence in Kenya, 1989–2004. *Journal of Peace Research, 49(1), 81-96.*

[46] Tworkoski, T. J., and Glenn, D. M. (2001). Yield, shoot and root growth, and physiological responses of mature peach trees to grass competition. *Hort Science, 36(7), 1214-1218.*

[47] Verdoodt, A., Mureithi, S. M., and Van Ranst, E. (2010). Impacts of management and enclosure age on recovery of the herbaceous rangeland vegetation in semi-arid Kenya. *Journal of Arid Environments, 74(9), 1066-1073.*

[48] Vicente-Serrano, S. M., Beguería, S., and López-Moreno, J. I. (2010). A multiscalar drought index sensitive to global warming: the standardized precipitation evapotranspiration index. *Journal of Climate, 23(7), 1696-1718.*

3

Evaluation of Some Performance Traits and Carcass Characteristics of *Archachatina marginata* Snails Fed Plant Wastes

Olubukola Omolara Babalola[1, 2]

[1]Department of Biological Sciences, Landmark University, Omu-Aran, Kwara State, Nigeria
[2]Department of Science Technology, Federal Polytechnic, Ado-Ekiti, Ekiti State, Nigeria

Email address:
babalola.olubukola@lmu.edu.ng, olubukolababalola2@gmail.com

Abstract: There is inadequate information on quality feedstuffs for large scale production and all year round availability of snails in Nigeria. This study evaluated the performance, carcass analysis and sensory evaluation of cooked meat of growing *Archachatina marginata* snails fed plant wastes as a sole feed ingredient. 120 growing snails of mean weight of 132.91±2.13g were randomly allotted to 4 dietary treatments of pawpaw leaves (PL), whole lettuce (WL), lettuce wastes (LW) and cabbage wastes (CW). Each treatment was replicated thrice with 10 snails per replicate in a completely randomized design. The feeding trial lasted 6 months. Treatment effect on shell length and width was significant (P<0.05) with snails on LW recording highest while no significant differences were observed in the shell thickness gain (P>0.05). The highest dressing percentage of 43.2% was obtained for snails on LW while the lowest value of 35.19% was recorded for snails on PL. The treatments had no appreciable effect on the nutrient composition and sensory quality of the snail meat. The highest dry matter digestibility of 83.50% was recorded in snails on CW which was statistically similar to those on LW (83.33%) while the least value of 78.33% was recorded in snails on PL. The weight gain and feed per gain followed the same pattern as the dry matter digestibility. It can be concluded that growing snails can utilize lettuce waste as well as cabbage waste as sole feed thereby increasing the feed data base for snail production in the Tropics.

Keywords: Pawpaw leaves, Lettuce, Cabbage, *Archachatina marginata* Snails, Feed per Gain

1. Introduction

Large scale snail farming is needed in order to meet the animal protein need in human diet. There is a dearth in the supply of conventional feed concentrates which has greatly affected animal production in the tropics. This low level of livestock production in the tropics cannot meet the needs of the rapidly growing human populations. There is therefore the need to source for cheaper alternative sources of animal protein.

Snails are invertebrates with a soft body and a covering of hard shell. It is one of the micro livestock that has recently attracted attention among agriculturists in Nigeria as an aftermath of alarm raised by Food and Agricultural Organization (FAO) on animal protein deficiency among Nigerians [1, 4]. It has small body size and is easy to handle and manage. It is found in a cool environment, in gardens, vegetable plantation, refuse heap, orchards, etc. They require humid environment and thrive well on decay materials [3, 7, 12].

Snail meat is tender and tastes good. It is highly nutritious and when eaten serve as a special delicacy in the diet. Several studies have been conducted on snails in the last three decades. Ajayi et al. (1978) indicated that snail meat is particularly rich in protein, iron, calcium and phosphorous [2]. Snail meat has a protein content of about 18-20% [7, 8, 9] which compare quite well with protein contents of conventional meat such as beef (18%), mutton (18%) and poultry (20%) [13] Imevbore and Ademosun (1988) reported a fat content of 1.36% which is lower than the corresponding values of 9.6%, 21.4% and 23% obtained for egg, mutton, and duck respectively [14, 16[]. The study also showed that snail meat is low in saturated fatty acids (28.71%) and cholesterol (20.28mg/100g fresh sample) when compared with beef, goat meat, mutton, pork, broiler meat and fish. The low contents of fat and cholesterol make snail meat a good antidote for vascular diseases such as heart attack, cardiac arrest, hypertension and stroke [5]. Soup prepared with snail meat is a good source of iron for pregnant and nursing mothers [19].

Unfortunately, in spite of the obvious nutritional value of snail in human diet, no significant effort has been made at its large scale production as with other livestock like cattle, goat, sheep, and poultry. The main source of supply to the consumers is through marketers who gather them from the wild making the supply relatively higher during the rainy season than the dry season. FAO (1986) encouraged raising one's own snails, a practice referred to as 'snail farming' with the advantage of continuous supply of fresh snail meat whenever this is desired and sale of excess to other consumers [12].

For effective performance, nutrition in snail production cannot be underestimated. African giant land snails (A. marginata) are naturally herbivores. They feed mostly during the night because they are nocturnal animals. However experience has shown that they can eat at any time of the day if served with their delicacy in a cool, humid environment. Their conventional feed comprises of fresh leaves/shoots (pawpaw, lettuce, cabbage, cassava, cocoyam, African spinach, waterleaf); Ripe fresh fruits (pawpaw, banana, plantain, mango) and household/agro wastes (poultry litter, rice bran, palm kernel meal) etc. [8, 11]. These are mostly of plant origin and there is possibility of scarcity during the dry season. Sourcing for these feeds in the urban areas may be very difficult. There is therefore the need to source for acceptable feed that is available all year round.

Feed accounts for at least 70% of total cost of livestock production. The high cost of producing animal products due to the exorbitant prices of feed ingredient has forced animal nutritionists to explore the use of agricultural byproducts hitherto referred to as wastes as feed resources in order to reduce cost of production [18].

There is paucity of information on quality feedstuffs for large scale production of snail. Hamzat (2004) evaluated the use of kola testa, a byproduct of kola fruit, for feeding snails in Nigeria [15]. Lettuce waste, an inedible foliage after harvesting was found to be cherished by snails. It is succulent and available all year round. Lettuce has been reported to contain water (94g); energy (18kcal); protein (1.3g); fat (0.3g); carbohydrate (3.5g); fibre (1.9g) and ash (0.9g) per 100g of edible portion. It also contains (mg) Ca (68); Fe (1.4); Mg (11); P (25); Cu (0.044) per 100g of edible portion [21]. There is a dearth of information on the use of lettuce waste by snails, hence this study was embarked upon to evaluate the performance of African giant land snail fed lettuce waste in comparison with cabbage waste and pawpaw leaves.

2. Materials and Method

One hundred and twenty (120) growing A. marginata snails

of mean weight of 132.91±2.13g were used for the experiment. The snails were randomly allotted into 4 dietary treatments of pawpaw leaves, whole lettuce, lettuce wastes and cabbage wastes. Each treatment was replicated thrice with 10 snails per replicate in a completely randomized design. The snails were reared in wooden cages of $0.5 \times 0.5 \times 0.5m^3$ compartments. Feed and water were supplied ad libitum. Egg shell powder was added to the soil weekly to supply calcium. Feed intake and weight gain were measured on a daily and weekly basis respectively. Shell length and width were measured with vernier caliper while micrometer screw gauge was used to measure the shell thickness. Other parameters determined were mortality and feed conversion ratio. The feeding trial lasted six (6) months.

2.1. Digestibility Trial

Three (3) snails per replicate were put inside cages demarcated into different compartments devoid of soil but lined with foam. The snails were fed with the same diet fed during the feeding trial. Daily feed intake and excreta voided were recorded for each treatment. The daily excreta for each treatment was dried in the oven at 600C and dry matter determined. The trial lasted 10 days, including 3 days for acclimatization and 7 days for excreta collection.

2.2. Carcass Analysis

Nine growing snails per treatment (3 per replicate) were used at the end of the feeding trial for carcass analysis. The snails were starved overnight and their weights taken. They were killed by striking iron rod on their shell after which the visceral, shell, haemolymph and foot were separated. Parameters determined were: dressing percentage, visceral to live weight percent, shell to live weight percent and haemolymph per live weight percent.

2.3. Chemical Analysis

Proximate composition of the experimental diets as well as that of the foot of the snails was carried out [6]. Parameters analysed were dry matter, crude protein, crude fibre, ash and ether extract.

2.4. Organoleptic Evaluation of the Cooked Meat

The snail meat from each treatment was washed with alum and cooked separately in pots containing 3g of salt dissolved in 300mls of water at 1000c for 20minutes. A twelve –member taste panellist was set up. They were trained prior to serving of the meat. The snail meat from each treatment was served in individual plates and given to the panellist. They were also served with drinking water to rinse their mouth after tasting each treatment of the meat. There was partitioning in between the panellists in such a way that there was no interaction with one another. Questionnaires were given to the panellist for rating of the samples according to the method of Larmond (1977) [17]. The ratings were based on a 9 point hedonic scale of 1(dislike extremely) and 9(like extremely). The evaluation was based on colour, taste, flavour, tenderness, and overall

2.5. Data Analysis

All data were subjected to analysis of variance while the treatment means were separated "using Duncan multiple range test (SAS 2003) [20]. All snails in the cage by replicate represent the experimental unit.

3. Results

3.1. Proximate Composition of Test Ingredients.

The proximate composition of experimental diets is as shown on Table 1. The crude protein of PL was significantly higher than that of the other test diets. Crude fibre and ash follow the pattern WL >PL >LW >CW with CW recording the highest Nitrogen Free Extract (NFE) while PL had the lowest.

Table 1. Proximate composition of the snail diet (% dry matter).

Nutrients	Pawpaw Leaf (PL)	Whole Lettuce (WL)	Lettuce Waste (LW)	Cabbage Waste (CW)
Dry matter	25.43	5.96	7.04	10.10
Crude protein	33.25	11.20	7.35	9.80
Crude fibre	7.26	8.96	6.32	5.48
Ether extract	0.78	0.56	0.27	0.23
Ash	10.86	11.65	9.67	6.94
Nitrogen free extract	47.85	67.63	76.39	77.55
Gross energy (kcal/kg)	3.25	3.16	3.23	3.33

3.2. Feeding and Growth Performance

The results obtained for the feeding and growth performance for growing A. marginata snails is as presented in Table 2. The mean dry matter feed intake showed that there were significant differences among the treatment means (P<0.05). The mean weekly feed intake of 6.45, 8.79, 8.18 and 7.81g were recorded for snails placed on PL, WL, LW and CW respectively. The highest mean weekly feed intake was recorded in WL (8.79g) while the lowest was recorded in PL (6.45g).

The weights gained by the experimental snails were affected by the dietary treatments (P<0.05). Snails on CW recorded the highest mean weekly weight gain of 3.55g which was statistically similar to that of LW (3.50g) while those on PL recorded the least weekly weight gain of 2.35g.

Treatment effect on monthly shell length gain was significant. It was observed that the highest mean monthly shell length gain of 3.85mm occurred in snails on LW while the least value of 3.21mm was recorded in snails on PL. The mean monthly shell length gain of snails on WL and CW were similar (P>0.05). There were also significant differences in the mean monthly shell width gain (P<0.05) with snails on WL recording the highest value of 3.26mm which was statistically similar to that of LW (3.13mm). Snails on PL recorded the lowest monthly shell width gain of 2.56mm. No significant differences were observed in the mean monthly shell thickness gain (P>0.05). The values ranged between 0.21 and 0.24mm.

The result of the dry matter digestibility showed significant differences amongst the treatments. The highest digestibility of 83.50% was recorded in snails on CW which was statistically similar to those on LW (83.33%). The least digestibility of 78.33% was recorded in snails on PL.

The best feed per gain of 2.20 was obtained in snails on CW which was similar to that of snails on LW (2.34) while snails on PL recorded a value of 2.74. No mortality was recorded in all the treatments.

Table 2. Performance characteristics of growing snails fed the experimental diets.

Parameters (mean values)	Pawpaw Leaf (PL)	Whole Lettuce (WL)	Lettuce Waste (LW)	Cabbage Waste (CW)	SEM
Weekly dry matter feed intake (g)	6.45d	8.79a	8.18b	7.81c	0.32
Initial weight (g)	135.08	132.15	131.91	132.50	0.38
Final weight (g)	191.48b	213.99a	215.97a	217.60a	1.95
Weekly weight gain (g)	2.35c	3.41b	3.50a	3.55a	0.57
Total weight gain (g)	56.40c	81.84b	84.06a	85.10a	2.05
Monthly shell length gain (mm)	3.21b	3.60a	3.85a	3.65a	0.20
Monthly shell width gain (mm)	2.56d	3.26a	3.13b	2.86c	0.09
Monthly shell thickness gain (mm)	0.21a	0.23a	0.21a	0.24a	0.01
Mortality (%)	0.00	0.00	0.00	0.00	
Dry matter digestibility (%)	78.33c	81.18b	83.33a	83.50a	0.71
Feed per gain	2.74a	2.58b	2.34c	2.20d	0.10

a, b, c, d: means along the same row with different superscripts are significantly different (p<0.05)
SEM – Standard Error of Means
n = 3 per diet

3.3. Carcass Analysis

Table 3 presents results of foot yield, visceral, shell and haemolymph components of growing snails. The highest foot weight was recorded for snails on PL. There were significant differences (P<0.05) in the dressing percentages of the snails.

The highest dressing percentage of 43.20% was obtained for snails on LW while the lowest value of 35.19% was recorded for snails on PL. The mean weight of the shell followed the same trend as the dressing percentage. The shell to live weight for snails PL and LW were similar and were significantly higher than the values obtained for snails on WL and CW.

Table 3. Carcass evaluation of growing snails (A. marginata) fed the experimental diets.

Parameters (mean values)	Pawpaw Leaf (PL)	Whole Lettuce (WL)	Lettuce Waste (LW)	Cabbage Waste (CW)	SEM
Number of snails	9	9	9	9	
Live weight (g)	190.80[b]	211.70[a]	210.90[a]	213.20[a]	1.98
Foot (edible portion) (g)	67.14[d]	85.17[b]	91.11[a]	82.38[c]	2.48
Visceral (g)	42.42[a]	38.50[b]	39.60[b]	41.12[a]	2.07
Shell (g)	50.84[b]	51.40[b]	54.10[a]	51.10[b]	1.68
Haemolymph (ml)	30.40[b]	36.63[a]	26.09[c]	38.60[a]	1.26
Dressing (%)	35.19[d]	40.23[b]	43.20[a]	38.64[b]	1.09
Shell/live weight (%)	26.65[a]	24.28[b]	25.65[a]	23.97[b]	0.75
Visceral/live weight (%)	22.23[a]	18.19[b]	18.78[b]	19.29[b]	1.01
Haemolymph/live weight (%)	15.93[b]	17.30[a]	12.37[c]	18.10[a]	1.39

a,b,c,d: means along th same row with different superscripts are significantly different (p<0.05), SEM – Standard Error of Means

3.4. Nutrient Composition of the Meat

The percentage crude proteins in all the treatments were similar with values ranging from 17.82 to 18.53% (Table 4). The percentage ash contents ranged between 2.14 and 2.33 and the differences among the treatments were not significant.

The values obtained for the fat content were also similar and ranged between 2.11 and 2.44. There was however significant differences in the nitrogen free extract with the highest value of 3.85% obtained for snails on CW while the lowest (2.80%) was recorded for snails on WL.

Table 4. Nutrient composition (g/100g fresh meat) of snail meat from growing snails fed the experimental diets.

Nutrient (%)	Pawpaw Leaf (PL)	Whole Lettuce (WL)	Lettuce Waste (LW)	Cabbage Waste (CW)	SEM
Dry matter	26.33	25.89	25.32	26.01	0.11
Moisture content	73.67	74.11	74.68	73.99	0.13
Crude protein	18.53	18.32	17.95	17.82	0.08
Ash	2.26	2.33	2.14	2.23	0.02
Ether extract	2.31	2.44	2.21	2.11	0.03
Nitrogen free extract	3.23[b]	2.80[b]	3.02[b]	3.85[a]	0.06

a,b: means along the same row with different superscripts are significantly different (p<0.05)
SEM – Standard Error of Means

3.5. Organoleptic Evaluation of Cooked Meat

The dietary treatments had no significant effect on the colour, taste, flavour, texture and overall acceptability of the snail meat (Table 5).

Table 5. Organoleptic properties of snail meat from growing snails fed the experimental diets.

Properties	Pawpaw Leaf (PL)	Whole Lettuce (WL)	Lettuce Waste (LW)	Cabbage Waste (CW)	SEM
Colour	7.34	7.58	7.92	7.39	0.07
Taste	8.65	8.31	8.14	8.62	0.06
Flavour	8.11	8.34	8.94	8.03	0.10
Texture	8.62	8.57	8.62	8.42	0.02
Overall acceptability	8.45	8.21	8.50	8.55	0.04

SEM – Standard Error of Means

4. Discussion

The findings of this study were similar to those obtained in an earlier experiment reported by Babalola and Akinsoyinu, 2010 for snailets fed the same set of experimental diets. The zero mortality recorded in all the treatments indicates that growing snails are more resilient than their snailets counterpart which recorded some mortality [10]. This means that any of the feed could be used in feeding growing snails without adverse effect.

Snails on LW recorded the highest shell weight and also high shell per live weight. This may be as a result of the high mineral content of LW most especially calcium which supported shell growth [21]. LW also enhanced the highest dressing percentage. The organoleptic properties of the snail meat were similar, an indication that the feed had no appreciable effect on the meat quality of the snails.

Snailets are better converter of the experimental feeds than their growing counterparts as the values obtained for the feed conversion ratio of snailets were generally lower than those obtained for the growing snails. One can therefore suggest the feeding of these plant wastes to snailets and as they grow older supplementing with compounded ration to meet their energy needs.

It can be concluded that lettuce wastes contain high nutrients which favour snail growth and development as evidenced in the total weight gain, feed per gain, shell weight and dressing percentage and incorporation of the dried lettuce waste into the feed could enhance better growth of snails and increase in the supply of animal protein in Nigeria and so prevent these animals from going into extinction.

Acknowledgement

The author is grateful to Landmark University, Omu-Aran for providing a conducive environment to embark on the write-up.

References

[1] O.M. Adesope, Attitudes of household in a Niger Delta zone towards snail meat consumption. In: Ukachukwu S.N. et al. (eds). Animal production in the new millennium. Challenges and options. Zaria: NSAP secretariat. 2000.

[2] S.S. Ajayi, O.O. Tewe, C. Moriaty and M.O. Awesu, Observation on the biological and nutritive value of the African giant snail, Archachatina marginata. East Africa Wildlife J. Vol. 16 pp 85-95. 1978.

[3] O. Akinnusi, Snail farming – low investment, high profit business. Livestock Echo April-June pp14-23. 1997.

[4] O. Akinnusi, Snail rearing – Case study Abeokuta, Ogun State, Nigeria. Proceedings of 5th Annual Conference of Animal Science Association of Nigeria, Port Harcourt Nigeria, Sept. 19-22. 2000.

[5] O. Akinnusi, Introduction to snails and snail farming. Triolas publishing company, Abeokuta, p.70. 2002.

[6] Association of Official Analytical Chemist (A.O.A.C). Official methods of analysis, Washington DC. 2005.

[7] M.O. Awesu, A biology and management of the African giant land snail (A.marginata). M.Phil. Thesis, University of Ibadan, Nigeria. 1980.

[8] I.A. Ayodele and A.A. Asimalowo, Essentials of snail farming. Agape Printers, Ibadan. Pp 7-37. 1999.

[9] O.O. Babalola and A.O. Akinsoyinu, Proximate composition and Mineral Profile of Snail meat from different Breeds of Land Snail in Nigeria. Pakistan Journal of Nutrition Vol. 8(12): 1842-1844. 2009.

[10] O.O. Babalola and A.O. Akinsoyinu, Performance, carcass analysis and sensory evaluation of cooked meat of snailets of African giant land snail (Archachatina marginata) fed pawpaw leaves, whole lettuce, lettuce waste and cabbage waste as sole feed ingredient. African Journal of Agricultural Research Vol. 5(17): 2386-2391. 2010

[11] O.O. Babalola and E.E. Owolabi, Comparative evaluation of performance of snails (Archachatina marginata) fed milk leaf (Euphorbia heterophylla) as against pawpaw leaf (Carica papaya) and Concentrate as sole feed. International Journal of Research in Applied, Natural and Social Sciences Vol. 2(11), pp 137-144. 2014.

[12] FAO, Farming snails by FAO. Better Farming Series, 3/33 Rome, Italy. 1986.

[13] FAO, Proximate composition of FOODS. In: Improving nutrition through home gardening. A training package for preparing field worker in Africa. Food and Nutrition Division, FAO. United Nations, Rome. Appendix 2. 2001.

[14] Food and Agricultural Organisation (FAO), Trade book. Vol. 23. Rome Italy. 1969.

[15] R.A. Hamzat, Utilization of Testa of kola (Cola nitida) in the feeding of African giant land snail (Archachatina marginata) in Southwestern Nigeria. PhD Thesis, University of Ibadan. 2004.

[16] E.A. Imevbore and A.A. Ademosun, The nutritive value of the African giant land snail, Archachatina marginata. J. Animal. Prod. Res. Vol. 8 (2), 76-87. 1988.

[17] E. Larmond, Laboratory methods for sensory evaluation of food. Research Branch. Canadian Department of Agriculture 1637 pp. 50-59. 1977.

[18] O.G. Longe and Fagbenro-Bryon, Composition and physical characters of some fibrous waste and by-products for pig feed in Nigeria. Biets, Trop. Landwirthsch Vet. Med. Vol. 28 pp. 199-205. 1990.

[19] A.J. Omole, How to start and manage snail farming. A paper presented at the workshop organized by Petroleum Staff Training Programme for Retiree at Petroleum Training Institute, Effunrun-Warri, Delta State on March 13-15th 2001.

[20] S.A.S, Version 8 Edition, Statistical Analysis System Institute Inc. Carry, N.C. USA. 2003.

[21] USDA, USDA Nutrient Database for standard Reference, Release 12. 1998.

4

Photorhabdus Luminescens: Virulent Properties and Agricultural Applications

Elizabeth Gerdes[1], Devang Upadhyay[1, *], Sivanadane Mandjiny[2], Rebecca Bullard-Dillard[3], Meredith Storms[4], Michael Menefee[5], Leonard D. Holmes[1]

[1]Sartorius Stedim Biotechnology Laboratory, Biotechnology Research and Training Center, The University of North Carolina at Pembroke, Pembroke, USA

[2]Department of Chemistry and Physics, The University of North Carolina at Pembroke, Pembroke, USA

[3]School of Graduate Studies and Research, The University of North Carolina at Pembroke, Pembroke, USA

[4]College of Arts & Sciences, The University of North Carolina at Pembroke, Pembroke, USA

[5]Thomas Family Center for Entrepreneurship, The University of North Carolina at Pembroke, Pembroke, USA

Email address:

danny.uncp@gmail.com (D. Upadhyay)

Abstract: *Photorhabdus luminescens* is a gram-negative, bioluminescent bacterium from the family *Enterobacteriaceae* which has been found in countries across the globe. It is part of a symbiotic relationship with, and resides in the gut of the entomopathogenic nematode, *Heterorhabditis bacteriophora*. *Photorhabdus luminescens* produces many virulence factors, toxin complexes and antimicrobial compounds which allow it to kill insect hosts while simultaneously protecting itself and its symbiotic partner from other bacteria. Due to its virulent properties and the ability to infect a wide range of insect hosts, the *Photorhabdus luminescens-Heterorhabditis bacteriophora* relationship is a promising candidate for agricultural use as a mass produced biological control agent. The use of Photorhabdus luminescens has been deemed safe towards humans, animals, non-target insects, plants, as well as the environment. The entomopathogenic nematode is also exempt from registration regulations in most countries. The significance of *Photorhabdus luminescens* is the potential for an insecticide that allows different species of insect pests to be effectively controlled by a single natural product rather than multiple chemical products. The purpose of this review is to provide readers with an overview of the safety of *Photorhabdus luminescens* to the community and environment, to inform readers of the virulence factors associated with the bacteria, and to outline the potential the product possesses as a mass produced biological control agent.

Keywords: *Photorhabdus Luminescens*, *Heterorhabditis Bacteriophora*, Symbiosis, Bioluminescence, Biocontrol Agent

1. Introduction

Photorhabdus luminescens is an insect pathogenic, gram-negative [Figure 1], bioluminescent bacterium [Figure 2] formerly known as *Xenorhabdus luminescens* belonging to the family *Enterobacteriaceae*. The carcass of the insect host becomes bioluminescent as the bacterial infection progresses [1]. The biological purpose of bioluminescence is not yet clear, however, it has been speculated to be an indicator of insect virulence [1,2]. *Photorhabdus luminescens* lives in the gut of the entomopathogenic nematode (EPN) *Heterorhabditis bacteriophora* by means of a specialized and free-living form known as the infective juvenile (IJ) [3]. This particular combination is toxic to a wide range of insects such as, *Pyralidae*, *Sphingidae* and *Aleyrodidae* families, thus making the *P. luminescens* - *H. bacteriophora* pairing a candidate for agricultural applications [4].

P. luminescens is 5 microns in length and 1 micron in width when viewed with an electron microscope [5,6]. Important properties of the *P. luminescens* bacterium are its ability to: (1) produce and secrete high-molecular-weight toxin complexes that kill the host quickly (within 24 h to 48 h) [21]; (2) synthesize enzymes to convert the insect body into nutrients; (3) antibiotic production to inhibit competing bacteria; (4) pigmentation and (5) bioluminescence [2,6,7,8,9]. The bacterium lives a bi-phasic lifecycle: a nematode-symbiotic phase where the bacteria enter and reproduce in the nematode gut, and an insect-pathogenic phase in which the bacteria release different toxins in order to kill insect host [6,10]. The nematodes itself converts through 4 stages within the insect which will be discussed more thoroughly below [6].

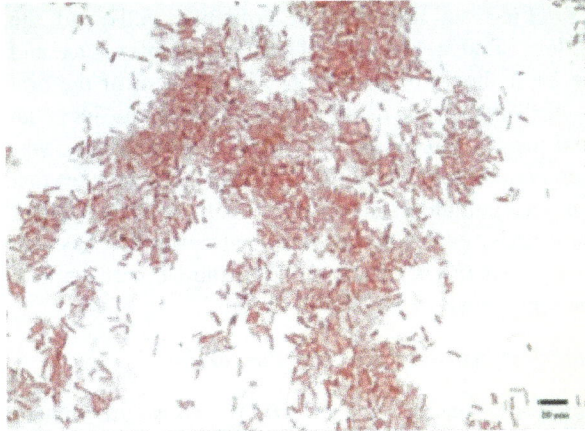

Figure 1. Gram stain of Photorhabdus luminescens. From Inman, F. L. et al. [5].

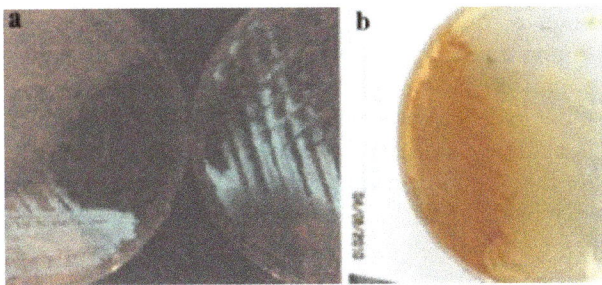

Figure 2. a: Phase I bioluminescence. b: Pigmentation on rich media. From Singh, S. et al. [2].

2. Bacterial Relationship with *Heterorhabditis Bacteriophora*

The symbiotic or mutualistic relationship between *Photorhabdus luminescens* and *Heterorhabditis bacteriophora* is required for nematode viability [11]; however, the relationship is not obligate for both partners as *P. luminescens* can be cultured anoxically. In nature, *P. luminescens* releases toxins into the open bloodstream of the insect host causing the death of the insect. Insect tissue is converted by *P. luminescens* enzymes for food utilization for bacteria and the nematodes. The nematodes reproduce and move onto the next insect host. In exchange, *H. bacteriophora* allows the bacteria to enter and live in its gut until it enters the insect host by means of the insect cuticle or gut with the buccal tooth [12]. The relationship is highly specific in that *Heterorhabditis bacteriophora* can only pair with *Photorhabdus luminescens* [6].

3. Symbiotic Life Cycle

The life cycle of *P. luminescens* and *H. bacteriophora* is

divided into four stages [Figure 3, Table 1]. Stage I, the bacteria is located in the midgut of the nematode while it searches for and enters an insect host [12]. In Stage II, the bacteria are released into the hemolymph of the insect and secrete toxins resulting in the death of the insect host [13]. It has been shown that as few as one to ten cells of *P. luminescens* can be fatal to insect hosts [12]. In Stage III, the nematodes reproduce within the insect while the bacteria produce antibiotics, exoenzymes, and crystal proteins to repel competing bacteria and convert the insect corpse into nutrients. In this stage, the nematode's ability to reproduce is dependent on the presence of the *P. luminescens*; therefore the relationship is both obligatory and specific. The entirety of stage II and III is complete within approximately 48 hours after initial infection of the insect host [6]. Stage IV is the final stage where the development of new nematodes occurs as well as the colonization of the nematode intestines by the bacteria [6].

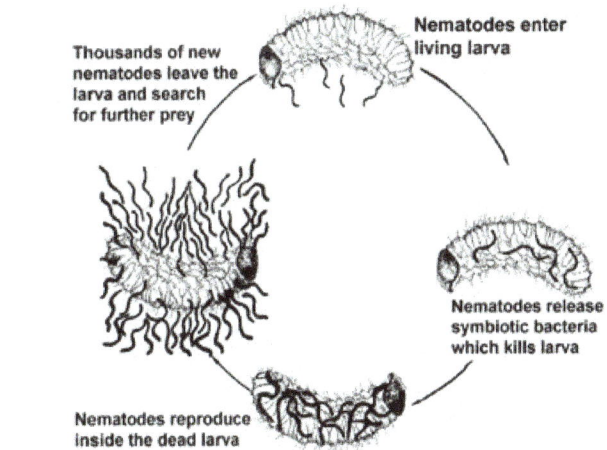

Figure 3. The life cycle of Photorhabdus luminescens. From Ehlers, R. [3].

Table 1. Stages of the symbiotic relationship between Photorhabdus luminescens and Heterorhabditis bacteriophora. From Frost, S. and Clarke, D. [12].

Stage	Bacteria Life Cycle	Nematode Life Cycle
I	Bacteria retained in nematode gut	Infective juvenile in the soil Search for insect host Infective juvenile enters into insect haemocoel
II	Bacteria released into haemolymph Production of virulence factors Death of insect	Recovery in the haemocoel
III	Bacteria in stationary phase Production of antibiotics, exoenzymes, crystal protein Bioconversion of insect	Nematode reproduction
IV	Colonization of the intestine of infective juveniles	Development of new infective juveniles

4. Primary and Secondary Cells

In *Photorhabdus luminescens*, there is a phenomenon known as phase variation in which secondary cells are formed [3]. Primary cells (Phase I cells) are carried by Infective Juvenile (IJ) stages of nematodes and possess numerous phenotypic traits. The secondary cells (Phase II cells) are the variant cells in which the traits of the primary cells are present at low levels or not present at all [6,14]. Traits lost or altered in the secondary cells include the loss of dye-binding ability, less pigmentation, fewer antibiotics produced and almost undetectable bioluminescence. Regardless of the loss of these traits in Phase II cells, it is interesting to note that the Phase II are virulent; but to a lesser extent than the Phase I cells [15]. Secondary cells also produce a protease inhibitor which accounts for the lack of protease activity compared to primary cells [6]. The presence of protease is important because it is widely believed that protease is actively involved in the breakdown of proteins in insects which would allow the bacteria and its symbiotic nematode to be able to reproduce inside the insect before moving on to the next [6].

5. Virulence Factors

The toxins produced by *Photorhabdus luminescens* are classified into four different groups: (1) the "makes caterpillars floppy" (mcf) toxins [10,16]; (2) the toxin complexes (Tcs) [17]; (3) the *Photorhabdus* insect related (Pir) proteins and (4) the *Photorhabdus* virulence cassettes (PVC) [25].

5.1. The Mcf Toxin

The mcf toxins cause cells to adopt a phenotype similar to those undergoing apoptosis [Figure 4]. Cells infected with the mcf toxins expressed changes in the morphology within 6 hours post infection. Infected cells disintegrate due to multiple cellular disruptions in the cell membrane [10]. Further studies show when injected into the insect midgut, the mcf toxins produce signs of severe infection in as little as 12 hours. The midgut of the insect is a primary organ tasked with osmoregulation and the destruction of this system would cause the floppiness associated with the mcf toxins [18].

Figure 4. *M. sexta 24 h after injection of mcf toxin. Caterpillar on left has lost body turgor. From Daborn, P. J. et al. [10].*

5.2. Toxin Complexes

Photorhabdus luminescens produces toxin complexes (Tc). These are further categorized into toxin complexes *a* through

d (*tca*, *tcb*, *tcc*, *tcd*). These toxins are high molecular-weight molecules with oral insecticidal abilities. It has been suggested that the Tc toxins are fusion proteins for several different functions [19]. The toxin complex *tca* is active after oral delivery or injection and the median lethal dose (LD) is 875 ng per square centimeter [20]. The ingestion of *tca* causes swelling and blebbing of columnar cells and the eventual loss of the cell nuclei [21]. Toxin complexes *tca* and *tcd* have been shown to be responsible for most of the oral toxicity in *Manduca sexta* caterpillars. When the loci that form both the *tca* and *tcd* complexes were deleted there was no longer any oral toxicity [20,22,23]. This demonstrates that the *tca* and *tcd* complexes are responsible for a large majority of the oral toxicity to *Manduca sexta* caterpillars. It has also been shown that *tca* is rapidly toxic to the Colorado potato beetle and the sweet potato whitefly [24].

5.3. Photorhabdus Insect Related (Pir) Toxins

Photorhabdus insect related (*Pir*) toxins are binary proteins encoded by *PirA* and *PirB* [18,25]. There is little genetic similarity between *PirA* and any other known substance, however, *PirB* is highly similar to the genetic composition of the *Cry2A* insecticidal toxin suggesting a possible conserved structure and/or function for these different proteins [25]. Significantly these proteins express traits similar to the δ-endotoxins of *Bacillus thuringiensis* (Bt) which is currently used as a crop pesticide. The similarity allows for the possibility of using *Photorhabdus luminescens* and its symbiotic nematode, *Heterorhabditis bacteriophora* as an alternative to Bt for crop pesticides [25].

5.4. Photorhabdus Virulence Cassettes (PVC) and Lipases

Lipase activity has been shown to be present mainly in *Photorhabdus luminescens* primary cells, but secondary cells produce low levels of lipase activity as well. The *lip-1* lipase gene in *P. luminescens* has been cloned from secondary cells and shown to be secreted in an inactivated form displaying entomotoxic ability towards the greater wax moth, *Galleria mellonella* [6]. *Photorhabdus* virulence cassettes (PVCs) are phage-like loci responsible for an "anti-feeding" effect in insects. *P. luminescens* strain TT01 contains many copies of the prophage-like loci that are each tasked with encoding different proteins. PVCs of *Photorhabdus luminescens* have no proven antibacterial abilities. However, the injection of PVCs destroys the hemocytes of the insect which causes cytoskeleton damage. It has been suggested that PVCs may be important in the formation of virulent properties against multiple types of insects [18].

6. Antimicrobial Compounds

In order to allow for the reproduction of the bacteria and its symbiotic nematode, *P. luminescens* produces antimicrobial compounds which prevent other bacteria, unrelated as well as related, from colonizing the insect host carcass while nematode reproduction is taking place [8,26].

One of the main antimicrobial compounds found to be produced and secreted by *Photorhabdus luminescens* are carbapenems, members of the *β-lactam* class of antibiotics. Carbapenems possess a broad potency to both gram-negative and gram-positive bacteria [27]. It has been suggested these carbapenems are produced by a cluster of 8 different genes referred to as *cpmA* to *cpmH*. It is clear that there are more antimicrobial compounds being produced by *P. luminescens* due to the observation of inhibition zones in tests involving *cpm* mutations. Also, inhibition zone studies have shown that these carbapenem-like molecules are not effective against gram-positive bacteria [Figure 5] [27]. A known antibiotic, 3,5-dihydroxy-4-isopropylstilbene [Figure 6], has been isolated from a strain of *P. luminescens* and it was shown that this compound was produced rapidly during the first day of culture and then declined until it was hardly detectable in the sample [28]. This test proved that 3,5-dihydroxy-4-isopropylstilbene possesses a strong antifungal potential for agricultural and medical applications. It is also fairly easy to manufacture, making it potential supplement for current pesticides and insecticides employed in agricultural [26].

Indicator strain	TT01	PL2101	Indicator strain	TT01	PL2101
Escherichia coli			Enterobacter cloacae T39180297		
Enterobacter cloacae 39177636			Pseudomonas aeruginosa		
Klebsiella pneumoniae			Xanthomonas maltophilia		

Figure 5. *Activity of P. luminescens shown by antibiosis.*

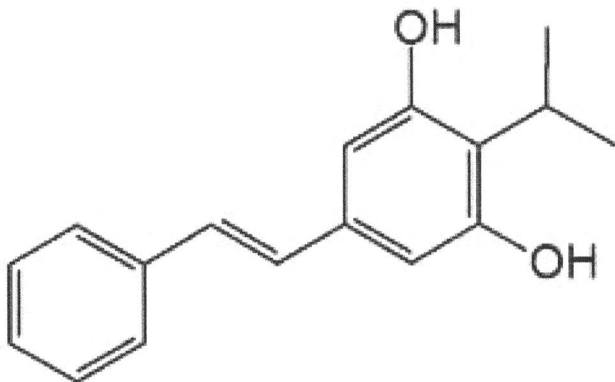

Figure 6. *Structure of 3,5-dihydroxy-4-isopropylstilbene. From Li, J. et al. [28].*

After 3 days, LB plates were spotted with *Photorhabdus luminescens* strain TT01 and *Photorhabdus luminescens* strain PL2101 were inoculated with various indicator strain cultures in soft agar. Growth inhibition around a spot shows production of antibiotics that the indicator strain is sensitive to the antibiotics produced. This figure shows that the inhibition zones of the TT01 strain were larger than those of the PL2101 strain under similar conditions. The zones seen in the PL2101 strain are likely due to other antibiotics produced by the *Photorhabdus luminescens* bacteria. From Derzelle, S. et al. [27].

7. Agricultural Applications

The fear that insect pests may develop resistance to the widely used pesticide, *Bacillus thuringiensis* (Bt), has encouraged scientists to explore alternative pesticides and insecticides. This exploration has brought them to the *Photorhabdus luminescens* bacteria and their symbiotic nematode, *Heterorhabditis bacteriophora* as a potential alternative for insects that have developed a resistance to Bt. This is a highly probable solution to the problem of insect resistance because, other than the few documented cases of opportunistic cultivation of an open wound, the use of this bacteria-nematode relationship as a pesticide has shown no adverse consequences to humans, wildlife, or plants and therefore, is a highly attractive option [29,30]. Another approach that some scientists have experimented with is plant mutations that would allow plants to be able to defend themselves against pests. But, it was determined that it does not provide the same levels of protection that the use of direct methods, such as Bt [31].

There are many positive qualities of the *Photorhabdus luminescens-Heterorhabditis bacteriophora* relationship as an insecticide which include: 1) its wide range of insect hosts; 2) safety to humans, non-target insects, animals, plants, and the environment and 3) its exemption to registration in many countries [3].

7.1. The Wide Range of Insect Hosts

The *Photorhabdus luminescens-Heterorhabditis bacteriophora* relationship has virulence towards a large number of different insects when tested in the laboratory. Due to the wide variety of susceptible insects, this bacteria-nematode relationship has developed the ability for versatile penetration routes that can be taken depending on the insect's defenses which could be a high defecation rate, low carbon dioxide output to minimize attention, cocoons, the quarantine of infected insects, and other grooming or evasive behaviors [32].

7.2. Safety to Humans, Non-Target Insects, Animals, Plants, and Environment

The *H. bacteriophora* nematodes infected with *P. luminescens* can be used as an insecticide to prevent insect populations of certain pests from growing too large. *P. luminescens - H. bacteriophora* provide a safe alternative for Bt pesticides. The symbionts show no negative side effects towards the environment, which means that use will not harm the soil or the water supply. Traditional chemical inputs can seep into soil, thus contaminating ground water. The symbiont will not harm wildlife because this combination resides in the soil where most wildlife will not come into contact with them [3].

7.3. Exemption from Registration in Most Countries

Nematodes with their bacteria symbionts have been found in soils across the world. The *P. luminescens - H.*

bacteriophora symbiotic pair has been recovered from soils in countries with continental and Mediterranean climates and has been isolated from soil in Turkey [32]. Most European countries, Australia, and the United States do not require the registration of indigenous nematodes while other countries require similar procedures to that of chemical pesticides [3,32].

8. Application Challenges

On the other hand, the negative aspects include (1) the inability to tolerate environmental changes and (2) the potential problems related to mass production and transportation.

8.1. Tolerance to Environmental Changes

Having a higher tolerance for heat would allow this bacteria-nematode combination to be stored for longer periods of time and in a wider variety of locations than is currently possible [33]. The ability for lower temperature tolerance would allow the nematodes and their symbiotic bacteria to be used in colder climates. It has been shown that modified strains of the *Heterorhabditis bacteriophora* can successfully survive at temperatures as high as 39 degrees Celsius and as low as 6 degrees Celsius [33], However, *Photorhabdus luminescens* cannot be stored at such varied temperatures. *Photorhabdus luminescens* can be stored in the gut of their symbiont nematodes in refrigerated tanks with an oxygen supply for long periods of time (between 3 and 6 months) or they can be used immediately. This storage method requires refrigeration to protect quality and that is where scientists and businesses run into the problem of mass storage. Scientists have begun researching ways to alter the bacteria in order to allow for a higher heat tolerance as well as a lower cold tolerance for easier transportation and storage [33].

8.2. Challenges with Mass Production and Transportation

The most cost-effective means of mass production of this *Photorhabdus luminescens*-*Heterorhabditis bacteriophora* combination is in liquid media using bioreactors because it offers the ability to produce a high yield and the media is relatively inexpensive when compared to solid media components. However, the initial costs of liquid fermentation are exponentially higher due to the advanced equipment required for the process. Many developing countries use *in-vivo* methods because it is reliable and it produces high quality nematodes with Phase I bacteria. The *in-vivo* method is problematic for mass production, because it is costly and labor intensive [32]. Larger companies most often use liquid fermentation because it is the most profitable since the profit increases as the production costs decrease [34]. The down side is that this method is highly susceptible to contamination and the bacteria must be kept in their Phase I during production and transportation in order to remain virulent [3].

9. Conclusion

Discoveries to date have not yet produced a comprehensive list of the agricultural uses and benefits of the *Photorhabdus luminescens* and *Heterorhabditis bacteriophora* relationship but, the relationship does have a promising future in the agricultural industry that has yet to be fully defined. The use of this bacteria-nematode combination as a biological control agent offers a product that is safe, effective, cost effective, and easy to apply to crops. This will save time and money without allowing potentially harmful chemicals to be introduced into the environment and become potential hazards to animals and humans. There are many promising abilities held by this bacteria-nematode combination which make it a highly viable alternative: (1) The wide range of insect hosts; (2) being exempt from regulation in many countries and (3) the unique virulence factors associated with the Photorhabdus luminescens bacterium. There are still a few unresolved negative issues such as the bacteria's intolerance to extreme temperatures, the issues involved in storage, and the high cost of mass production. Finally, more field-testing must continue to promote wider acceptance of this bio-control product. Once these issues are addressed and resolved, the viability of the bacteria-nematode relationship as biological control agents will be widely recognized.

Acknowledgments

Financial support was provided, in part, by the: Farm Bureau of Robeson County, North Carolina; University of North Carolina at Pembroke (UNCP) Office of the Provost and Academic Affairs; UNCP Department of Chemistry and Physics; North Carolina Biotechnology Center (NCBC) and UNCP Thomas Family Center.

References

[1] Daborn, P., Waterfield, N., Blight, M. A. and Ffrench-Constant, R. H. (2001) Measuring Virulence Factor Expression by the Pathogenic Bacterium *Photorhabdus luminescens* in Culture and during Insect Infection. Journal of Bacteriology, 183(20): 5834-5839.

[2] Singh, S., Moreau, E., Inman, F. and Holmes, L. D. (2011) Characterization of *Photorhabdus luminescens* Growth for the Rearing of the Bacterial Nematode *Heterorhabditis bacteriophora*. Indian Journal of Microbial, 52(3): 325-331.

[3] Ehlers, R. (2001) Mass production of entomopathogenic nematodes from plant protection. Applied Microbiology and Biotechnology, 56: 623-633.

[4] Kooliyottil. R, Upadhyay, D., Inman, III F., Mandjiny, S. and Holmes, L.D. (2013) A Comparative Analysis of Entomoparasitic Nematodes *Heterorhabditis bacteriophora* and *Steinernema carpocapsae*. Open Journal of Animal Science, 3(4): 326-333.

[5] Inman, F. L. III, Singh, S. and Holmes, L. D. (2012) Mass Production of the Beneficial Nematode *Heterorhabditis bacteriophora* and Its Bacterial Symbiont *Photorhabdus luminescens*. Indian Journal of Microbial, 52(3): 316-324.

[6] Frost, S. and Clarke, D. (2002) Bacteria-Nematode Symbiosis. In Gaugler R. (Ed.) Entomopathogenic Nematology (pp 57-77).

[7] Marokhazi, J., Waterfield, N., LeGoff, G., Feil, E., Stabler, R., Hinds, J., Fodor, A. and Ffrench-Constant, R. H. (2003) Using a DNA Microarray To Investigate the Distribution of Insect Virulence Factors in Strains of *Photorhabdus* Bacteria. Journal of Microbiology, 185(15): 4648-4656.

[8] Ffrench-Constant, R. H., Waterfield, N., Burland, V., Perna, N. T., Daborn, P. J., Bowen, D. and Blattner, F. R. (2000) A Genomic Sample Sequence of the Entomopathogenic Bacterium *Photorhabdus luminescens* W14: Potential Implications for Virulence. Applied and Environmental Microbiology, 66(8): 3310–3329.

[9] Schmidt, T. M., Kopecky, K. and Nealsoni, K. H. (1989) Bioluminescence of the Insect Pathogen *Xenorhabdus luminescens*. Applied and Environmental Microbiology, 55(10): 2607-2612.

[10] Daborn, P. J., Waterfield, N., Silva, C. P., Au, C. P. Y., Sharma, S. and Ffrench-Constant, R.H. (2002) A single Photorhabdus gene, makes caterpillars floppy (mcf), allows *Escherichia coli* to persist within and kill insects. Proceedings of the National Academy of Sciences, 99 (16): 10742-10747.

[11] Patterson, W., Upadhyay, D., Mandjiny, S., Bullard-Dillard, R., Storms, M., Menefee, M. and Holmes, L. D. (2015) Attractant Role of Bacterial Bioluminescence of *Photorhabdus luminescens* on a *Galleria mellonella* Model. American Journal of Life Sciences, 3(4): 290-294.

[12] Ciche, T. and Ensign, J. C. (2002) For the Insect Pathogen Photorhabdus luminescens, Which End of a Nematode Is Out? Applied and Environmental Microbiology, 69(4): 1890-1897.

[13] Guo, L., Fatig III, R. O., Orr, G. L., Schafer, B. W., Strickland, J. A., Sukhapinda, K., Woodsworth, A. T. and Petell, J. K. (1999) *Photorhabdus luminescens* W-14 Insecticidal Activity Consists of at Least Two Similar but Distinct Proteins. The Journal of Biological Chemistry, 274(14): 9836-9842.

[14] Bowen, D. J. and Ensign, J. C. (1998) Purification and Characterization of a High-Molecular-Weight Insecticidal Protein Complex Produced by the Entomopathogenic Bacterium *Photorhabdus luminescens*. Applied and Environmental Microbiology, 64(8): 3029–3035.

[15] Dowds, B. C.A. and Peters, A. (2002) Virulence Mechanisms. In Gaugler R. (Ed.) Entomopathogenic Nematology (pp 79-98).

[16] Derzelle, S., Turlin, E., Duchaud, E., Pages, S., Kunst, F., Givaudan, A. and Danchin, A. (2003) The PhoP-PhoQ Two-Component Regulatory System of *Photorhabdus luminescens* Is Essential for Virulence in Insects. Journal of Microbiology, 186(5): 1270-1279.

[17] Brillard, J., Duchaud, E., Boemare, N., Kunst, F. and Givaudan, A. (2002) The PhlA Hemolysin from the Entomopathogenic Bacterium *Photorhabdus luminescens* Belongs to the Two-Partner Secretion Family of Hemolysins. Journal of Bacteriology, 184(14): 3871-3878.

[18] Ffrench-Constant, R. H., Dowling, A. and Waterfield, N. R. (2007) Insecticidal toxins from *Photorhabdus* bacteria and their potential use in agriculture. Toxicon, 49: 436-451.

[19] Waterfield, N., Bowen, D. J., Fetherston, J. D., Perry, R. D. and Ffrench-Constant, R. H. (2001) The tc genes of *Photorhabdus*: a growing family. Trends in Microbiology, 9(4): 185-191.

[20] Bowen, D., Rocheleau, T. A., Blackburn, M., Andreev, O., Golubeva, E., Bhartia, R. and Ffrench-Constant, R. H. (1998) Insecticidal Toxins from the Bacterium *Photorhabdus luminescens*. Science Magazine, 280: Pages 2129-2132.

[21] Blackburn, M., Golubeva, E., Bowen, D. and Ffrench-Constant, R. (1998) A Novel Insecticidal Toxin from *Photorhabdus luminescens*, Toxin Complex a (Tca), and Its Histopathological Effects on the Midgut of Manduca sexta. Applied and Environmental Microbiology, 64(8): 3036-3041.

[22] Ffrench-Constant, R. H. and Bowen, D. J. (2000) Novel insecticidal toxins from nematode-symbiotic bacteria. Cellular and Molecular Life Sciences, 57: 828-833.

[23] Pinheiro, V. and Ellar, D. J. (2007) Expression and insecticidal activity of *Yersinia pseudotuberculosis* and *Photorhabdus luminescens* toxin complex proteins. Cellular Microbiology, 9: 2372-2380.

[24] Blackburn, M. B., Domek, J. M., Gelman, D. B. and Hu, J. S. (2005) The broadly insecticidal *Photorhabdus luminescens* toxin complex a (Tca): Activity against the Colorado potato beetle, *Leptinotarsa decemlineata*, and sweet potato whitefly, *Bemisia tabaci*. Journal of Insect Science, 5(32): 1-11.

[25] Rodou, A., Ankrah, D. O. and Stathopoulos, C. (2010) Toxins and Secretion Systems of *Photorhabdus luminescens*. Toxins, 2: 1250-1264.

[26] Bondi, M., Messi, P., Sabia, C., Baccarani C. M. and Manicardi, G. (1998) Antimicrobial Properties and Morphological Characteristics Of Two *Photorhabdus luminescens* strains. Microbiologica, 22: 117-127.

[27] Derzelle, S., Duchaud, E., Kunst, F., Danchin, A. and Bertin, P. (2002) Identification, Characterization, and Regulation of a Cluster of Genes Involved in Carbapenem Biosynthesis in *Photorhabdus luminescens*. Applied and Environmental Microbiology, 68(8): 3780–3789.

[28] Li, J., Chen, G., Wu, H. and Webster, J. M. (1995) Identification of Two Pigments and a Hydroxystilbene Antibiotic from *Photorhabdus luminescens*. Applied and Environmental Microbiology, 61(12): 4329–4333.

[29] Peel, M. M., Alfredson, D. A., Gerrard, J. G., Davis, J. M., Robson, J. M., McDougall, R. J., Scullie, B. L. and Akhurst, R. J. (1999) Isolation, Identification, and Molecular Characterization of Strains of *Photorhabdus luminescens* from Infected Humans in Australia. Journal of Clinical Microbiology, 37(11): 3647–3653.

[30] Morgan, J.A.W., Kuntzelmann, V., Tavernor, S., Ousley, M.A. and Winstanley, C. (1997) Survival of *Xenorhabdus nematophilus* and *Photorhabdus luminescens* in water and soil. Journal of Applied Microbiology, 83: 665-670.

[31] Ferry, N., Edwards, M. G., Gatehouse, J. A. and Gatehouse, A. M.R. (2004) Plant-insect interactions: molecular approaches to insect resistance. Current Opinion in Biotechnology, 15: 155-161.

[32] Hazir, S., Kaya, H. K., Stock, S. P. and Keckun, N. (2003) Entomopathogenic Nematodes (*Steinernematidae* and *Heterorhabditidae*) for Biological Control of Soil Pests. Turkish Journal of Biology, 27:181-202.

[33] Ehlers, R., Oestergaard, J., Hollmer, S., Wingen, M. and Strauch, O. (2005) Genetic selection for heat tolerance and low temperature activity of the entomopathogenic nematode-bacterium complex *Heterorhabditis bacteriophora-Photorhabdus luminescens*. BioControl, 50: 699-716.

[34] Upadhyay, D., Kooliyottil, R., Mandjiny, S., Inman, III F. and Holmes, L. (2013) Mass production of the beneficial nematode *Steinernema carpocapsae* utilizing a fed-batch culturing process. EScience Journal of Plant Pathology, 02(01): 52-58.

[35] Mahajan-Miklos, S., Rahme, L. G. and Ausubel, F. M. (2000) Elucidating the molecular mechanisms of bacterial virulence using non-mammalian hosts. Molecular Microbiology, 37: 981-988.

[36] Eleftherianos, I., Marokhazi, J., Millichap, P.J., Hodgkinson, A.J., Sriboonlert, A., Ffrench-Constant, R.H. and Reynolds, S.E. (2006) Prior infection of Manduca sexta with non-pathogenic *Escherichia coli* elicits immunity to pathogenic *Photorhabdus luminescens*: Roles of immune-related proteins shown by RNA interference. Insect Biochemistry and Molecular Biology, 36: 517-525.

5

Ecology of Basal Stem Rot Disease of Oil Palm (*Elaeis guineensis* Jacq.) in Cameroon

Afui Mathias Mih[1, *], **Tonjock Rosemary Kinge**[2]

[1]Department of Botany and Plant Physiology, Faculty of Science, University of Buea, South West Region, Cameroon
[2]Department of Biological Sciences, Faculty of Science, University of Bamenda, North West Region, Cameroon

Email address:
afuimih@yahoo.com (A. M. Mih), rosemary32us@yahoo.com (T. R. Kinge)

Abstract: Basal stem rot (BSR) disease caused by species of *Ganoderma* is of immense importance in oil palm production. Although much is known on the occurrence of this devastating disease, fundamental studies on the ecology in oil palm in plantations are rather limited. This study sought to determine the incidence, severity, distribution and spread pattern of BSR disease in oil palm plantations and relate disease parameters to climatic and edaphic factors. Surveys were carried out for two years on two–hectare plots in each of five oil palm estates of the Cameroon Development Corporation. Data for disease incidence and severity in each estate were recorded. Disease spread patterns were generated from Arc GIS version 9.3 using GIS coordinates of diseased plants. A correlation between disease parameters and soil physicochemical properties and multivariate analyses were done. Typical BSR disease symptoms were observed including unopened spear leaves, skirt–like appearance of leaves, basidiocarp formation, bole creation and death of the palm. The disease incidence ranged from 5.4% in 16-year old palms at Bota to 39.0% in palms of the same age in Mungo were about 50% of infected plants had extreme severe symptoms. Although principal component analysis showed that six soil properties account for variation in BSR disease incidence and severity, only fine sand content was positively correlated (P≤0.05) with disease incidence and severity, while C/N ratio was negatively correlated. This study has established the occurrence and spread of basal stem rot disease in five oil palm plantations in South Western Cameroon.

Keywords: Epidemiology, Incidence, Severity, Spread Pattern, *Ganoderma*

1. Introduction

Oil palm, *Elaeis guineensis* Jacq is an important oil crop in Cameroon where over 230,000 tons of crude palm oil is produced in about 190,000ha [1]. In the past, production was based on the Dura type palm whose production was low, but with the development of plantation agriculture, the more productive Tenera hybrid is widely cultivated. This hybrid produces up to eight times the amount of oil produced by other vegetable oil seeds like soybean and sunflower [2]. The inflorescence is also tapped for the much cherished palm wine.

Generally, oil palm suffers from relatively few important diseases in each of the different environments where it has been planted commercially. In Southeast Asia, basal stem rot (BSR) disease caused by species of *Ganoderma* is the most important disease of oil palm [1]. Although in several African countries, vascular wilt caused by *Fusarium oxysporum* f.s. *elaeidis* was thought to be the only disease causing serious problems in some plantations, BSR disease has become one of the major diseases of oil palm, especially in Cameroon [4]. The disease was first recorded in Malaysia where it was initially considered a disease of older palms as it occurred in palms of over 25 years [5]. The BSR disease has also been recorded in Malaysia, Indonesia, Nigeria, Ghana, Zaire, Angola, Tanzania, North Mozambique, Papua New Guinea and Cameroon [5,6]. The causal agent of basal stem rot disease in Malaysia was first identified as *G. lucidum* (W. Curt.) Karst [5]. At least seven species of *Ganoderma* have been associated with BSR of oil palm in Malaysia, Indonesia, Papua New Guinea and Cameroon including *G. boninense* Pat., *G. miniatocinctum* Steyaert, *G. chalceum* (Cooke) Steyaert, *G. tornatum* (Pers.) Bers., *G. zonatum* Murill, *G. xylonoides* Steyaert, *G. ryvardense* Tonjock and Mih and *G. lobenense* [7, 5, 6, 8, 9]. BSR disease has been found to infect oil palms as young as 1 to 2 years of age, and is serious on palms aged 4 – 5 years of age, particularly in replanted areas [10]. In new oil palm planted from jungle or old rubber plantations, BSR incidence of 25% has been recorded after 25 years while in that planted from old coconut plantations, an incidence of 60% occurred after 16 years [11], whereas oil palm to oil palm under planting has resulted in 33% infection after 15 years. The highest disease incidence is in coastal areas [5, 12]. In Malaysian coastal areas, a 50% loss of yield was recorded from 80% disease incidence on 13 year old plantings [13]. A survey has also reported typical levels of disease incidence of 30% on 13 – year old palms in both inland and peat soils [14]. In North Sumatra (Indonesia), by the time of replanting (25 years) 40 – 50% of palms are lost in some fields with the majority of standing palms showing disease symptoms. Where oil palm stumps were left in the ground at replanting then more serious palm losses due to *Ganoderma* have been observed in some fields with up to 25% incidence that occurred within 7 years [15].

The natural infection with *Ganoderma* in oil palm occurs as a result of contact between healthy roots and diseased tissues left buried in the soil [16]. Subsequent spread occurs by root to root contact once a few palms are infected [17].

In young palms, the external symptoms of basal stem rot normally comprise a one sided yellowing, or mottling of the lower fronds, followed by necrosis [10]. The newly unfolded leaves are shorter than normal and chlorotic, and additionally the tips maybe necrotic. As the disease progresses within the plant, the diseased palm may take on

an overall pale appearance, with retarded growth and the spear leaves remain unopened [10]. Affected leaves die, necrosis sets in, beginning with the oldest leaves and extending progressively upwards through the crown. Dead dessicated fronds droop at the point of attachment to the trunk or fracture at some point along the rachis and hang down to form a skirt of dead leaves.

Although there have been sporadic reports of basal stem rot in Cameroon, there is no comprehensive study on the ecology of the disease that could guide its management. The objectives of this work was therefore to determine the incidence, severity, distribution and spread pattern of basal stem rot disease in south western Cameroon and see how the disease relates to climatic and edaphic factors.

2. Materials and Methods

2.1. Establishment of Sampling Plots

The ecology of basal stem rot was studied in five oil palm estates belonging to the Cameroon Development Corporation (CDC) in south western Cameroon (Fig. 1). The means of various environmental parameters are shown on Table 1.

In each estate, four 2ha plots were mapped out, each of the same age as shown on Table 2. Location of plot was done through a stratified random sampling technique.

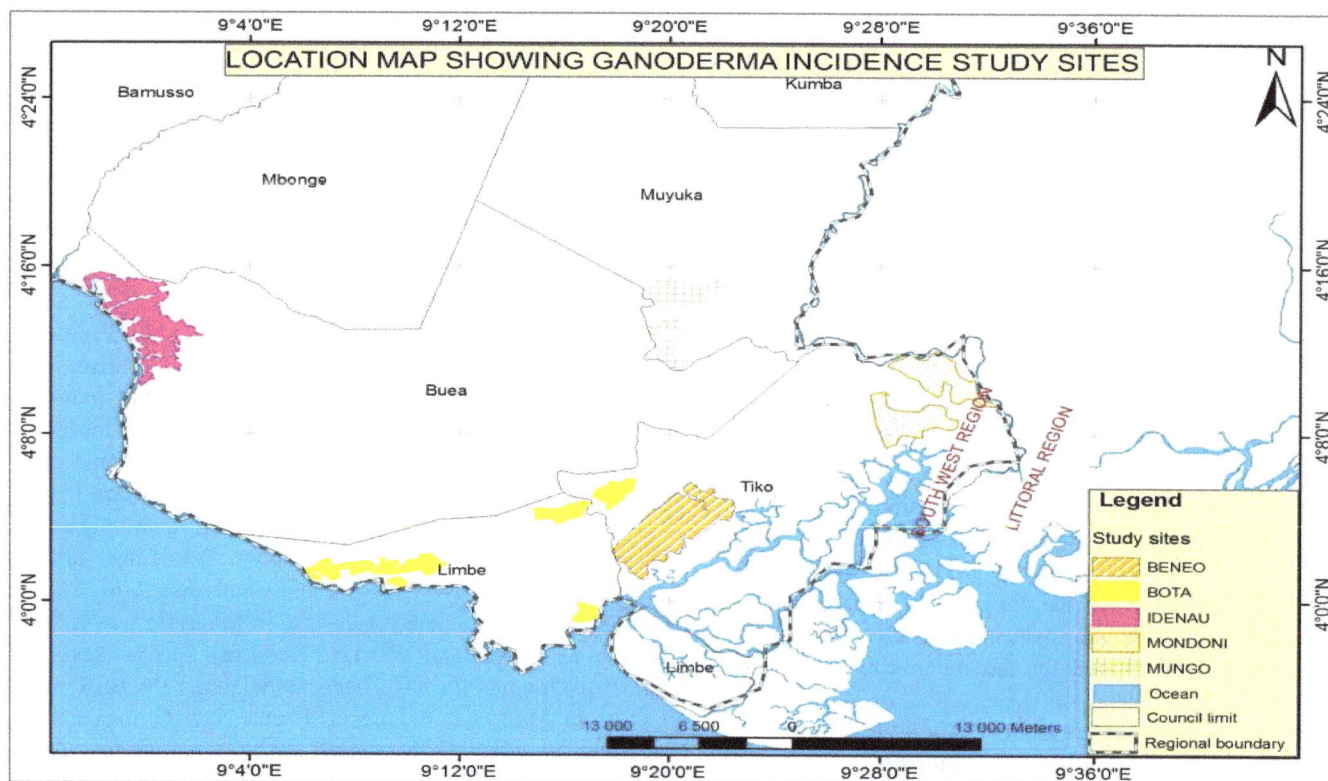

Fig. 1. Survey sites of basal stem ro disease of oil palm in south western Cameroon.

Table 1. Mean environmental parameters for five oil palm estates surveyed for basal stem rot disease in south western Cameroon.

Descriptors	Sites				
	Beneo	Bota	Idenau	Mondoni	Mungo
Location(Lat (N)	N 05° 38 28.6'	N04° 00 908'	N04° 14.949'	N04° 10.483'	N04° 14.813'
Long (E)	E04° 50 91.2'	E009° 09.659'	E009° 00.059'	E009° 25.361'	E009° 23.895'
Altitude (m)	62–72	50–72	42–80	24–39	44–60
Max.Temp (°C)	28.4	30	31	33	32
Min. Temp (°C)	22.1	20.6	22	23.3	22.9
Rainfall (mm)	2803	4927.6	6899.7	2306.1	1636.8
Rainfall days	145.7	182.8	221.5	135.8	100.4
Rel. humidity (%)	84.9	86.9	85.5	84.8	85.4
Sunshine days	309	280	280	297	261
Sunshine hours	1427	1133	1129	1334	1122
Soil	Clay loam	Volcanic	Volcanic	Sandy loam	Sandy

Table 2. Age of sampling plots for basal stem rot disease in five oil palm estates of south western Cameroon.

Estate	Year of planting	No. of plots	No. of plants
Beneo	1995	1	286
	1973	1	286
Bota	1996	2	572
Idenau	1978	2	572
Mondoni	1973	1	286
	1995	1	286
Mungo	1995	1	286
	1996	1	286

2.2. Disease Scoring

Each plant in the sampling plots was observed for symptoms of BSR during the wet (June – August) and dry (October – December) seasons of 2010 and 2011 respectively, and scored for BSR severity on a scale of 0 – 4 as shown on (Table 3) according to the method of Abdullah *et al.* [18].

Table 3. Scoring scale for basal stem rot disease of oil palm.

Severity class	Description
0	Healthy looking plants with green leaves without appearance of fungal mycelium on any part of plants.
1	Appearance of white fungal mass on any part of plants, with or without chlorotic leaves and unopened spear leaves at the centre.
2	Appearance of basidiomata on any part of plants with chlorotic leaves, skirt–like appearance of the leaves resulting in collapse of the lower leaves.
3	Formation of well-developed basidiocarp and bole creation.
4	Death of the plant and creation of bare land.

Adapted from Abdullah *et al.* [18]

The disease incidence was calculated as follows;

$$I = \frac{No.of\ plants\ with\ score\ 1-4}{No.of\ plants\ observed} x100\% \qquad (1)$$

Where I= Incidence

To assess the severity of the BSR disease, the disease severity index (DSI) was calculated using the method of Abdullah *et al.* [18] thus:

$$DSI = \frac{\sum AB}{4\sum B} x100\% \qquad (2)$$

Where:

A – Disease class (0, 1, 2, 3 or 4)

B – Number of plants showing that disease class per estate.

The Geographical Positioning System (GPS) point of each symptomatic plant was recorded. The data were processed using arc GIS version 9.3, to generate the disease spread pattern map.

2.3. Soil Sampling and Analysis

For each of the estates surveyed, five core soil samples were collected at a depth of 0 –10 cm, bulked, mixed thoroughly, air dried and sieved through a 2mm sieve. Sub samples were then analysed for physicochemical properties according to the method of Anderson and Ingram [29] at the soil science laboratory of IRAD Nkolbisson, Yaounde, Cameroon. Water holding capacity (WHC) was determined in the Life Sciences Laboratory at the University of Buea. Five replicate samples per site (Mungo, Mondoni, Beneo, Bota and Idenau) were each saturated with tap water and weighed, (w_1). The weighed samples were each dried at 70°C to a constant weight, w_2. The WHC was estimated as follows: $WHC = \frac{w_1-w_2}{w_2} x1000\ mL\ kg^{-1}$ (Assuming the density of H2O = 1 g/mL) The relationship between BSR disease and other environmental parameters was determined by multivariate analyses (correlation and principal component analysis, PCA) using MINITAB version 16.

3. Results and Discussion

Both asymptomatic and symptomatic plants of various levels of severity were observed in the field. Symptoms observed in the field are shown on Fig. 2. These ranged from presence of multiple unopened spear leaves at the centre, to skirt–like appearance of the leaves. Generally, the symptomatic palms had a pale appearance. Other symptoms observed were, production of fruiting bodies, bole creation on the base of the trunk, and finally death of the palms. The symptoms were typical of those described for the disease [19]. These are different from symptoms of vascular wilt of oil palm in that there is dryness of the lower leaves, the breaking of the rachis at about one third the length from the trunk, the hanging of the dry leaves along the trunk for the typical or acute form, the narrowing of the trunk at the top taking a "pencil – point" appearance, and the cracking of the trunk resulting from deterioration of the vessels for the chronic form [20, 4]. During the first year of observation, the incidence ranged from 4.6% in Beneo estate to 38.0% in 16 year old palms at the Mungo estate (Table 4). The incidence was generally high in young palms (≤16years) when compared to old palms (>30years) which recorded an incidence range of 14.2 – 20.7%. The second year of observation showed a general trend of increasing incidence from 6.8% in Bota to as high as 40.0% in Mungo. Considering the two years of observation, the average disease incidence was highest at Mungo in palms that are at about their peak production age. Except for Mungo, the incidence of basal stem rot was generally low when compared to

reports from other parts of the world on palms of comparable age. For example, values of 85% have been reported in 25 year old palms in Malaysia [21] and 30% in 13 years old palms in Malaysia [14]. The basal stem rot severity index was highest in Mungo where the mean for the period of observation was 80%, rated as very severe (Fig. 3). This is supported by the fact that the proportion of symptomatic plants with a score of five was highest there (Fig. 4). The high severity indices are an indication of the threat of this disease to production. Other observations on small holder fields in the Mungo area have similarly been shown to be highly infected, resulting in death of palms [1]. Basal stem rot disease spread showed a cluster pattern of spatial dispersion with only Bota showing a sparse pattern (Fig. 5). The disease had a dense cluster pattern in Mungo, Mondoni, Beneo, and Idenau estates. The spread pattern of BSR disease in this study was typical of soil–borne diseases which occurred in patches [22].

Fig. 2. *Field symptoms of BSR disease of oil palm in plantations of the CDC. A) Asymptomatic plant, B) Unopened spear leaves at the centre, C) Skirt–like appearance, D) Basidiocarp formation, E) Bole creation, F) Death of palm.*

Table 4. *Incidence of basal stem rot disease in five oil palm plantations south western Cameroon.*

Location	Year of plantings	Age at first observation (years)	Incidence (%)		
			Mean (2010 dry and wet seasons)	Mean (2011 dry and wet seasons)	Mean (2010 and 2011)
Beneo	1995	15	4.6	9.4	7.0
	1973	37	20.7	21.3	21.0
Bota	1996	16	5.9	6.8	5.4
	1996	16	9.4	10.2	9.8
Idenau	1978	32	14.2	14.8	15.0
	1978	32	18.3	19.7	19.0
Mondoni	1995	15	16.8	18.4	17.6
	1973	37	17.5	19.1	18.3
Mungo	1996	16	38.0	40.0	39.0
	1995	15	31.4	36.6	34.0

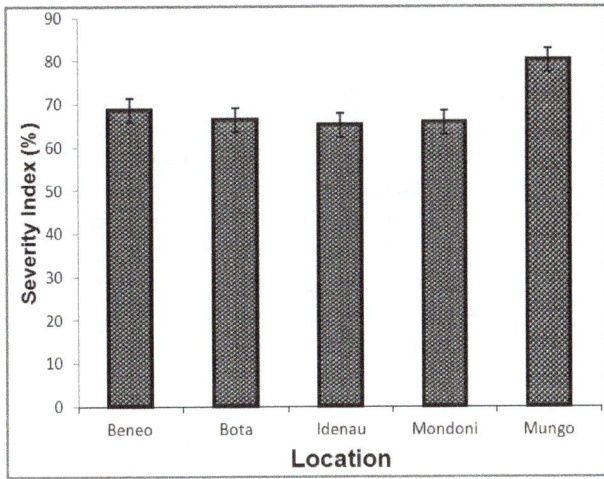

Fig. 3. *Mean disease severity index of five oil palm estates in south western Cameroon.*

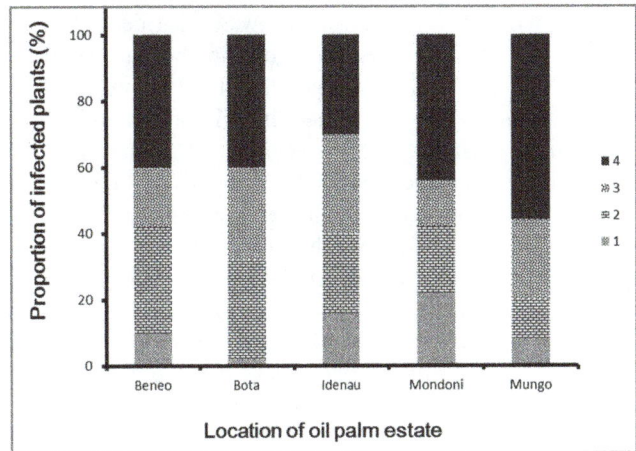

Fig. 4. *Proportion of diseased plants in each disease score (mean of two years) in five oil palm plantations in south western Cameroon.*

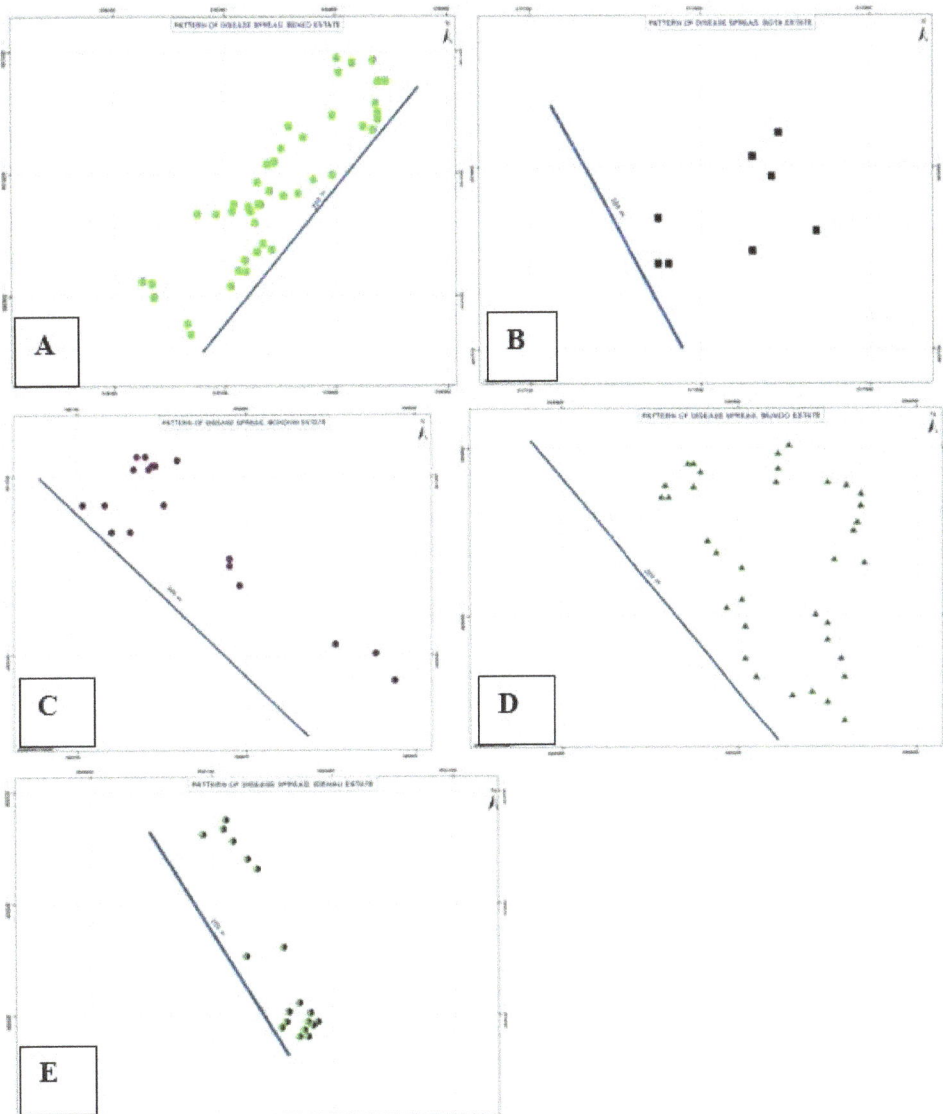

Fig. 5. *Spread pattern of BSR disease in oil palm plantations in south western Cameroon. A) Beneo estate, B) Bota estate, C) Mondoni estate, D) Mungo estate, E) Idenau estate.*

They typically appear in clusters or patches. The clustered pattern of infected plants was recorded at a number of sites and this was analogous to the pattern of basal stem rot reported by Rao *et al.* [14]. The mode of survival of *Ganoderma* as hyphae in oil palm residues would favour a clustered pattern of infected plants since dispersion of the residues of each plant would tend to overlap with that of other plants and a continuum of infested residue may result as the incidence of infected plants increases. Thus, under environmental conditions which favour a high incidence of infected plants, the pattern of infected plants would become regular. Soil physicochemical properties varied across the estates (Table 6). Of these properties, fine sand had a significant positive correlation with disease incidence and severity at 0.01 probability level, while C/N ratio had a significant negative correlation at 0.05% probability level. There was no significant correlation between any of the climatic factors and disease incidence. This does not preclude the indirect effect of climate through the vigour of the plant and soil. This observation is expected because the disease is soil borne and systemic. Results of the principal component analysis (PCA) showed that the first four principal components explained 100% of the total variation (Table 7). The PC_1 is strongly associated with fine silt, pH in KCl, organic carbon and total organic matter. PC_2 is strongly associated with pH in H_2O, base saturation and Sodium. PC_3 is strongly associated with total nitrogen, clay content, base saturation and fine sand, and PC_4 is strongly associated with coarse silt, moisture content, cation exchange capacity and water holding capacity (Table 8). The climatic data for Mungo was comparable to those of other sites. However, the edaphic factors were unique. The soils have a very high sand content, low clay content and consequently, low water holding capacity. The soil factors may partly contribute to this high incidence, given that there was a strong positive correlation between fine sand and incidence. Also, it may be due to the low nutrient status since there was a significant negative correlation between C/N ratio and disease incidence and severity. The low percentage of clay content of soil in Mungo estate might have been more conducive to the development of the basal stem rot disease than those with high clay contents. Soils low in clay content have been shown to be favourable to some soil borne diseases as is the case for damping–off of tomato seedlings caused by *Sclerotium rolfsii* in the Nigerian Savanna [23]. The incidence of basal stem rot would be much higher in future than the values obtained in the present survey because detection was based only on symptomatology. Plants at the early stages of infection generally do not manifest any symptoms. More elaborate techniques may be required to detect such plants [24].

Tengoua and Bakoume [4] recorded a 40% incidence level of basal stem rot in Mussaka palm plantation of the CDC planted between 1967 and 1969 but they actually projected a 60% incidence level of basal stem rot disease. However,

these were very old palms and their result was not comparable to the present result in terms of percentage disease incidence but it shows that this disease had been present in oil palm plantations of the CDC.

The high disease severity indices observed in the palms was expected. Plants with severity scores 3 and 4 constituted over 50% of the plants, thus resulting in the high disease indices observed. There was no management strategy put in place in the oil palm estates surveyed. The effects of the disease could easily be mitigated in Cameroon with proper management efforts. A preliminary survey in plantations of SOCAPALM where a rigorous eradication scheme was supplemented with trenching to curb spread revealed an extremely low incidence of less than 1%. However, in Malaysia a basal stem rot incidence of 80% and more were recorded in areas with attempted management strategies [14].

Fine sand was primarily the edaphic factor that influenced basal stem rot disease, having a significant positive correlation. High fine sand content (45.86%) found in Mungo estate allows for easy growth of roots thus enhancing contact and subsequent spread of basal stem rot. A very strong significant negative correlation was observed between incidence of basal stem rot and C/N ratio. Soils with low C/N ratios tend to release nutrients fast [25]. These nutrients easily leach away in the low water holding capacity soils thus starving plants and predisposing them to attack. Unsuitable soil conditions for plant development generally arise from lack of organic matter content in the soil [26, 27]. Higher content of easily decomposable organic matters might be associated with higher microbial activity and ultimately lead to the decline of *Ganoderma* species population. Thus higher population of *Ganoderma* may aggravate the disease situation in locations where soil reaction is more acidic and organic carbon content is less as in the case of Mungo estate. The lower disease incidence in other estates may be attributed to high organic carbon contents and weak acidity. Similar results were obtained by Sharma et al. [28] for ginger rhizome rot.

4. Conclusion

This study documents the first comprehensive study on the incidence, severity and associated factors of *Ganoderma* disease of oil palm in South Western Cameroon, after the initial report by Tengoua and Bakoume [4]. The incidence and severity of the disease shows that it is of increasing importance in oil palm production such that its management needs to be considered in the oil palm production plan, given that the disease is very devastating in Malaysia and other oil palm producing countries of the Indian sub – continent. The pattern of disease occurrence and distribution was attributed to variations in edaphic factors which are important in pathogen survival and host predisposition. Although there was no correlation between incidence and severity of BSR with climatic factors, principal component analyses showed the indirect influence of climate on the disease. Although no resistant clones are known for the disease, seed producers in Cameroon should screen for resistance in different progeny crosses. Also, there is need to monitor the disease on a continuous basis.

Acknowledgements

This research was funded by the Cameroon Development Corporation (CDC) and the University of Buea Research Grant No. 2008/A24 to the first author.

References

[1]   Nkongho R.N, Feintrenie L. and Levang P. (2014). *The Non-industrial Palm Oil Sector in Cameroon.* Working Paper 139, Banghor, Indonesia: CIFOR.

[2]   Jacqmard JC (2012). *Le Palmier a l'huile.* Montpellier – France, Edition Quae, CTA, Presses agronoque de Gembloux.

[3]   Naher N, Yusuf UK, Ismail A, Tan SG and Mondal MMS (2013). Ecological status of basal stem rot disease of oil palm. Australian Journal of Crop Science 7(11): 1723 – 1727.

[4]   Tengoua FF, Bakoume C (2005). Basal stem rot and vascular wilt, two threats for the oil palm sector in Cameroon. The Planter 81: 97 – 105.

[5]   Turner PD (1981) Oil Palm Diseases and Disorders. Kuala Lumpur, Oxford University Press, 281p.

[6]   Pilotti CA, Sanderson FR, Aitken AB, Armstrong W (2004). Morphological variation and host range of two *Ganoderma* species from Papua New Guinea. Mycopathologia 158: 251 – 265.

[7]   Steyaert RL (1967). Les *Ganoderma* palmicoles. Bulletin du Jardin Botanique Nationale Belgique 37: 465 – 492.

[8]   Kinge TR, Mih AM (2011) *Ganoderma ryvardense* sp. nov. associated with basal stem rot disease of oil palm in Cameroon. Mycosphere 2 (2): 179 – 188.

[9]   Kinge TR and Mih AM (2014). *Ganoderma lobenense* (Basidiomycete), a new species from oil palm (*Elaes guineensis*) in Cameroon. Journal of Plant Sciences 2 (5):242 – 245.

[10] Singh G (1990). *Ganoderma* the scourge of oil palms in the coastal areas. *In*: Arrifin, D. and Jalani, S. (editors) Proceedings of *Ganoderma* workshop, 11 September 1990 pp113 – 131. Palm Oil Research Institute of Malaysia, Bangi, Selangor, Malaysia.

[11] Singh G (1991). *Ganoderma* the scourge of oil palms in the coastal areas. The Planter 67:421 – 444.

[12] Khairudin H (1990). Results of four trials on *Ganoderma* basal stem rot of oil palm in Golden Hope Estates. *In*: Ariffin, D. and Jalani, S. (eds) Proceedings of *Ganoderma* Workshop, 11 September 1990, pp 113 – 131. Palm Oil Research Institute of Malaysia, Bangi, Selangor, Malaysia.

[13] Lim TK, Chung GF, Ko WH (1992). Basal stem rot of oil palm caused by *Ganoderma* infection in the field. Mycopathologia 159: 101 – 107.

[14] Rao V, Lim CC, Chia CC, Teo KW (2003) Studies on *Ganoderma* spread and control. The Planter 79: 367 – 83.

[15] Subagio A, Foster HL (2003). Implications of *Ganoderma* disease on loss in stand and yield production of oil palm in North Sumatra. *In*: Proceedings of the MAPPS Conference (Aug 2003). Kuala Lumpur.

[16] Sanderson FR, Pilotti CA, Bridge PD (2000). Basidiospores: their influence on our thinking regarding a control strategy for basal stem rot. *In*: Flood, J., Bridge, P. D., Holderness, M. (Eds.), *Ganoderma* Diseases of Perennial Crops. CABI Publishing, Wallingford, UK, pp. 113 – 119.

[17] Flood J, Bridge PD, Holderness M (2000). *Ganoderma* Diseases of Perennial Crops. CABI Publishing, Wallingford, UK. pp32.

[18] Abdullah F, Ilias GNM, Nelson M, Nur Ain Izzati MZ, Umi KY (2003). Disease assessment and the efficacy of *Trichoderma* as a biocontrol agent of basal stem rot of oil palms. Research Bulletin Science Putra 11: 31 – 33.

[19] Abdul RJH, Ahmad K, Ramdhan AS, Idris S, Abdul R, Aminul R, Fauzi I (2003). Mechanical Trunk Injection for Control of Ganoderma MPOB Information Series. MPOB TT No. 215.

[20] Hartley CWS (1988). The Oil Palm. Third edition, Harlow, England: Longman.

[21] Gurmit S (1990). *Ganoderma*– The scourge of oil palms in the coastal areas. P7–35. *In*: Proceedings of the *Ganoderma* Workshop, 11 September. (edited by Ariffin D and Jalani S) Palm Oil Research Institute of Malaysia. P1 – 142.

[22] Agrios GN (2005) Plant Pathology 5th edition. Elsevier Academic Press. 901pp.

[23] Wokocha RC (1987). Effect of Soil type on the damping–off of tomato seedlings caused by *Sclerotium rolfsii* in the Nigerian Savanna. Plant and Soil 98(3): 443 – 444.

[24] Ishaq I, Alias MS, Kadir J and Kasawani I. (2014). Detection of basal stem rot disease at oil palm plantations using sonic tomography. Journal of sustainable science and Management 9(2): 52 – 57.

[25] Myers RJK, Palm CA, Cuevas E, Gunatilleke IUN, Brossard M (1993). The synchronization of nutrient mineralization and plant nutrient demand. *In*: Woomer P. L., Swift, M. J., (Eds.). The biological management of tropical soil fertility. TSBF–John Wiley and Sons, Chichester, UK.

[26] Boyle M, Frakenburge, WT, Stolyz LH (1989). The influence of organic matter on soil aggregation and water infiltration. Journal of Production Agriculture 2: 290 – 299.

[27] Tejada M, Hernandez MT, Garcia C (2006). Application of two organic amendments on soil restoration: Effects on the soil biological properties. Journal of Environmental Quality 32: 1010 – 1017.

[28] Sharma BR, Dutta S, Roy S, Debnath A, Roy DM (2010). The effects of soil physico–chemical properties on Rhizome rot and wilt disease complex incidence of ginger under hill agro–climatic region of west Bengel. Plant Pathology Journal 26(2): 198 – 202.

[29] Anderson JM, Ingram JSI (1993) Tropical Soil Biology and Fertility: a handbook of methods Second edition. CAB International, the Camrian News, Aberstwyth, United Kingdom, 221pp.

6

Effect of a Hormone Containing Nitrobenzene in Combination with Fertilizers on Early Flower Induction of *Ixora coccinea* Hybrids Under Outdoor and Shaded Conditions

Hapu Arachchige Ruwani Kalpana Jayawardana[1], Mohamed Cassim Mohamed Zakeel[2, *], Channa De Zoysa[3]

[1]Department of Botany, Open University of Sri Lanka, Nawala, Sri Lanka
[2]Department of Plant Sciences, Faculty of Agriculture, Rajarata University of Sri Lanka, Anuradhapura, Sri Lanka
[3]Serendib Horticulture Technologies (Pvt) Ltd., Kalagedihena, Sri Lanka

Email address:
ruwanikal@hotmail.com (H. A. R. K. Jayawardana), zakeelag48@yahoo.com (M. C. M. Zakeel), dezoysa.channa@gmail.com (C. D. Zoysa)

Abstract: *Ixora* or Jungle Geranium (*Ixora coccinea*) is an invaluable plant in tropical landscapes for its beautiful foliage, easiness of growing and fabulous flowers freely formed and artfully presented. Sri Lanka exports *Ixora* hybrids as planting materials after rooting the *Ixora* cuttings on a coir dust media. Exporting potted *Ixora* plants with flowers is more beneficial than that without flowers as flowers increase product quality and gain customer attraction. In commercial floriculture venture, flower induction at young age using plant growth regulators and fertilizers is a popular practice. An experiment was conducted to investigate the most effective combination of a plant growth regulator and fertilizers for early flower initiation of *Ixora* hybrids; Vulcanus, Chanmai, Nora grant and Kontiki under outdoor and shaded conditions. Six-month-aged healthy plants of *Ixora* hybrids grown in containers were treated with a flowering hormone containing Nitrobenzene in combination with two fertilizer types: F1 – Bloom special and F2 – Krista K 44. Flowering hormone was sprayed in four concentrations; H1= 0.075% (V/V), H2 = 0.100% (V/V), H3 = 0.125% (V/V) and H4 = 0.150% (V/V) once in two weeks. Both fertilizers were applied as a liquid spray in same concentration (1 g/L of water) once a week. All the cultural practices were similarly applied to the plants under two light levels (outdoor light 'L1'and shaded conditions 'L2'). There were 16 treatment combinations and control, each with five replicates. Each treatment combination consisted of three plants. The number of flower-initiated plants per treatment combination was recorded as a percentage for five weeks. Percentage values after five weeks of treatments were transformed into log values and used for the analysis of variance. All the hybrids except Vulcanus showed a significant flower induction compared to the control. Outdoor light and shaded conditions were not significantly different to each other showing the potentiality of early flower initiation of hybrids under shaded condition. However, no significant difference in flower induction was observed among four levels of the flowering hormone and the two types of fertilizers ($p > 0.05$) in all four *Ixora* hybrids despite differences among the hybrids. Therefore, it could be concluded that the least expensive combination of hormone and fertilizer type could play a profitable and a positive role in early flower induction of *Ixora* at an age of six months.

Keywords: Floriculture, Ixora Coccinea, Nitrobenzene, Flowering Hormone, Flower Induction

1. Introduction

Ornamental crop culture, which was considered as an amateur or gardener activity, has now become an important and innovative business with a viable export potential [1]. Sri Lanka is recognized as one of the best quality production centers for floriculture products in the world as it is endowed with different climatic conditions caused by terrain enabling to develop floriculture products ranging from tropical to temperate flora. Floriculture industry has grown substantially during the last few decades to become the Sri Lanka's major foreign exchange generating venture [2].

Ixora (*Ixora coccinea*) hybrids are planted worldwide in tropical and subtropical climates [3]. There are over 400 species of *Ixora* and only a few of them are grown as landscape plants. One such species grown is *Ixora chinensis* or Chinese Ixora. This species offers many cultivars and many of which produce yellow to orange flowers. *Ixora coccinea* or Red Ixora is a popular species that bears orange-red flowers and reddish new growth. The *Ixora* hybrids, which are produced by crossing between species, account for the majority of *Ixora* plants used in landscapes. During the past many years, new *Ixora* hybrids differing in flower color, leaf size and plant height have appeared in the market due to the introduction and hybridization programs. *Ixora* is used in warmer climates for hedges and screens, foundation planting, massed in flowering beds and grown as a specimen shrub or small tree. *Ixora* is invaluable in tropical landscapes for its beautiful foliage, fabulous flowers that are freely formed and artfully presented and also for its easiness in management. In cooler climates, *Ixora* plants are grown in greenhouses or as a potted houseplant necessitating bright light.

Sri Lanka exports *Ixora* as planting materials after rooting of cuttings on coir dust media. Exporting potted *Ixora* plants with flowers is more beneficial than exporting young plants without flowers as flowers increase product quality and gain customer attraction. In Sri Lankan nurseries, *Ixora* hybrids are mostly grown in containers under shade nets. Therefore, natural flowering is very rare until the plants become matured. The regular pruning practice delays flowering of those plants as it removes emerging flower buds. Although a few plants initiate flowers, the flowering is not synchronized within the same aged plant groups.

Understanding the process of flower initiation is vital to growers as the plants have a great commercial and economic importance. Being able to control the flowering time, it enables growers to schedule crops to meet the demands of the market. However, it is a complex physiological process which requires a good deal of knowledge as different species respond to different stimuli to initiate flowering [4]. Plant growth substances change the flowering behavior of several plant species and their effects vary with plant species, age, concentration of the growth substance and temperature. Use of plant growth substances for flower induction -under environmental conditions which are not conductive to flowering- has considerable interest in agriculture. In addition to plant growth substances, growers should be able to control light and temperature as well in order to induce flowering. Naphthaleneacetic acid (NAA) and 6-Benzyladenine (BA) could play a big role in early flowering of *Ixora chinensis* [5].

As reported in [6], spraying of plant growth substances containing Nitrobenzene significantly increase the panicle production in wetland rice. Moreover, early flowering was achieved by 1% Nitrobenzene in Paprika cv.KtPl-19 [7]. Nitrobenzene is a combination of nitrogen and plant growth regulators extracted from sea weeds. Nitrobenzene, also known as Nitrobenzol or oil of mirbane, is an organic compound with the chemical formula $C_6H_5NO_2$. 'Elite' is a commercially available plant growth substance containing 20% (v/v) Nitrobenzene.

Although there are some literatures on induction of flowering in *Ixora* using different combinations of hormones and by controlling of light and temperature, no reported study is found to induce flowering of potted young *Ixora* plants. As the subject plants are very young, it was felt important to apply fertilizers in combination with the hormones to scaffold the growth of the plants. Since product quality of a flowering plant heavily depends on flowers, it is beneficial to induce flowering of potted young plants of *Ixora* by applying a combination of plant growth substances and fertilizers. This study investigates (1) the most effective combination of plant growth substances and fertilizers for early flower initiation of *Ixora* hybrids (Vulcanus, Chanmai, Nora grant and Kontiki) at outdoor and shaded conditions, (2) the effect of light and shade on early flower initiation and (3) a method of synchronization of flowering within the same batch of plants.

2. Materials and Methods

2.1. Study Area

This study was carried out at the Serendib Horticulture Technologies (Pvt) Ltd., Sri Lanka, with four *Ixora* hybrids namely Chanmai, Vulcanus, Nora Grant and Kontiki.

2.2. Experimental

Cuttings of *Ixora* hybrids were planted on coir dust beds and kept in propagators for four weeks for rooting. Rooted cuttings were then planted in pots (12 cm diameter) kept in a net house and the basal fertilizer (Osmocotte 5 g/pot) was given. The potted plants were allowed to grow under shade net for 3-4 weeks. Pinching was practiced at 2-months age to induce more shooting and thereby a compact nature. An additional fertilizer (Crop Master) was applied once a week for all plants since root initiation till the end of the experiment. Plants to be treated in normal light condition were trained and acclimatized for two weeks while other plants were maintained under the net house itself (shaded with 50% shade net).

Sixteen treatments with controls were applied by combination of four concentrations of Hormone containing Nitrobenzene; H1- 0.075% (V/V), H2- 0.1%, H3- 0.125% & H4- 0.15%, two types of Fertilizer; F1- Bloom special & F2- Krista K 44 and two light conditions; L1- outdoor light & L2- net house shade. Each treatment consisted of three *Ixora* plants and had five replicates. The hormone was sprayed on foliage once in two weeks and both fertilizers were applied as a liquid spray once a week at the same concentration (1 g/L of water). Flower buds initiation was recorded for five weeks.

2.3. Statistical Analysis

The percentage of flower induced plants at 5-weeks of treatment was transformed into log values and then used for the analysis of variance.

3. Results and Discussion

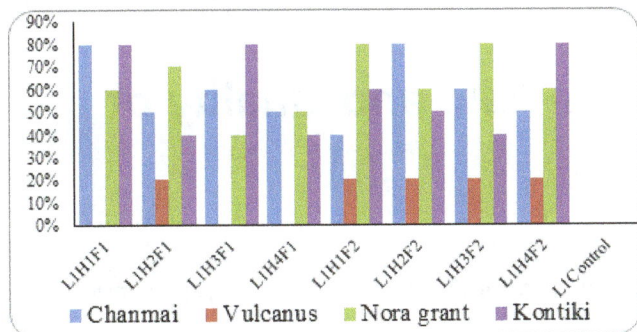

Figure 1. *Percentages of flower induced plants under each treatment for ixora hybrids at outdoor light condition*

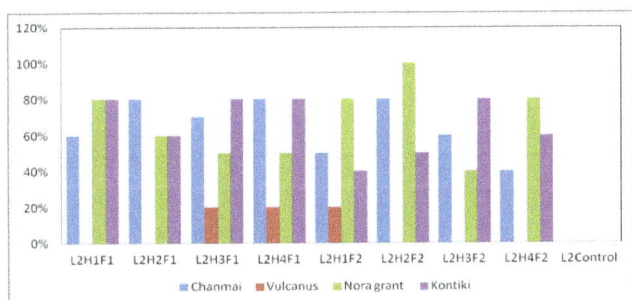

Figure 2. *Percentages of flower induced plants under each treatment for ixora hybrids at shade condition*

Figure 3. *Flower formation in Ixora coccinea hybrids. a: Chanmai; b: Vulcanus; c: Nora grant; d: Kontiki*

The results revealed that the plants in the control did not show any flower initiation at both light levels. However, all the treatment combinations showed more than 40% of flower initiation under both light conditions (Fig. 1 & 2) though the percentage of flower formation varied according to the variety (Fig. 3). The highest flower induction (100%) was observed in the hybrid Nora grant under shade condition (Fig. 2). The hybrid Vulcanus showed only 20% of flower initiation in few treatments while no flowers were observed in other treatments (Fig. 1 & 2).

Six month aged plants responded in flower initiation for growth regulator containing Nitrobenzene in combination with potassium supply by two fertilizers. The treatments significantly initiated flowers of all *ixora* hybrids except vulcanus compared to the control ($p > 0.05$). Hybrid Vulcanus was different in morphology with other three hybrids since it was a dwarf type chinese *ixora* hybrid. According to reference [8], the terminal bud of 6-month-old chinese *ixora* was in vegetative phase. Moreover, no significant difference was observed among four levels of the flowering hormone and between the two types of fertilizers ($p > 0.05$). In fact, least expensive combination (combination of least amounts of hormone and low cost fertilizer type) of hormone (containing Nitrobenzene) and the two types of fertilizers could play a profitable and a positive role in early flowering of *ixora hybrids* except vulcanus at six months age. Outdoor light condition and shade conditions were not significantly different showing the potential of early flower initiation of the hybrids even at shade condition with no effect by the sun loving character of *ixora* plants. However, as mentioned in reference [9] for Rosa hybrids Mercedes, the involvement of photoreceptor phytochrome for flowering response needs to be studied.

4. Conclusion

Least expensive combination of hormone and the two types of fertilizers could be successfully used for early flower initiation of six months aged *Ixora* hybrids except in Vulcanus either in light or shade conditions.

References

[1] Bhattacharjee S.K. and Chandra De L, (2003), Advanced commercial Floriculture, Aavishkar publishers, India, 802pp.

[2] Parthipan, S, (2001), Development of floriculture industry in up country, Department of Agriculture, Peradeniya, pp 30-44.

[3] Liogier, H. A, (1997), Descriptive flora of Puerto Rico and adjacent islands, Vol. 5. Editorial de la Universidad de Puerto Rico, San Juan, PR. 436 p.

[4] Keightley J,(2009), Aspects of Flowering, Quandio Publications, Great Britain.

[5] Lin Jinshui, Luanmei Lu and Yalan Zhang, (2010), Chinese Agriultural Science Bulletin.

[6] Aziz, M.A. and M.A Miah, (2009), Effect of "Flora" on the Growth and Yield of Wetland Rice.JARD. http://www.banglajol.info/index.php/ accessed on 20th May 2011.

[7] Kannan K, Jawaharlal M, and Prabhu M, (2009), Agricultural research communication centre, india.www.arccjournals.com / indianjournals.com.

[8] Chen L, Y.C. Chien and C.H. Min (2003), Inflorescence and Flower Development in Chinese Ixora, *Journal of American society for Horticultural science*.

7

Effect of GA$_3$, Girdling or Pruning on Yield and Quality of 'Parletta' Seedless Grape

Ismail Ali Abu-Zinada

Department of Plant Production & Protection, Faculty of Agriculture & Environment, Al-Azhar University-Gaza, Gaza Strip, Palestinian Territories

Email address:

isalznada@hotmail.com

Abstract: Grape vines of cultivar 'Parletta' were sprayed with GA$_3$ at 10 (G1), 15 (G2), 20 (G3), 30 (G4) or 40ppm (G5) when cluster was 7-12cm length (S1), at full bloom (S2) or when berry was 2-4mm (S3). Treatments were as the following combinations: Water spray (T0) as a control, G1S1 (T1), G1S1 + G2S2 (T2), G1S1 + G3S3 (T3), G1S1 + G5S3 (T4), G2S1 (T5), G2S1 + G2S2 (T6), G2S1 + G4S3 (T7), G2S1 + G5S3 (T8), G3S2 (T9), G3S2 + G4S3 (T10), G3S2 + G5S3 (T11), manual thinning by comb (T12), pruning to 50 eyes (T13) or girdling vine arms (T14). Treatments T6 and T1 produced the highest yield kg vine^{-1} where T7 and T6 had the highest number of clusters vine^{-1}. The heaviest clusters were produced in T13, T1 and T14. The treatments increased berry weight than control. Cluster length did not change significantly where the heaviest berries were produced in T13, T9, T5 and T3. The lowest cluster compactness was in T11, T9, T8, T14, T7 and T4. Shot berries % showed the lowest percentages in T8, T14, T4, T3 and T5. Fruit total soluble solids (TSS) increased in T5, T3, T12 and T13. Titratable acidity had the lowest values in T3, T14 while T12. TSS /acid ratio induced the highest ratio in T3, T12, T14 and T13.

Keywords: Acidity, Berry, Cluster, Grape, TSS

1. Introduction

Table grape (*Vitis vinifera*L.) is an important fruit and is traditionally produced in the Mediterranean region. The crop is also produced in all over the World during the last decades where the crop total production has increased, especially in Asia in China and India (O. I. V., 2007). Grape is used as dried berries (raisins), table grape fruit and juice (Mullins et al., 1992).

The local consumers have accustomed to consume traditional cultivars like Dapoki, Karishi, etc. where these grapes are of large berries and rain-fed. Nowadays, several seedless cultivars were imported where these varieties are grown in the open field or in greenhouses under drip-irrigation system (MoA, 2014).

Seedless grape is characterized by small berries, compacted cluster which are un favorite to the consumers. 'Perlette' is a seedless cultivar characterized by delicious early maturing fruits and edible soft leaves. The cultivar was produced in 1936 in California as a hybrid between Scolokertekhiralynoje 26 x Sultanina marble (English et al., 1990).

Gibberlic acidas GA$_3$ is a growth regulator that is widely used during fruit set stage to increase size of seedless berries. GA$_3$ had potential impact on grape quality where the impact depended on grape varieties (Rusjan, 2010). A positive correlation was observed between GA$_3$ application and amount of nutrients like N, P or K absorbed which enhanced the enlargement of grape berries and sink capacity of grape cluster to absorb water or nutrients such as potassium. (Zhenning et al, 2008).

Girdling is the removing of a small section of park (4 mm width) where it has been practiced for years to enlarge berries of table grape. Girdling resulted in an increase in carbohydrate concentrations above girdle and an increase weight per unit leaf area of grape vine (During, 1978). The application of GA$_3$ on grapevine mitigated the depressing effect of girdling on leaf gas exchange (Harrel and Williams, 1987 and Roper and Williams, 1989).

Spray of 25 or 50 ppm GA$_3$ on white 'Banaty' seedless grapevine significantly reduced each of TSS and the ratio between TSS/ acid (Wassel et al., 2007). Application of 50ppm GA$_3$ + CPPU [1-(2-chloro-4-pyridyl)-3-phenylure] at 2.5 ppm on 'Parletta' after 14 days of full bloom increased berry size, cluster weight and acidity without effect on TSS (Nilnond et al, 2010). Application of GA_3 or GA_3 + girdle produced heavier berries, increased berry diameter, heavier bunches, and increased number of berries punch[-1], TSS increased and decreased titratable acidity (Abu-Zahra and Salameh, 2012). Spray of 40 mg GA_3L^{-1} on cluster significantly increased cluster weight, insignificantly increased berry weight, significantly increased number of berries cluster[-1], insignificantly increased cluster length and significantly decreased TSS (Abu-Zahra, 2013). Spray of GA_3 at 20 mg L[-1] on flame seedless grape increased weight of the cluster and berry where GA3 had no significant influence on sugar content, TSS and acidity (Dimovska et al, 2014).

The current study aimed at improving the cluster and berries property of 'Parlette' cultivar by GA_3 spray on vine's canopy, manual thinning of cluster's berries or girdling of main vine arms.

2. Materials and Methods

2.1. Site and Season of the Experiment

The experiment was carried out in the North Governorate, Gaza Strip, Palestinian territories on grape (*Vitis vinifera* L.) in season 2012. 'Perlette' seedless cultivar of 9 years old vines were grown in a clay soil, drip-irrigated and spaced at 2 x 3 m.

2.2. Design and Treatments of the Experiment

The experiment was laid out in the completely randomized block design (CRBD) with 4 replications where each replicate-plot contained one tree. Treatments were as following in Table (1).

Table 1. Treatments applied.

Treatments code	Treatment stage		
	Spray when cluster 7 – 12 cm length	Spray at full boom	Spray after fruit set (2–4 mm diameter)
	GA₃ ppm		
T0	0	0	0
T1	10	0	0
T2	10	20	0
T3	10	0	30
T4	10	0	40
T5	15	0	0
T6	15	20	0
T7	15	0	30
T8	15	0	40
T9	0	20	0
T10	0	20	30
T11	0	20	40
T12	Manual thinning at fruitlets 2-4 mm stage		
T13	Retaining to 50 nodes at pruning time		
T14	Tree arms girdling at pruning time		

Vines were trained to the bilateral cordon system and pruned to short spurs (Alfonso et al., 2008) retaining an

average of 70 buds per vine. The other horticultural practices were carried out according to the recommendations of Ministry of Agriculture (MoA, 2014).

2.3. Data Collection

2.3.1. Yield Components

It was determined in each replicate as yield kg vine[-1] and clusters number vine[-1].

2.3.2. Cluster Properties

Eight clusters were randomly devoted to determine average of cluster weight (g), cluster length (cm), cluster compactness (Fig 1, consisted of 1-5 grads where 1 represented the least compacted cluster, grade 5 represented the most compacted where 3 represented the normal compactness) of cluster (El-Mahdi, 1960) and shot berries percentage.

2.3.3. Fruit Properties

Five clusters per each replicate were randomly devoted to determine average weight of 100 berry, titratable acidity as tartaric acid per 100 milliters of juice against 0.1 N NaOH (A.O.A.C., 1970), total soluble solids TSS using hand refractometer and TSS / titratable acidity ratio was calculated.

2.4. Statistical Analysis

Data were statistically analyzed using Duncan's multiple range test where means of similar letter/s are not significantly different at p = 0.05.

3. Results and Discussion

3.1. Yield Components

Yield as kg plant[-1] (Table, 2) did not change significantly by the different treatments than control. Treatment T6 (12.412) and T1 (12.373) respectively produced higher yield than the other treatments or control. Number of cluster plant[-1] was not affected significantly by the different treatments than control where T7 (50.3) and T6 (48.8) had higher number than the other treatments and the untreated control. Khan, 2009; Dimovska et al., (2006) reported that the effect of GA_3 is depending on variety of grapevine, the concentration and time of hormone application. Giberlic acid affects grape berry through different pathways: formation of flower cluster, berry set, berry enlargement, cluster length increase, thinning cluster berries and to prevent berry cracking (Korkutal et al., 2008).

3.2. Cluster and Berries Physical Properties

Cluster weight (Table, 3) showed insignificant changes due to the treatments compared to untreated control. Treatment T13 (332) and T1 (314) and T14 (308) respectively produced the heaviest clusters than the other treatments or control while T11 (172) and T9 (210) had the lightest cluster weight. Cluster length showed insignificant changes where no trend could be noticed. The average weight of 100 berries generally increased in the different treatments than control where treatments T13 (179), T9 (169), T5 (167) and T3 (161) produced the heaviest grape berries than the other treatments or control (145). GA$_3$ has a beneficial effect on cell division and cell enlargement, thus on a higher accumulation of sugar and water without changing pressure potential, which in consequence translate into larger berry and cluster size during harvest (Perez and Gomez, 200; Casonova et al., 2009). Cluster compactness (Figure. 1) showed indefinite trend at the scale 1-5. In this concern T11 (2.26), T9 (2.56), T8 (2.61), T4 (2.66), T7 (2.87) and T4 (2.66) respectively produced the lowest cluster compactness, however changes were insignificantly among all treatments and/ or control. Percentage of shot berries showed insignificant changes among the different treatments and control where no rend could be detected in this respect. Treatments T8 (6.80), T14 (7.28), T4 (8.21), T3 (8.24) and T5 (8.63) resulted in the lowest shot berries %. On the other hand T13 (14.92), T7 (13.04), T11 (12.73) and T10 (12.6) resulted in the highest percentage of small berries.

These findings in harmony with those of Abu-Zahra and Salameh.

3.3. Chemical Properties

Total soluble solids % (Table, 4) in general insignificantly increased whereas T5 (12.0), T3 (11.8), T12 (11.7) as well as T13 (11.7) had the highest TSS content. Titratable acidity of the grape juice also insignificantly affected however, T3, T14 and T12 respectively contained lower acidity content than the other treatments or control. Total soluble solids/ acid ratio also showed insignificant changes. The treatments T3 (26.3), T12 (25.6), T14 (24.3) and T13 (23.5) had the highest ratio than control and other treatments

The current study relatively came to the same results reported by Dimovsk et al., (2014).

Table 2. Yield components.

Treatments	Yield kg plant^{-1}	Number of clusters plant^{-1}
T0	11.968 a	47.0 abc
T1	12.373 a	38.0 abc
T2	9.881 a	40.5 abc
T3	11.319 a	44.3 abc
T4	10.545 a	41.0 abc
T5	9.383 a	34.8 bc
T6	12.412 a	48.8 ab
T7	10.292 a	50.3 a
T8	9.772 a	43.8 abc
T9	9.061 a	44.8 abc
T10	10.071 a	36.8 abc
T11	5.423 a	33.8 c
T12	8.445 a	33.5 c
T13	10.867 a	33.3 c
T14	11.505 a	41.0 abc

Means followed by same letter/s do not differ significantly at $p = 0.05$ according to Duncan's multiple range test.

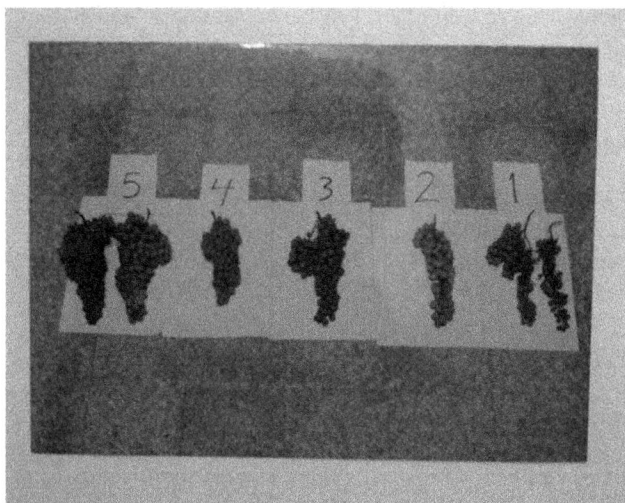

Figure 1. Cluster compactness (grade 1-5).

Table 3. Physical properties of cluster and berries.

Treatment	Cluster weight (g)	Cluster Length (cm)	average weight of 100 Berry (g)	Shot Berries %	Cluster Compactness (grade 1-5)
T0	283 abc	23.3 a	145 ab	10.98 a	3.25 ab
T1	314 ab	21.9 a	157 ab	10.72 a	3.51 a
T2	280 abc	23.9 a	153 ab	9.82 a	3.35 ab
T3	265 abc	21.6 a	161 ab	8.24 a	3.25 ab
T4	258 abcd	22.1 a	152 ab	8.21 a	2.66 ab
T5	276 abc	22.2 a	167 ab	8.63 a	3.09 ab
T6	266 abc	21.9 a	141 ab	9.64 a	3.55 a
T7	240 abcd	23.8 a	136 b	13.04 a	2.87 ab
T8	231 bcd	21.9 a	154 ab	6.80 a	2.61 ab
T9	210 cd	21.8 a	169 ab	9.90 a	2.56 ab
T10	249 abcd	21.5 a	163 ab	12.06 a	2.80 ab
T11	172 d	22.0 a	141 ab	12.73 a	2.26 b
T12	255 abcd	23.1 a	153 ab	9.12 a	3.22 ab
T13	332 a	23.1 a	179 a	14.92 a	3.69 a
T14	308 ab	22.0 a	158 ab	7.28 a	3.77 a

Means followed by same letter/s do not differ significantly at $p = 0.05$ according to Duncan's multiple range test.

Table 4. *Chemical properties of berries.*

Treatment	TSS%	Titratable acidity mg 100^{-1} ml	TSS /acid ratio
T0	10.8 a	0.53 ab	20.7 ab
T1	10.6 a	0.50 ab	21.3 ab
T2	11.2 a	0.51 ab	22.1 ab
T3	11.8 a	0.48 ab	26.3 a
T4	10.8 a	0.55 ab	19.9 ab
T5	12.0 a	0.55 ab	21.8 ab
T6	10.1 a	0.59 a	17.4 b
T7	11.1 a	0.54 ab	20.7 ab
T8	10.5 a	0.57 ab	18.8 ab
T9	11.2 a	0.54 ab	21.0 ab
T10	10.5 a	0.58 ab	18.3 ab
T11	9.9 a	0.51 ab	19.6 ab
T12	11.7 a	0.46 b	25.6 ab
T13	11.7 a	0.50 ab	23.5 ab
T14	11.4 a	0.47 ab	24.3 ab

Means followed by same letter/s do not differ significantly at $p = 0.05$ according to Duncan's multiple range test.

4. Conclusion

Seedless grapes in general are characterized by small berries where this creates a problem for consumers. The current study aimed at improving berry quality of cultivar 'Parletta' where 15 treatments were devoted to achieve the goal. Vines were treated by spraying with GA_3 at 3 stages and 5 levels, fruitlets manual thinning, pruned to 50 eyes or main arms girdling. Spraying with GA_3 at 15 ppm when cluster 7-12mm could be recommended for the potential economic benefits.

References

[1] Abu-Zahra, T.R. 2013. Effect of plant hormones application on fruit quality of 'Superior seedless' grape. Biosciences Biotechnology Research Asia, 10 (2): 527-531.

[2] A.O.A.C. 1970. Official and Tentative Methods of Analysis. Association of Official Agricultural Chemist, Washington, D.C.U.S.A.

[3] Dimoviska, V., Petropulos, V.I., Salamovska, A. and Ilieva, F. 2014. Flame seedless grape variety (*Vitis vinifera* L.) and different concentration of gibberelic acid (GA3). Bulgarian Journal of Agricultural Science, 20 (1): 137-142.

[4] During, H. 1978. Studies on environmentally controlled stomatal transpiration in grapevine. II. Effect of girdling and temperature. Vitis, 17: 1-9.

[5] English, J.T., Bledose, A.M., Marois, J. J. and Kliewer, W.M. 1990. Influence of grapevine canopy management on evaporative potential in the fruit zone. American Journal of Enology and Viticulture, 41 (2): 137-141.

[6] Harrel, D.C. and Williams, L.E. 1987. The Influence of Girdling and Gibberellic Acid Application at Fruitset on Ruby Seedless and Thompson Seedless Grapes. American Journal of Enology and Viticulture, 38 (2): 83-88.

[7] Ministry of Agriculture, 2013. Recommendations for fruits orchards, Gaza, Palestinian Territories.

[8] Ministry of Agriculture, 2014. Annual report. Gaza, Palestinian Territories.

[9] Mullins M.G., Bouquet A., Williams L.E. 1992. Biology of the grapevine. Cambridge University Press. ISBN-10: 0521305071.

[10] Organisation International de IaVigneet du Vin, (O.I.V.), 2003. Weltstatistiken / World statistics / Estadisticasmundial / Statistiquesmondial. Organization Intewrnational de Ia Vigneed du Vin. Paris: O.I.V.

[11] Rober, T.R. and Williams, L.E. 1989. Net CO_2 assimilation and carbohydrate partitioning of grapevine leaves in response to trunk girdling and gibberelic acid application. Plant physiology, 89 (4): 1136-1140.

[12] Rusjan, D. 2010. Impact of gibberellins (GA_3) on sensorial quality and storability of table grape (*Vitis vinifera* L.). Acta agriculturae Slovenica, 95 (2): 163-173.

[13] El-Mahdi, M.A. 1960. Physiological studies on maturity and storage of Thompson seedless (Banati) and Gharibi grapes. Ph. D. Thesis, Faculty of Agriculture., Egypt. Pp. 55-60.

[14] Wassel, A.H., Abd-Elhameed, M., Gobara, A. and Attia, M. 2007. Effect of some micronutrients, gibberelic acid and ascorbic acid on growth, yield and quality of Banaty seedless grapevines. African Crop Science Conference Proceedings vol. 8. pp. 547-553.

[15] Zhenming, N., Xuefeng X., Yi, W., Tianzhong L., Jin K. and Zhenhai H., 2008. Effect of leaf applied potassium, gibberelin and source-sink ratio on potassium absorption and distribution in grape fruits. Scientia Horticulturae. 115 (2): 164-167.

8

Comparative Effects of Foliar Application of Gibberellic Acid and Benzylaminopurine on Seed Potato Tuber Sprouting and Yield of Resultant Plants

Martin Kagiki Njogu[1], Geofrey Kingori Gathungu[1, *], Peter Muchiri Daniel[2, 3]

[1]Department of Plant Science, Chuka University, Chuka, Kenya
[2]Department of Plant Science and Crop Protection, University of Nairobi, Kangemi, Kenya
[3]MoA, Wambugu Agriculture Training Centre, Nyeri, Kenya

Email address:

gkgathungu@yahoo.com (G. K. Gathungu)

Abstract: Seed potato tubers planted immediately after harvest is characterized by delayed plant emergence, poor establishment and low yields. Gibberellic acid (GA) and Benzylaminopurine (BA) or their combinations cause dormancy breakage though little information is available on their combined application to dormancy termination. The effects of foliar application of GA and BA on potato tuber sprouting and subsequent yield were studied. Three potato varieties with different tuber dormancy period; 'Asante' (short dormancy), Dutch Robyjn (medium dormancy) and 'Kenya Sifa' (long dormancy) were planted at National Potato Research Centre, Tigoni and sprayed with a factorial combinations of 0, 50, 100, 300 ppm GA and 0, 50, 75, 100 ppm BA separately and combined at the rate of 1000 lts/ha spray volume towards the end of maturation. The resulting tubers were put in diffuse light storage (DLS) and data on number, length and vigour of sprouts recorded. Sprouted seed tubers were subsequently planted and evaluated for both growth characteristics and yields. The data collected was subjected to analysis of variance and significantly different means were separated using Fisher's protected least significant difference at p≤0.05. Higher rates of foliar application of GA+BA (300 ppm + 100 ppm) compared with the control (0 + 0) resulted in significant increase insprout length (cm), number of sprout/tuber, sprout vigour (score), and % sprouting from 3.24 to 7.02 and 3.84 to 9.03, 2.04 to 4.45 and 2.07 to 4.8, 1.7 to 3.06 and 1.63 to 3.23, 61.21to 86.67and 63.3 to 83.7 in Asante, 2.94 to 8.03 and 2.8 to 7.99, 1.84 to 5.24 and 1.87 to 4.76, 1.3 to 3.0 and 1.27 to 2.63, 50.61 to 92.7 and 52.7 to 85.7in Dutch Robyjn and 0.79 to 6.43 and 1.32 to 6.99, 0.61 to 3.49 and 0.79 to 3.33, 0.61 to 3.03 and 0.73 to 2.83, 22.12 to 85.76 and 28.3 to83.7 in Kenya Sifa after storage in 2008 and 2009 respectively. A combination of BA and GA resulted in significantly more growth than using only GA or BA alone at the same level. Similarly the subsequent tuber numbers per plant and yield (tons/ha) in resultant plants increased from 7.13 to 12.53 and 24.66 to 32.27, 6.93 to 10.47 and 16.73 to 23.37, and 5.63 to 9.6 and 17.53 to 30.13 in Asante, Dutch Robyjnand Kenya Sifa respectively. Combined application of GA + BA at varied rates can be used to improve sprouting characteristics of seed potato and yield of resultant plants.

Keywords: Potato, Giberrelic Acid, Benzylaminopurine, Seed Sprouting, Resultant Plants, Yield

1. Introduction

Potato tuber buds normally remain dormant throughout the growing season until several weeks after harvest [1] and the conditions that influence tuber formation and growth in potatoes may influence the duration of dormancy [2].Plant internal factors including plant hormones such as gibberellic acid, auxins, ethylene, cytokinins and abscissic acid have been known to affect potato sprouting [3]. There have been a number of studies of the effects of exogenous plant growth regulators (PGRs) on dormancy in potato where they have been applied to leaves shortly before harvest [4] or to whole tubers at harvest or during storage [5, 6, 7].The most consistent effects have been observed with gibberellins [8, 9].

A foliar spray of gibberellic acid, 3–6 days before haulm killing shorten potato tuber dormancy period and induced sprouting [10, 11]. When GA₃ was applied to the foliage of potato plants grown from true seeds towards the end of the vegetative cycle (60 days after transplantation), it induced rapid breakage of tuber dormancy, a reduction in specific weight, a higher rate of respiration and increased weight loss during storage [4]. The magnitude of the GA effect depends on the cultivar and storage temperature regime [11]. Most studies of foliar application of cytokinins have been based on potato yields [12, 13]. Foliar applications of 50 mg/L benzylaminopurine (BAP) and 50 mg/L gibberellic acid (GA) at early tuberization phase increased both tuber number and yields [12]. Dwelle[13] found that foliar application of a commercial sea weed extract "Cytex" containing cytokinins equivalent to 100ppm resulted to substantial increase in potato yields.

However, reports show that tuberization and sprouting of tubers are associated with high level of cytokinins[14]. Supplying cytokinins to tubers with innately dormant buds induce sprouting growth.It was found that cytokinins is the primary factor in the switch from innate dormancy to non-dormant state in the potato tuber buds but probably do not control the subsequent sprout growth [15].Early bioassay data suggested that increases in endogenous cytokinins accompanied dormancy break [3, 16]. It is proposed that cytokinins may be responsible for dormancy breakdown [15] while the gibberellins promote the subsequent sprout growth [14, 17]. Since potato tuber dormancy is thought to occur on or about the time of tuber initiation, it may be possible that application of gibberellic[4] acid and cytokinins[12] at late vegetative (tuber bulking) phase may shorten the dormancy period. However, there are no reports of the effect of combined use of GA and cytokinins on potato seed sprouting.

However, in Kenya there exists less information on the effects of exogenous application of gibberellic acid and benzylaminopurine on tuber sprouting characteristics and subsequent effect on yield of generated plants. This study was aimed at investigating the effects of foliar application of a factorial combination of Gibberellic acid and Benzylaminopurine in breaking dormancy and subsequent sprouting in three cultivars of different dormancy period.

2. Materials and Methods

2.1. Establishment of the Experiment

The experiment was conducted at Kenya Agricultural & Livestock Research Organisation (KALRO) formally Kenya Agricultural Research Institute (KARI), National Potato Centre, Tigoni from September 2008 to October 2009. Three potato varieties with different tuber dormancy period [18]; 'Asante' (short dormancy), Dutch Robyjn (medium dormancy) and 'Kenya Sifa' (long dormancy) were planted in staggered manner so that all varieties flowered at the same time. The experiment was laid out as a randomized complete block design (RCBD) with a split plot layout where the main blocks were the different rates of combinations of GA and cytokinins and the sub plots were the different genotypes. Each experimental plot was 4.5 m by 0.9 m. Plots and blocks were separated by 1 m and 1.5 m respectively. Di-ammonium

phosphate (DAP) fertilizer was applied during planting at the rate of 500 kg/ha. The potatoes were sprayed with a factorial combinations of concentrations of 0, 50, 100, 300 ppm Gibberellic acid (GA) and 0, 50, 75, 100 ppm Benzylaminopurine (BA) at the rate of 1000 litres/ha spray volume towards the end of tuberization phase (75 days after emergence). Dehaulming was done 6 days after treatment application [11] and harvesting was done 14 days thereafter. Untreated varieties were used as control.

2.2. Data Collection

2.2.1. Tuber Dormancy and Sprouting Characteristics

At harvest, twenty uniform potato tubers from each treatment were randomly selected and put in paper trays, labeled and put in diffuse light store (DLS). Data on dormancy period, number, length and vigour of sprouts and subsequent yields were collected, recorded and analyzed. Tuber sprouting was defined as when a tuber had at least one visible sprout of at least 2mm long [3].For dormancy period, the buds of all seed tubers from each treatment were observed after every week. The number of sprouted tubers was counted and recorded. Sprouting was got as a percentage of the number of sprouted tubers in a sample. A sample was considered to have broken dormancy when its sprouting was 80% and dormancy period was given by the duration from when the sample was treated to time when sample tuber dormancy was broken [6]. For the number and length of sprouts, five tubers from each treatment were picked at random after every week and the number of sprouts per tuber and the length of the longest sprout per tuber were noted.

Sprout vigour was determined as a 5 point rating score based on sprout base thickness and sprout length where;1= Very low vigour (where half or more of the tubers in a treatment sample had produced sprouts of at least 1mm base diameter and 2mm long), 2=low vigour (where half or more of the tubers in a treatment sample had produced sprouts of at least 2mm base diameter and 3mm long), 3= medium vigour (where half or more of the tubers in a treatment sample had produced sprouts of at least 3mm base diameter and 4mm long), 4=high vigour (where half or more of the tubers in a treatment sample had produced sprouts of at least 4mm base diameter and 4mm long) and 5=very high vigour (as described in score 4 but had green colouration, firm and had no defects).

2.2.2. Field Evaluation of Sprouted Tubers

After data collection, the sprouted tubers were conventionally planted at the same station for field yield evaluation. Pest and disease management was done using karate® and ridomil®. Supplementary irrigation was also done. The experiment was laid in a complete randomized block design with three replications. In the field, data on %germination (score), number of stems per plant and number of leaflets per plant were taken after every two weeks and recorded.

Germination score was determined as a 4 points rating whereby; 1 means $\leq 25\%$ germination, 2 means $> 25\% \leq 50\%$ germination, 3 means $> 50\% \leq 75\%$ germination and 4 means $>$

75% germination. After germination, 5 plants per treatment per genotype were selected at random, labeled and the number of stems per plant and leaflets per plant were counted and recorded. At harvest, the tuber were graded into three grades (Chatts, Seed, and Ware) based tuber size where; Chatts = tubers < 25mm diameter, Seed = tubers ≥ 25 mm ≤ 55 mm and Ware = tuber > 55mm.

2.3. Data Analysis

The data collected was subjected to Analysis of variance (ANOVA) using the PROC ANOVA procedure of Genstat (Lawes Agricultural Trust Rothamsted Experimental station 2006, Version 9). Where the treatment was significant, difference among the treatment means were compared using the Fisher's protected LSD test at p≤0.05 probability level.

3. Results

3.1. Sprouting

Potato tuber sprouting exhibited significant differences among treatments and genotypes in both seasons compared with the control (Tables 1 and 2). Duration to commencement of sprouting of tubers from plants treated with GA alone or its combination with BA decreased with increased rate of GA concentration in all genotypes in both seasons except for 50ppm GA which was not significantly different with the control.GA at 50ppm level had no significant difference in all genotypes in both seasons compared with the control. However, at 100ppm level duration to sprouting was reduced by 1 week in Asante, 3 weeks in Dutch Robyjn and Kenya Sifa with respect to control. Visible sprouts were observed during the first week after harvest in Dutch Robyjn and Kenya Sifa at 300ppm GA while all the three genotypes sprouted in the second week at the same rate in the second season. When GA and BA were sprayed in combination, duration to sprouting was the same as that of GA alone at the same rate of concentration in all genotypes in both seasons. Benzylaminopurinealone exhibited no significant difference in sprouting at all rates in all genotypes in both seasons compared with the control. Kenya Sifa took the longest time (7weeks) to sprout at low rates of GA (0-50ppm) while Asante sprouted during the 4th week at the same rate. At higher rates (300ppm) all varieties took the same time (2 weeks) except in 1st season where visible sprouts were observed during the 1st week in Kenya Sifa.

Table 1: *Effects of foliar application of rates of combination of gibberellic acid and benzylaminopurine on sprouts length, number of sprouts per tuber, sprouts vigour and % sprouting of potato genotypes Asante, Dutch Robyjn and Kenya Sifa (Year 2008).*

Treatments	Genotypes											
	Asante				Dutch Robyjn				Kenya Sifa			
	Sprout length (mm)	Sprouts/ tuber	Sprouts vigour	%Sprouting	Sprout length (mm)	Sprouts/ tuber	Sprouts vigour	%Sprouting	Sprout length (mm)	Sprouts/ tuber	Sprouts vigour	%Sprouting
Control	3.24a*	2.04a	1.70a	61.21a	2.94a	1.84a	1.30a	50.61a	0.79a	0.61a	0.61a	22.12a
0G.50C	3.53b	2.11a	1.73a	64.24b	3.09ab	2.03b	1.36a	51.21a	0.86a	0.67a	0.67ab	24.85b
0G.75C	3.67bc	2.30b	1.73a	65.76bc	3.21bc	2.17bc	1.36a	53.33b	0.98ab	0.70ab	0.76b	32.76c
0G.100C	3.84c	3.38b	1.91b	67.27c	3.30bc	2.32c	1.33a	54.24bc	1.13bc	0.74abc	0.82c	31.82c
50G.0C	3.84c	2.38b	1.91b	66.33cd	3.31bc	2.18c	1.39a	53.03b	1.24cd	0.79bc	0.88cd	33.33cd
50G.50C	3.81c	2.44b	1.94bc	66.33cd	3.40cd	2.34d	1.67b	55.15c	1.27cd	0.88cd	0.88cd	32.13c
50G.75C	4.11d	2.56cd	2.03c	67.76d	3.57d	2.42de	1.61b	56.36d	1.30cd	0.98d	0.97d	43.24d
50G.100C	4.16d	2.68d	2.15d	73.94e	3.90e	2.55ef	1.67b	60.00e	1.47d	1.34e	1.00e	36.36e
100G.0C	4.61e	2.99e	2.27e		5.41f	3.63fg	2.39c	77.58f	3.26e	1.65f	1.55f	49.70f
100G.50C	4.84ef	3.10f	2.46f	74.85e	5.53f	3.77gh	2.39c	76.06f	3.35e	1.92g	1.52f	53.33g
100G.75C	4.99f	3.19f	2.58g	75.15e	5.84g	3.91hi	2.30c	81.52g	3.81f	2.16h	1.76g	57.88h
100G.100C	5.24g	3.89g	2.58g	73.95e	6.02g	4.02i	2.29c	80.91g	4.08g	2.29h	2.03h	59.09h
300G.0C	6.15h	4.14g	2.94h	80.00f	6.82h	4.71j	2.51d	89.70h	5.26h	2.90i	2.58i	72.73i
300G.50C	6.30h	4.04g	2.91i	84.85g	7.12i	4.97k	2.67e	89.70h	5.67i	3.15j	2.76j	79.39i
300G.75C	6.62i	4.05g	3.03j	85.76h	7.67j	5.01k	2.79f	92.12i	6.06j	3.35k	2.76j	84.55j
300G.100C	7.02j	4.45h	3.06j	86.67i	8.03k	5.24L	3.00g	92.73i	6.43k	3.49k	3.03k	85.76j
LSD$_{0.05}$ Treatment	0.12	0.05	0.04	0.90	0.12	0.05	0.04	0.90	0.12	0.05	0.04	0.90
LSD$_{0.05}$ Genotype	0.45	0.21	0.18	3.36	0.45	0.21	0.18	3.36	0.45	0.21	0.18	3.36
CV%	6.8	4.2	5.5	2.8	6.8	4.2	5.5	2.8	6.8	4.2	5.5	2.8

*Means followed by the same letters along the column are not significantly different (p≤0.05).LSD $_{0.05}$ =least significant difference at 5% probability level; CV% =percent coefficient of variation and xG.yC means combination of x parts per million (ppm) gibberellic acid and y ppm Cytokinins (benzylaminopurine).

3.2. Sprout Length

Sprout length showed significant difference following treatment with different rates of GA, BA or their combinations and genotypes in both seasons compared with the control. When GA was applied alone, it exhibited significant ascending difference of sprouts length with increasing doses of GA in all varieties. However, when BA was applied alone, variation amongst different rate was genotype dependent with significance observation from 50 ppm for Asante, 75ppm for Dutch Robyjn and 100ppm for Kenya Sifa with respect to control(Tables 1 and 2).At lower doses of GA (0-50ppm), Asante genotype produced longer sprouts than Dutch Robyjn but at higher doses (100-300ppm) GA, it was the vice versa. Kenya Sifa recorded the shortest sprout in all treatments. There were higher sprouts length observed when a combination of GA and BA was employed than when each hormone was applied alone at the same rates. It was noted that BA exhibited significant difference after each increase rate at highest rate of GA (300ppm) unlike at lower rates of GA. There was a linear increase of sprout length with duration of storage.

3.3. Sprouts per Tuber

The number of sprouts per tuber was significantly different amongst treatments and genotypes in both seasons (Tables 1 and 2). The number of sprouts per tuber was significantly higher with increased rate of GA in all genotypes while the same occurred in BA treated tubers except in Kenya Sifa where increasing the rate of BA caused no significant difference when applied alone. However, the number of sprouts per tuber increasingly and significantly varied with increase in rate of combination of GA and BA in all genotypes. At lower doses of GA (0-50ppm), Asante recorded higher number of sprouts per tuber than Dutch Robyjn but this was the vice versa at higher doses of GA (100-300ppm) (Tables 1 and 2).

Table 2: *Effects of foliar application of rates of combination of gibberellic acid and benzylaminopurine on sprouts length, number of sprouts per tuber, sprouts vigour and % sprouting of potato genotypes Asante, Dutch Robyjn and Kenya Sifa (Year 2009).*

Treatments	Genotypes											
	Asante				Dutch Robyjn				Kenya Sifa			
	Sprout length (mm)	Sprouts/ tuber	Sprouts vigour	%Sprouting	Sprout length (mm)	Sprouts/ tuber	Sprouts vigour	%Sprouting	Sprout length (mm)	Sprouts/ tuber	Sprouts vigour	%Sprouting
Control	3.84a	2.07a	1.63a	63.3a	2.80a	1.87a	1.27a	52.7a	1.32a	0.79a	0.73a	28.3a
0G.50C	3.99ab	2.15a	1.77a	64.3a	2.91ab	1.99b	1.27a	52.0a	1.36a	0.87ab	0.77a	30.3a
0G.75C	4.07b	2.34b	1.73a	64.3a	2.99b	2.03b	1.33a	52.3a	1.41a	0.94b	0.83a	35.0b
0G.100C	4.15bc	2.41b	1.93a	63.3a	3.13b	2.16c	1.30a	54.7a	1.46ab	0.99bc	0.90b	35.7b
50G.0C	4.28cd	2.41bc	1.97a	63.3a	3.51c	2.39d	1.43b	61.7b	1.47ab	1.02c	0.90b	36.3b
50G.50C	4.36d	2.47cd	2.00ab	64.3ab	3.60c	2.46de	1.73c	64.0bc	1.58bc	1.10c	0.90b	36.3b
50G.75C	4.52e	2.58de	2.10bc	65.3ab	3.86d	2.51e	1.70c	65.3c	1.61c	1.15cd	1.07c	38.7b
50G.100C	4.70f	2.67e	2.17bcd	66.3b	3.97d	2.64f	1.77c	69.3d	1.68c	1.22d	1.07c	37.7b
100G.0C	5.67g	3.32f	2.43cde	75.7c	5.32e	3.76g	2.40d	80.7e	3.38d	1.91e	1.60d	54.7c
100G.50C	5.88h	3.39f	2.53de	75.7c	5.51f	3.85g	2.40d	80.0e	3.53d	2.04ef	1.60d	58.3d
100G.75C	6.17i	3.55g	2.60de	76.1c	5.70g	4.04h	2.37d	82.0e	3.76e	2.13f	1.76e	61.7e
100G.100C	6.38j	3.75h	2.70ef	75.7c	5.92h	4.16i	2.43d	81.0e	3.98f	2.28g	2.17f	65.0f
300G.0C	8.31k	4.42i	3.03fg	82.0d	7.19i	4.39j	2.50de	84.3f	6.11f	3.11h	2.53g	75.0g
300G.50C	8.60l	4.55j	3.17g	84.0d	7.48j	4.55k	2.60e	83.3ef	6.38g	3.22h	2.70h	79.3h
300G.75C	8.92m	4.65j	3.20g	83.0d	7.72k	4.60k	2.73e	85.0f	6.71h	3.33i	2.53g	83.0i
300G.100C	9.03m	4.80k	3.23g	83.7d	7.99l	4.76l	2.63e	85.7f	6.99i	3.33i	2.83h	83.7i
$LSD_{0.05}$ Treat.(T)	0.166	0.105	0.138	3.05	0.166	0.105	0.138	3.05	0.166	0.105	0.138	3.05
$LSD_{0.05}$ Genotype	0.078	0.065	0.055	1.22	0.078	0.065	0.055	1.22	0.078	0.065	0.055	1.22
CV%	4.1	5.8	6.8	2.8	4.1	5.8	6.8	2.8	4.1	5.8	6.8	2.8

*Means followed by the same letters along the column are not significantly different (p≤0.05).$LSD_{0.05}$ =least significant difference at 5% probability level; CV% =percent coefficient of variation and xG.yC means combination of x parts per million (ppm) gibberellic acid and y ppm Cytokinins (benzylaminopurine).

Kenya Sifa had the least number of sprouts per tuber in all treatments in both seasons.

3.4. Sprout Vigour

Sprouts vigour score varied significantly among treatments and genotypes in both seasons (Tables 3 and 4). Vigour score increased with duration of storage and with rates of application of GA, BA or their combination. Asante genotype had the highest vigour score in all treatments while Dutch Robyjn recorded higher vigour score than Kenya Sifa at lower dose of GA (0-100ppm) alone or in combination with BA but at 300ppm GA, the Kenya Sifa scored higher than Dutch Robyjn. At higher doses of GA (300ppm), Dutch Robyjn produced longer but slender sprouts.

3.5. Germination Score

Onset of germination was evident in all treatments in all genotypes by the second week after planting. However, there was significant difference in % germination score among treatments and genotypes (Table 5). At lower rates of GA applications, there was no significant difference in % germination score for Asante (0-50ppm) and Dutch Robyjn (0ppm) even with increased rates of concentration of BA. Kenya Sifa exhibited significant variation in the same rate bracket of GA with increase of BA concentration. Tubers treated with 300ppm GA recorded the highest % germination score in all genotypes while at the same rate no significant difference was observed with increase in BA concentration rates in all genotypes. Asante had the highest %germination score in all treatments. Dutch Robyjn recorded higher germination than Kenya Sifa at lower rates of GA (0-50ppm) while the opposite occurred at higher rates (100-300ppm) of GA applications.

3.6. Stems per Plant

Variation in the number of stems per plant differed significantly between treatments and genotypes (Table 5). Gibberellic acid treated tubers produced significantly more stems per plant with increase in the rate of application in all genotypes. Plants treated with BA alone showed no significant variation with increased rates of BA for Asante and Dutch Robyjn genotypes. However, increase in BA concentration exhibited significant difference in stems per plant when applied in combination with GA. A combination of BA and GA gave significantly more stems than when each hormone was applied alone. Asante genotype had the highest number of stems per tuber in all treatments while Kenya Sifa had the lowest.

3.7. Leaflets per Plant

The number of leaflets per plant differed significantly among treatments and genotypes.Increase in concentration of GA or BA when applied alone or in combination resulted in significant increase in the number of leaflets per plant. However, results for the combination were significantly higher than when each hormone was applied alone at the same rate. The number of leaflets per plant increased with time in each treatment in Asante and Dutch Robyjnupto the 8th week after which no increase was observed (Table 5).

However, Kenya Sifa exhibited linear increase in the number of leaflets per plant throughout the 12 weeks. The plants in the control had the lowest leaflets while those treated with a combination of 300ppm GA and 100ppm BA gave the highest in all genotypes. Asante gave the most number of leaflets per plant while Kenya Sifa produced the lowest in all treatments (Table 5).

3.8. Tubers per Plant

Number of tubers per plant varied significantly between treatments and genotypes (Table 5). When GA was applied alone, increase in concentration caused significant upward variation in all the genotypes compared with the control and previous rate except Kenya Sifa where concentration of 50ppm caused no significant difference with respect to control. Increase in the rate of concentration of BA resulted in significant increase in number of tuber per plant in all genotypes except in Kenya Sifa. When GA and BA were used in combination, variation of BA concentration rate had no significant difference at high level of concentration of GA (100-300ppm) in all genotypes. At low concentration rare of GA (0-50ppm), the number of tuber per plant in Asante and Dutch Robyjn were not significantly different except at when BA was 100ppm at 50ppm GA. When GA concentration was 100ppm and 300ppm, Asante genotype significantly outweighed Dutch Robyjn in the number of tubers per plant. Kenya Sifa had the lowest number of tubers in all treatments.

Table 3: *Effects of sequence of application of gibberellic acid and benzylaminopurine on sprouts vigour of potato tubers stored under diffuse light conditions (2008).*

Treatment	Genotypes								
	Asante			Dutch Robyjn			Kenya Sifa		
	Sprouts vigour score			Sprouts vigour score			Sprouts vigour score		
	2Wks	6 Wks	10 Wks	2 Wks	6 Wks	10 Wks	2 Wks	6 Wks	10 Wks
Control	0a*	2a	3a	0a	1a	3a	0a	0a	2a
0G.50C	0a	2a	3.33b	0a	1a	3a	0a	0a	2.33b
0G.75C	0a	2a	3.33b	0a	1.33a	3a	0a	0a	2.33b
0G.100C	0a	2a	4c	0a	1a	3a	0a	0a	2.67c
50G.0C	0a	2a	4c	0a	1.33a	3a	0a	0a	2.33b
50G.50C	0a	1.67a	4c	0a	2b	3a	0a	0a	2.33b
50G.75C	0a	2ab	4c	0a	1.33a	3.33b	0a	0a	3d
50G.100C	0a	2.33b	4c	0a	1.33a	3.67c	0a	0a	3d
100G.0C	0a	2.33bc	4c	1b	2b	4d	0a	1b	3d
100G.50C	0a	2.67cd	4.33d	1b	2.33bc	4d	0a	1.67c	3d
100G.75C	0.33b	3de	4c	1b	2.33bc	4d	0a	2cd	3d
100G.100C	0a	3de	4.33d	1b	2b	4d	0a	2.33d	3.33e
300G.0C	1c	3de	5e	1b	2.67cd	3.67c	1b	3e	3.33e
300G.50C	1c	3de	5e	1.33c	3d	4d	1b	3e	3.67f
300G.75C	1c	3de	5e	2d	2.67cd	4d	1b	2.67de	3.67f
300G.100C	1c	3.33e	5e	2d	3d	4d	1b	3e	3.67f
LSD$_{0.05}$ Treat.(T)	0.12	0.37	0.28	0.12	0.37	0.28	0.12	0.37	0.28
LSD$_{0.05}$ Genotype (G)	0.05	0.13	0.13	0.05	0.13	0.13	0.05	0.13	0.13
CV%	17.8	12.1	4.7	17.8	12.1	4.7	17.8	12.1	4.7

*Means followed by the same letters along the column are not significantly different (p≤0.05). LSD $_{0.05}$ =least significant difference at 5% probability level; CV% =percent coefficient of variation and xG.yC means combination of x parts per million (ppm) gibberellic acid and y ppm Cytokinins (benzylaminopurine).

Table 4: Effects of sequence of application of gibberellic acid and benzylaminopurine on sprouts vigour of potato tubers stored under diffuse light conditions (2009).

Treatment	Genotypes								
	Asante			Dutch Robyjn			Kenya Sifa		
	Sprouts vigour scores after n weeks			Sprouts vigour scores after n weeks			Sprouts vigour scores after n weeks		
	n=2	n=6	n=10	N=2	n=6	n=10	n=2	N=6	N=10
Control	0a	3.47a	3.73a	0a	3a	3.33a	0a	2a	2.33a
0G.50C	0a	3.33a	3.73a	0a	3a	3.47a	0a	2.33b	2.33a
0G.75C	0a	3.33a	3.67a	0a	3a	3.33a	0a	2.33b	2.47ab
0G.100C	0a	3.47a	4b	0a	3.33b	3.27a	0a	2.67cd	2.67bc
50G.0C	0a	4b	4.13b	0a	3a	3.33a	0a	2.33b	2.47ab
50G.50C	0a	3.93b	4b	0a	3.33bc	3.53ab	0a	2.47bc	2.47ab
50G.75C	0a	4b	4b	0a	3.47cd	3.47ab	0a	2.93d	2.73cd
50G.100C	0a	4b	4b	0a	3.67d	3.67b	0a	3d	2.933d
100G.0C	0.67c	4b	4.14bc	1.33c	4e	4c	0a	3d	3d
100G.50C	0.67c	4b	4.33c	1b	4e	4c	0a	3d	3d
100G.75C	0.69c	4.33c	4.27bc	1b	4e	4c	0a	3.33e	3.33e
100G.100C	0.67c	4.33c	4.33c	1b	4e	4c	0a	4f	4f
300G.0C	1.63c	4.67d	4.73d	1b	4e	4c	1b	4f	4f
300G.50C	1.33b	4.67d	5d	1b	4.13e	4.13cd	1b	4f	4f
300G.75C	1.33b	4.73d	5d	1b	4e	4.33d	1b	4f	4f
300G.100C	1.63c	4.57cd	5d	1b	4.33	4.33d	1b	4.33	4.13f
LSD$_{0.05}$ Treat.(T)	0.210	0.298	0.282	0.210	0.298	0.282	0.210	0.298	0.282
LSD$_{0.05}$ Genotype (G)	0.99	0.136	0.138	0.99	0.136	0.138	0.99	0.136	0.138
CV%	28.0	5.0	4.7	28.0	5.0	4.7	28.0	5.0	4.7

*Means followed by the same letters along the column are not significantly different (p≤0.05).LSD $_{0.05}$ =least significant difference at 5% probability level; CV% =percent coefficient of variation and xG.yC means combination of x parts per million (ppm) gibberellic acid and y ppm Cytokinins (benzylaminopurine).

Table 5: Effects of foliar application of rates of combination of gibberellic acid and benzylaminopurine on subsequent germination, number of stems per plant and number of leaflets per plant of potato genotypes Asante, Dutch Robyjn and Kenya Sifa (2009).

Treatments	Genotypes											
	Asante				Dutch Robyjn				Kenya Sifa			
	%Germ. Score	Stems/ plant	Leaflets /plant	Tubers/ plant	%Germ score	Stems/ plant	Leaflets /plant	Tubers/ plant	%Germ score	Stems/ plant	Leaflets /plant	Tubers/ plant
Control	3.50a	1.90a	204.0a	7.13a	2.94a	1.82a	159.1a	6.93a	3.00a	1.74a	131.6a	5.63a
0GA.50BA	3.50a	1.98a	208.2b	7.37ab	2.94a	1.81a	165.1b	7.13a	3.11b	1.81ab	137.2b	5.70a
0GA.75BA	3.50a	2.20b	213.0b	8.60bc	3.00a	1.81a	171.1c	7.67ab	3.11b	1.90bc	142.4c	6.00ab
0GA100BA	3.50a	2.34c	218.4c	8.01bc	3.07bc	1.89a	178.7d	8.43b	3.13b	2.00cd	147.2d	6.17ab
50GA.0BA	3.50a	2.41c	220.4c	9.00c	3.02b	1.90a	181.1d	8.63b	3.22c	2.01cd	153.7e	6.13ab
50GA.50BA	3.50a	2.59d	226.4d	9.00c	3.11c	2.16b	189.3e	8.47b	3.22c	2.08de	160.2f	6.67abd
50GA.75BA	3.50a	2.73e	229.0d	10.95d	3.28d	2.20b	195.6f	8.47b	3.22c	2.11def	165.8g	6.87abd
50GA.100BA	3.56a	2.82e	235.6e	10.83d	3.28d	2.40c	200.5g	8.33b	3.28d	2.18efg	171.4h	6.80abd
100GA.0BA	3.61b	3.11f	245.6f	11.10d	3.31d	2.46c	205.2h	9.33c	3.39e	2.23fgh	185.2i	7.10bd
100GA.50BA	3.67b	3.14f	251.7g	11.73de	3.39e	2.69d	219.4i	8.60b	3.50f	2.28ghi	192.9j	7.93de
100GA.75BA	3.78c	3.30g	257.4h	11.27de	3.39e	2.60d	224.6j	10.20cd	3.56g	2.33hi	197.7k	7.60bde
100GA.100BA	3.78c	3.32g	262.4i	12.20de	3.56f	2.69d	231.2k	10.07cd	3.61g	2.39ij	204.9l	8.73e
300GA.0BA	3.83c	3.52h	272.4j	11.73de	3.56f	2.93e	239.6l	10.43d	3.67h	2.49jk	218.6m	8.80e
300GA.50BA	3.83c	3.63h	275.2j	11.70de	3.56f	2.97e	246.6m	10.93d	3.67h	2.62kl	229.6n	8.93e
300GA.75BA	3.83c	3.90i	282.6k	12.37e	3.56f	3.19f	246.8m	10.33d	3.67h	2.60kl	236.7o	8.67e
300GA.100BA	3.85c	3.96i	287.1k	12.53e	3.56f	2.98e	258.2n	10.47d	3.67h	2.72l	245.3p	9.60e
LSD$_{0.05}$Genotype	0.06	0.08	4.69	0.44	0.06	0.08	4.69	0.44	0.06	0.08	4.69	0.44
CV%.	1.00	3.4	6.7	12.1	1.00	3.40	6.7	12.1	1.00	3.40	6.7	12.1

*Means followed by the same letters along the column are not significantly different (p≤0.05).LSD $_{0.05}$ =least significant difference at 5% probability level; CV% =percent coefficient of variation and xG.yC means combination of x parts per million (ppm) gibberellic acid and y ppm Cytokinins (benzylaminopurine).

Table 6: *Effects of foliar application of rates of combination of gibberellic acid and benzylaminopurine on subsequent yields of potato genotypes Asante, Dutch Robyjn and Kenya Sifa (Year 2009).*

| Treatments | Yields in tons per hectare | | | | | | | | | | | |
| | Asante | | | | Dutch Robyjn | | | | Kenya Sifa | | | |
	Chatts	Seed	Ware	Total	Chatts	Seed	Ware	Total	Chatts	Seed	Ware	Total
Control	0.93a	10.83a	12.90f	24.66a	1.60a	9.63a	5.50i	16.73a	0.73a	5.33a	11.27ef	17.53a
0GA.50BA	1.06a	11.73ab	12.37ef	25.26ab	1.70a	10.27ab	5.17i	17.13a	0.92ab	6.07ab	10.80cdf	17.80ab
0GA.75BA	1.07a	12.63bc	12.37ef	26.07b	1.53a	10.37ab	4.97gi	16.87a	0.73a	6.40b	10.67cde	17.80ab
0GA100BA	1.33b	13.23cd	12.20e	26.26b	2.03b	10.63bc	4.47fg	17.13a	1.03bc	6.90b	10.40abcd	18.33abc
50GA.0BA	1.47bc	14.13d	11.73de	27.33cd	2.00b	11.30cd	4.13ef	17.43a	1.00bc	7.97c	10.40abcd	19.07bc
50GA.50BA	1.40bc	15.80e	11.27cd	28.47de	2.10c	11.07c	4.43fg	17.60a	1.13cd	7.93c	9.93ab	18.99bc
50GA.75BA	1.53c	16.07ef	10.87dc	28.47de	1.90b	12.10d	3.87e	17.87a	0.93bc	8.47c	10.20abc	19.60cd
50GA.100BA	1.57cd	16.90f	10.70dc	29.17ef	2.33d	13.40e	3.70e	19.43b	1.33def	8.97c	9.83a	20.43d
100GA.0BA	1.67cde	17.93g	10.37d	29.97fg	2.47de	15.60f	2.90d	20.97c	1.20d	13.63f	10.57bcd	25.40ef
100GA.50BA	1.73cde	19.30h	9.30c	30.33fg	2.43de	15.67f	2.70cd	20.80c	1.27de	14.17f	10.97df	26.41fg
100GA.75BA	1.77def	19.63hi	9.60c	31.00gh	2.30d	16.30f	2.60bcd	21.20cd	1.40e	14.40f	10.93df	26.73g
100GA.100BA	1.83ef	20.30i	9.20bc	31,33hi	2.47de	17.37g	2.50bcd	22.34d	1.33def	15.47g	11.13ef	27.63g
300GA.0BA	1.73cde	21.33j	8.80ab	31.86i	2.40de	18.57hi	2.10abc	23.32e	1.50fg	16.50h	11.37f	29.37h
300GA.50BA	1.93f	21.67j	8.47a	32.03i	2.57e	18.90i	2.20abc	23.29e	1.60fh	16.67h	11.07ef	29.34h
300GA.75BA	1.88f	21.40j	8.60ab	31.88i	2.60e	18.80i	1.97ab	23.37e	1.73h	16.87h	11.13ef	29.73h
300GA.100BA	2.10g	21.97j	8.20a	32.27i	2.57e	19.20i	1.90a	23.37e	1.93i	17.20h	11.00def	30.13h
LSD$_{0.05}$ Genotype	0.13	0.46	0.30	0.60	0.13	0.46	0.30	0.60	1.13	0.46	0.30	0.60
CV%	19.1	7.9	2.7	6.1	19.1	7.9	2.7	6.1	19.1	7.9	2.7	6.1

*Means followed by the same letters along the column are not significantly different (p≤0.05).LSD $_{0.05}$ =least significant difference at 5% probability level; CV% =percent coefficient of variation andxG.yC means combination of x parts per million (ppm) gibberellic acid and y ppm Cytokinins (benzylaminopurine). Chatts= tubers< 25mm diameter, Seed= tubers ≥25mm≤ 55mm and Ware=tuber >55mm

3.9. Yields of Generated Plants

Both treatments and genotypes showed significant variation in total yields compared with the control (Table 6). When GA singly applied, there was a significantly difference in total yields in all genotypes with increased hormone application rates. However for BA treated plants, effects were only significant at higher application rates (75-100ppm) for Asante but total yields showed no significant difference in Dutch Robyjn and Kenya Sifa.

When a combination of GA and BA was employed, significant total yields was only observed at higher dose (100ppm) of BA at each rate of GA except at 300ppm where no significant difference was observed compared to that of GA treated plant at that rate in all genotypes. Asante gave the highest total yields while Dutch Robyjn gave the lowest in all treatments.

Tuber grades (tuber sizes) differed significantly between treatments and genotypes compared with the control (Table 6). The potato seed and the quantity of chatts increased with increase in rates of rate of concentration of GA in all genotypes. A decrease in yields of ware was registered with increased concentration of GA when applied alone or in combination with BA. However, application of BA alone only reflected significant difference at 100ppm in Asante and Dutch Robyjn while variation of ware yield in Kenya Sifa was not sequential and did not vary greatly with concentration of both GA and BA. Yields of chatts were highest in Dutch Robyjn while Kenya Sifa had the lowest grades of both chats and seed in all treatments. Asante yielded most seed grade in all treatments. Dutch Robyjn gave the lowest ware yields while at lower levels of GA concentration (0-50ppm), Asante outweighed Kenya Sifa. However, at higher GA concentration (100-300ppm), Asante recorded lower ware yields (less than

10tons/ha) while Kenya Sifa steadily maintained yields above 10 tons/ha (Table 6).

4. Discussions

4.1. Effects of Foliar Application of Gibberellic Acid and Cytokinins on Subsequent Potato Tuber Sprouting

In this study, foliar application of GA alone or GA+BA resulted in significant decrease in duration of subsequent tuber dormancy period and sprouting and this varied among the genotypes. However, BA alone had no effect on duration of tuber dormancy and sprouting. These results are in agreement with [11] findings that a foliar spray of 300ppm- 375ppm gibberellic acid, 3–6 days before haulm killing shorten potato tuber dormancy period and induced sprouting. The results were also in agreement with [4] findings that when GA was applied to the foliage of potato plants grown from true seeds towards the end of the vegetative cycle, it induced rapid breakage of tuber dormancy. Alexopoulos [5] also observed that exogenous application of GA on potato plant foliage drastically increased the concentration of endogenous gibberellins and simple reducing sugars of the resulting tubers leading to visible sprouting. These results also agreed with[3] findings that application of low doses of cytokinins had no effect on potato dormancy.

This study revealed that foliar spray of GA, BA or GA + BA increased tuber sprout length, number of sprouts per tuber and sprout vigour in a dose dependent manner and also varied with genotype. The number of sprouts per tuber, sprout length and vigour were significantly higher with increased rate of GA in all genotypes. However, when BA was applied alone, variation of tuber sprout length and number for different rate was genotype dependent with significant observation from

50ppm for Asante, 75ppm for Dutch Robyjn and 100ppm for Kenya Sifa with respect to control. Results from various researchers while working on effects of foliar spray of individual hormones on different variable found similar findings. The growth length and vigour of potato tubers sprouts was greatly advanced by a foliar spray with gibberellic acid [10], the magnitude of the GA effects depended on the cultivar [11] and increased level of GA was found to exhibit premature sprouting [14]. The number and length of sprouts of tubers treated with a combination of GA and BA were higher than those of tubers treated with each hormone at the same rate and varied with increase in rate of combination of GA and BA in all genotypes. At each rate of BA, both sprouts length and number of sprouts per tuber increased with increased rate of GA concentration. However, it was noted that BA exhibited significant difference after each increase rate at highest rate of GA (300 ppm) unlike at lower rates of GA. Vigour score increased with duration of storage and with rates of application of GA, BA or their combination. Asante tubers produced the strongest sprouts while Kenya Sifa gave the weakest. However, at higher doses of GA (300ppm), Dutch Robyjn produced longer but slender sprouts.

Despite the involvement of BA in cell division, the number of cell may not necessarily results to visible growth [19, 20]. The results from this study indicate that GA alone effected duration of dormancy termination and sprouts growth. This may indicate that GA is involved in both cell division and elongation. The above results may also be physiologically explained by several research findings that GA is involved in synthesis of α-amylase enzyme [19, 21] involved in breakdown of starch to glucose and fructose and facilitation of the movement of cytokinins to the buds enhancing cell division [22].

4.2. Effect of Foliar Application of Gibberellic Acid and Cytokinins on Subsequent Potato tuber Yields

Foliar application of GA alone caused significant increase in subsequent germination, stems per plant, leaflets per plant and number of tubers per plant and yields at harvest with increased rate of application in all genotypes. Asante had the highest germination score in all treatments. Dutch Robyjn recorded higher germination than Kenya Sifa at lower rates of GA (0-50ppm) while the opposite occurred at higher rates (100-300ppm) of GA applications. At low concentration rate of GA (0-50ppm), the number of tuber per plant in Asante and Dutch Robyjn were not significantly different. These results are in agreement with Abd[23] findings who observed thatfoliar application of GA enhances vegetative growth, length of plant, average number of shoots, leaves number, fresh and dry weight of shoots and gives more yields. They also agree with Alexopoulos[22] findings that foliar application of GA in plants derived from TPS caused an increase in the number of tubers per plant and increased the yields.

Spraying of BA alone caused significant increase in germination in Dutch Robyjn and Kenya Sifa genotypes but not in Asante, whereas application of 100ppm BA resulted in increased stems/tuber, leaflets/plant and tubers/plant except in Kenya Sifa for tubers/plant. This was in tandem with

Dwelle[24] and Blunden [25] who while working separately found thatfoliar application of 'Cytex' product, a commercial aqueous seaweed extract equivalent to 100ppm cytokinins activity and synthetic cytokinin 'kinetin' produced a significant increase in the yield of potatoes. However, at lower rates of GA applications, there was no significant difference in germination score for Asante (0-50ppm) and Dutch Robyjn even with increased rates of concentration of BA.

Plants treated with a combination of BA and GA gave significantly more stems per plant, leaflets per plant than when each hormone was applied alone. Asante genotype gave the highest stems per plant and leaflets per plant and total yields while Kenya Sifa gave the lowest except the total yield which was given by Dutch Robyjn. These findings were in agreement with Caldiz[12] findings who also observed that foliar applications of a combination of benzylaminopurine and gibberellic acid under both field and glasshouse conditions increased both tuber number and tuber yields. However,yields of ware decreased with increased concentration of GA when applied alone or in combination with BA but in BA treated plants, yields of ware tuber differed only at 100ppm in Asante and Dutch Robyjn. Production of ware yield did not vary with concentration of GA, BA or their combination in Kenya Sifa. Both the number and weight of seed and chatts tubers increased with increase in rates of concentration of GA alone or its combination with BA in all genotypes.These observations agree with Stuik[26] results who similarly observed that increase of GA application on potato plant increased the number of tubers and total yields but production shifted the tuber size distribution towards the smaller grades. It was observed that not all sprouts gave rise to stems. Increase in the numbers of sprouts per tuber may have resulted in increased stems per plant, leaflets per plant and faster rate of canopy cover. This may have resulted to higher amount of intercepted radiation increasing number of tubers per plant and total yields [27].

5. Conclusions and Recommendations

This study revealed that foliar application of GA alone or a combination of GA and BA late in the growth cycle results in significant decrease in duration of subsequent tuber dormancy period and sprouting and this vary among the genotypes. Foliar spray of GA, BA or GA + BA increase tuber sprout length, number of sprouts per tuber and sprouts vigour in a dose dependent manner but varies with genotype. A combination of BA and GA results in significantly more growth than using only GA or BA alone at the same level.It seems that BA and GA has dose dependent synergistic effect on potato tuber sprouting and sprouts growth.

Foliar application of GA alone or GA+BA causes significant increase in subsequent germination, stems per plant, leaflets per plant and number of tubers per plant at harvest with increased rate of application in all genotypes. GA enhances vegetative growth; average number of stems, leaflets number and more yields. More sprouts per tuber may result in more stems per plant. Consequently, more stems per plant results more leaves, and the ground cover taking place at a faster rate of ground cover, higher amount of intercepted radiation and assimilation and hence higher total yields. Foliar application of gibberellic acid or a combination of GA and BA late in the growth cycle may be of practical value in cases where tubers are required for planting soon after harvest. However, higher doses of hormones are recommended for foliar application than when the hormones were directly sprayed on tubers

Acknowledgements

The authors appreciate assistance by International Potato Centre (CIP), Nairobi in importation of True Potato Seed (TPS) from Lima, Peru; Centre Director, KALRO, National Potato Research Centre, Tigoni and Principal, Njabini Agricultural Training Centre, Nyandarua, for providing land resource and other facilities essential for the field work.

References

[1]   Demo P., Akoroda M.O., El-Bedewy R. and Asiedu R. (2004). Monitoring storage loses of seed potato (*Solanumtuberosum* L.) tubers of different sizes under diffuse light conditions. Proceedings of 6th triennial congress of the African Potato Association (APA). 5-10 April, 2004. Agadir, Moroco. Pg. 363-370.

[2]   Burton W.G., Van E. A. and Hartmans K.J. (1992). The physics and physiology of storage. In: Harris P. M., ed. The potato crop. London: Chapman and Hall; 608–727.

[3]   Suttle J.C. (2004). Involvement of endogenous gibberellins in potato tuber dormancy and early sprout growth: a critical evaluation. Journal of Plant Physiology 161:157-164.

[4]   Alexopoulos A. A., Akoumianakis A., Olympio C. M. And Passam H. C. (2007). The effect of time and mode of application of gibberellic acid and inhibitor of gibberellic acid biosynthesis on the dormancy of potato tubers grown from true potato seed. Journal of the science and agriculture 87(10): 1973-1979.

[5]   Alexopoulos A. A., Akoumianakis A. K., Vemmos S. M and Passum H. C. (2006).The effect of postharvest application of gibberellic acid and benzyl adenine on duration of dormancy of potato produced by plant grown from true potato seeds. Post harvest biology and technology 46(1): 54-62.

[6]   Shibairo S. I., Demo P., Kabira J.N., Gildemacher P., Gachago E., Menza M., Nyankanga R.O., Cheminingwa G.N. and Narla R.D. (2006). Effects of Gibberellic Acid (GA3) on Sprouting and Quality of Potato Seed Tubers in Diffuse Light and Pit Storage Conditions. Journal of Biological Sciences 6 (4): 723-733.

[7]   Suttle J.C. (1996).Dormancy in tuberous organs: problems and perspectives. In: Lang GA, ed. Plant dormancy: physiology, biochemistry and molecular biology, Wallingford, UK: CAB International, 133–146.

[8]   Craufurd P.Q., Summerfield R. J., Asiedu R. and Vara P.V. (2001).Dormancy in yams (*Dioscorea spp.*). Experimental Agriculture 37: 147–181.

[9]   Ile E. I. (2004). Control of tuber dormancy and flowering in yam (*Dioscorearotundata*Poir.) tuber. PhD. Thesis, The University of Reading, Reading, United Kingdom.

[10] Ittersum M. K.(1992). Dormancy and growth vigour of seed potato. Wageningen Agricultural University dissertation No. 1556 Wageningen University.

[11] Ittersum M. K.and ScholteK. (1993). Shortening dormancy of seed potatoes by a haulm application of gibberellic acid and storage temperature regimes. American Journal of Potato Research 70(1): 7-19

[12] Caldiz D. O., (1996). Seed potato (Solanumtuberosum L.) yield and tuber number increase after foliar applications of cytokinins and gibberellic acid under field and glasshouse conditions. Plant Growth Regulation20 (3): 185-188

[13] Dwelle, R.B. (1985). Photosynthesis and Photoassimilate Partitioning. Potato Physiology. Academic Press, Inc. Orlando, Florida: 35-38

[14] Suttle J.C. and Banowetz, J. (2000). Comparing potato tuberization and sprouting opposite phenomena. Cited on American journals of Potato Research, July/Aug. 2004.

[15] Turbull C. G and Hanke D. E. (1985). The control of bud dormancy in potato tubers. Springer Link journals 165(3): 359-365.

[16] Banas A., M. Bielinska-Czarnecka, and J. Klocek.(1984). Activity of endogenous cytokinins in potato tubers during dormancy and sprouting 23:213-218.

[17] Mikitzel. L.J. and Fuller N. (1995). Dry Gibberellic Acid Combined With Talc or Fir Bark Enhances Early Stem and Tuber Growth of Shepody Potato. American Potato Journal. 72: 545-550.

[18] National Potato Council of Kenya (NPC). (2015). Potato variety catalogue. 2015. Eagle Creations, Nairobi, Kenya. 50p. http://www.npck.org/images/potato%20variety%20catalogue%202015.pdf

[19] Arteca R.N. (1996). Plant growth substance. Principles and application: 148-156. Chapman and Hall, New York.

[20] Vreugdenill D. (2004). Comparing potato tuberization and sprouting: Opposite phenomena? American Potato Research 81: 275-280.

[21] Vivanco J.M. and Flores H.F. (2000). Control of root formation by plant growth regulators. In: Basra A.S. (ed). Plant growth regulators in agriculture and horticulture. Their role and commercial use: 1-16.

[22] Alexopoulos A. A., Akoumianakis A. K. and Passum H. C. (2006). The effects of time and mode of application of gibberellic acid of the growth and the yields potato plant derived from true potato seeds. Journal of science of food and agriculture Vol. 86:2189-2195.

[23] Abd El-Aal F.S., Shaheen A.M. and Fatma A.R. (2008). The effect of foliar application of gibberellic acid and soil dressing of NPK at different levels on the plant productivity of potatoes *(Solanumtuberosum L)*. Research Journal of Agriculture and Biological Sciences, 4(5): 384-391.

[24] Dwelle R. B. and Hurley P. J.(1984). The effects of foliar application of cytokinins on potato yields in southeastern Idaho American journal of Potato Research 61 (5): 293-299.

[25] Blunden G and Wildgoose P.B. (2006). The effects of aqueous seaweed extract and kinetin on potato yields. Journal of the science of food and agriculture 28 (2): 121-125.

[26] Stuik P.C., Kramer G. and Smit N.P. (1989). Effects of soil application of gibberellic acid on yields and quality of tuber of Solanumtuberosum L. cv.Bintje. Journal of potato research 32(2): 203-209.

[27] Wiersema S.G. (1989). Comparative performance of three small seed tuber size and standard size seed tubers planted at similar densities. Potato Res. 32:81-89.

9

Humic Acid to Decrease Fertilization Rate on Potato (*Solanum tuberosum* L.)

Ismail Ali Abu-Zinada[1, *]**, Kamal Soliman Sekh-Eleid**[2]

[1]Department of Plant Production & Protection, Faculty of Agriculture & Environment, Al-Azhar University-Gaza, Gaza Strip, Palestinian Territories

[2]Department of Plant Production & Protection, Faculty of Agriculture & Environment, Al-Azhar University-Gaza, Gaza Strip, Palestinian

Email address:

isalznada@hotmail.com (I. A. Abu-Zinada), kamaleid@yahoo.com (K. S. Sekh-Eleid)

Abstract: The current study aimed at decreasing fertilizers applications on potato (*Solanum tuberosum* L.) cultivar Spunta. Plants were subjected to six treatments as follows: unfertilized control (T0), fertilization program (166-N+ 80-P_2O_5+ 80-k_2O + 30M^3 cattle manure ha[-1]) of Ministry (MAP) of Agriculture (T1), 100% MAP + 20kg humic acid (HA) ha[-1] (T2), 100% MAP + 15 kg HA ha[-1] (T3), 50% MAP + 20 kg HA ha[-1] (T4), 50% MAP+15 kg HA ha[-1] (T5). Vegetative growth increased after the different fertilization applications than control where T1, T4 and T3 had the longest plants; T1, T2 and T4 emerged the significant highest number of main stems; T1 and T4 produced the significant highest leaf area and T4 in both seasons and T3 and T2 in the first season had the significant heaviest plant fresh weight. Yield components in general significantly increased where T2 and T3 produced the significant highest tubers number plant[-1]; T3 yielded the significant highest tubers weight plant[-1]; T4, T1 and T2 significantly had the highest number of tubers >60 mm diameter plant[-1]. Tuber physical properties were also significantly and positively affected as compared to control where T3 and T4 resulted in the significant longest tubers. T5, T3 and T1 had the significant widest tubers and T3 and T4 produced the significant heaviest tubers. It could be recommended under similar conditions to add 83-N+40-P_2O_5+ 40-k_2O + 15M^3 cattle manure ha[-1] + 20kg humic acid ha[-1].

Keywords: Humic Acid, Potato Yield, Tuber Properties, Vegetative Growth

1. Introduction

Potato is the world's fourth largest food crop where it plays an important role as a staple food in the Mediterranean Basin countries. The crop occupied an overall area about 1 million hectares which produced 28 million tons of tubers (FAO, 2011). Potato locally is considered as one of the most important vegetable crops where the crop total cultivated area reached about 1380 hectares constituting 52% of the vegetables area (MoA, 2010).

Humic acid contains many elements and it acts as an amendant to improve soil fertility. This increases the availability of nutrients and consequently it increases plant growth and yield. Humic acid particularly is used to ameliorate or reduce the side effect of chemicals. The acid application increased organic matter in soil which improved plant growth and yield (Chen and Aviad., 1990; David, et al., 1994; Hartwigson and Evans., 2000; Hafez., 2003; Erik, et al., 2000; El-Desuki., 2004). Humic substances are able to capture more moisture content that will increase the water use efficiency in the sandy soil. This may be attributed to the swelling and retention of water by the amended soil (Suganya and Sivasamy, 2006). Humic acid efficiently improves soil fertility and crop productivity (Chen and Aviad, 1990; Rajpar et al., 2011). Humic acid affects chemical and biological properties of soil as well as morpho-physiological processes of a plant (Ohta et al., 2004).

Vegetative growth, yield and tuber quality as well as the tuber nutritive value of potato significantly increased with humic acid level increase where no significant differences were noticed between 1 and 2 kgfed[-1] (Mahmoud and Hafez, 2010). Humic acid application led to positive changes in vegetative growth, leaf area index due to increase in root growth and nutrients availability (El-Hefny, 2010). Tuber yield increased by 16.47% after addition of humic substances compared to the recommended rate solely. These substances + 75% of the recommended NPK fertilizer was beneficent (Selim et al., 209). Soil application of humic acid significantly increased plant growth, photosynthetic pigments, total and marketable yield and tuber root quality (El-Sayed Hameda et al., 2011). Humic acid at 0, 10, 20 and 30 cm L^{-1} of irrigation water enhanced potato growth parameters, yield and tuber physical and chemical properties. The highest dose of the acid resulted in highest plant vigor increase, the heaviest tuber yield and the best tuber properties (Rizk et al., 2013). Soil application of humic acid did not affect tuber size, total yield or other chemical composition of tubers. However, 80g m^{-2} increased incidence of tubers with hollow heart (Suh et al., 2014).

This study aimed at investigation the effect of Humic acid application levels and fertilization rate on growth and productivity of potato crops.

2. Material and Methods

2.1. Location and Season

This trial was carried in the two seasons of 2013 and 2014 at a farm of the privates sector in Gaza Strip, Palestine. Spunta cultivar which is the most favorite and cultivated variety was used in this experiment.

2.2. The Experiment Layout and Treatments

It was arranged in the Randomized Complete Block Design (RCBD), with four replications. The plot area was 26.25 m^2 containing 150 plants which were spaced at 70 x 25 cm. Six fertilization treatments were used as they follow in table (1).

Table 1. Treatments applied.

Treatments content	Treatment code
Unfertilized control	T0
100% Ministry of Agriculture recommended (166-N+ 80-P_2O_5+ 80-k_2O + 30M^3 cattle manure ha^{-1}) program	T1
100 % recommended program + Humic acid 20 kg ha^{-1}	T2
100% recommended program + Humic acid 15 kg ha^{-1}	T3
50% recommended program + Humic acid 20 kg ha^{-1}	T4
50% recommended program + Humic acid 15 kg ha^{-1}	T5

Humic acid at15 or 20 kg ha^{-1} was applied on soil by spraying before sowing the cut-seeds.

2.3. Data Recorded

2.3.1. Vegetative Growth
A random sample of five plants plot^{-1} were devoted after 70 days of planting to determine plant length (cm), number of stems plant^{-1}, leaf area (cm^2) and plant fresh weight (g).

2.3.2. Yield Components
They were determined after 100 days of sowing as tubers weight plant^{-1}, number of tuber plant^{-1} and number of tuber of diameter >60 mm plant^{-1}.

2.3.3. Tuber Quality
At harvesting time (after 100 days of sowing), a random samples of 20 tubers plot^{-1} were devoted to determine the physical properties as an average of tuber weight (g), tuber length (cm) and tuber diameter (cm)

2.4. Statistical Analysis

Data were statistically analyzed according to Steel and Torrie (1980), where means comparison was carried out using Duncan's multiple range test. Means followed by the same letter/s within each columns are not significantly different at $p= 0.05$.

3. Results and Discussion

3.1. Vegetative Growth

Data reported in Table (2) show the different measurements of vegetative growth. It is clear that plant height generally showed a significant increase during the two seasons than control. Treatments T1 and T4 gave the significant longest plants in both seasons where T3 had the longest plants in the second season only. Number of stems plant^{-1} increased in the fertilizers applications than control during the two seasons where T1and T2 had the highest significant increase in both seasons. In this concern, no significant changes were noticed among fertilizers treatments in both seasons. Leaf area increased after fertilizers applications than untreated control in the two seasons, where in general this increase was significantly in the first season only. In this concern, T1 produced the significant highest area in both seasons where T4 in the first season and T3 in the second one came to the second position. Plant fresh weight showed that the different fertilizers treatments had heavier plants than control in the two seasons where treatment T4 resulted in the significant heaviest plant in both seasons. In this issue, T2 and T3 respectively came to the second significant rank in the first season and the same trend insignificantly was also true in the second experiment. No significant differences were detected between T4 and the other fertilization treatments in the two seasons.

Humic substances such as humic acid is the main component (65-70%) of soil organic matter which enormously increases plant growth by increasing permeability of cell membrane, respiration, phosynthesis rate, oxygen and supplying root cell growth (Russo and Berlyn, 1990). The increase in plant height and stems number of potato is due to humic acid nutrient providing. These elements involve in plant bioactivities and finally encourage plant growth induction (Abdel-Mawgood et al., 2007 and Taha, 2011). This increment in vegetative growth may be attributed to the enhancing effect of humic acid on the availability of nutrients and the role of potassium in plant nutrition which in turn increased the vegetative growth of potato plants (Mahmoud and hafez, 2010). Humic acid increases the soil prosperity which improves root growth and to increase shoot system (Gracia et al., 2008). Humic acid is beneficial to shoot and root growth by playing a role in nutrient's uptake in vegetable crops (Dursun et al., 2002;

Cimrin and Yilmaz, 2005). Humic substances have a direct action on plant growth by influencing metabolic processes such as nucleic acid synthesis, ion uptake and regulation of hormone levels (Serenella et al., 2002).

These results were in harmony with those reported on potato by El-Hefny, (2010) Mahmoud and Hafez (2010), and El-Sayed Hameda et al., (2011) and Rizk et al., (2013).

Table 2. Effect of humic acid and fertilization levels on potato vegetative growth.

Treatments	Plant height (cm)		Main stems no.		Leaf area (cm^2)		Plant fresh weight (g)	
	First	Second	First	Second	First	Second	First	Second
T0	28.0 c	39.9 c	1.83 b	1.80 b	139 d	163 b	186 b	296 b
T1	36.0 a	51.9 ab	3.66 a	2.35 a	275 a	290 a	333 ab	372 ab
T2	33.3 ab	50.2 b	3.33 a	2.30 ab	193 c	210 ab	433 a	453 ab
T3	32.0 b	57.4 a	3.00 ab	2.15 b	187 cd	272 ab	342 a	458 ab
T4	36.6 a	53.2 ab	3.10 ab	2.25 b	258 ab	232 ab	445 a	508 a
T5	31.6 bc	50.9 b	2.66 ab	2.15 b	219 bc	204 ab	314 ab	389 ab

Means of same letter/s in a column don't differ significantly at $p= 0.05$ (Duncan's multiple range test).

3.2. Yield Components

The different components of yield i.e., number of tubers of plant^{-1}, weight of tubers plant^{-1} and tuber number size >60 mm are reported in Table (3). It is evident that number of tubers plant^{-1} increased by the different fertilization treatments than control in both seasons. In this respect, T4, T2 and T3 respectively in the first season and T2 in the second one resulted in the highest significant increase. In addition, T3 came to the second position in second season only with no significant differences among T3 and the other fertilization treatments. Tubers weight plant^{-1} in general significantly increased due to the fertilization treatments as compared to control in the two seasons. Treatment T3 in the first season and all the fertilization applications in the second season showed the highest significant increase. Generally, no significant differences were observed among fertilization treatments in the two seasons. Number of tubers plant^{-1} (diameter >60 mm) increased in both seasons where the increase was significantly higher than that of control in T4 in

the first season and T4, T1 and T3 in the second one. No significant changes were noticed among all fertilization treatments in the first season while T3 was significantly lower than T4, T1 and T2 in the second season. Humic acid and nitroxin leads to increase plant yield through positive physiological effect such as impact on metabolism of plant cells and increasing the concentration of leaf chlorophyll (sure, et al. 2012). Humic acid is a promising natural resource that can be used as an alternative to synthetic fertilizers to increase crop production. It exerts either a direct effect, such as on enzymatic activities and membrane permeability, or an indirect effect, mainly by changing the soil structure (Biondi, et al., 1994).

Yield components in this trial came to the same trend of the results reported on potato by El- Selim et al., (2009), Hefny., (2010), Mahmoud and Hafez., (2010), El-Sayed Hameda et al., (2011) and Rizk et al., (2013). On the other hand, soil application of humic acid did not affect potato yield components (Suh et al., 2014).

Table 3. Effect of humic acid and fertilization level on yield components.

Treatments	Tuber number plant^{-1}		Tubers weight g plant^{-1}		Tuber number plant^{-1} (diameter>60 mm)	
	First	Second	First	Second	First	Second
T0	5.6 b	5.5 b	753 c	632 b	2.43 b	2.20 c
T1	7.3 ab	7.2 ab	955 abc	1004 a	3.80 ab	4.85 a
T2	9.0 a	8.5 a	1133 ab	1016 a	3.96 ab	4.60 a
T3	8.6 a	7.8 ab	1177 a	1209 a	4.20 ab	3.80 b
T4	9.3 a	7.3 ab	893 bc	1084 a	4.86 a	5.17 a
T5	7.6 ab	7.3 ab	1024 ab	1115 a	3.73 ab	4.47 ab

Means of same letter/s in a column don't differ significantly at $p= 0.05$ (Duncan's multiple range test).

3.3. Tuber Properties

The physical traits i.e., tuber length, tuber diameter and average of tuber weight are reported in Table (4). Tuber length in the different fertilization treatments was significantly higher than that of the respective control in the first season. The significant longest tuber was obtained by treatments T3 and T4 respectively in comparison with other fertilizers applications in the first season where the different fertilization treatments also showed longer tuber than control

in the second season. This increase was significantly the highest in T3 only where T4 came to second position in this concern. Tuber diameter generally showed significant increase in all fertilization treatments compared to control in both seasons. In this issue, T5 in the first season and T3, T5 and T1 in the second one had the significant widest tubers. Average of tuber weight in general produced higher significant increase in fertilization treatments than control in the first season. This increase was significantly in T3 and T4 in the second seasons only where T3 produced the heaviest tubers in the first seasons.

The current results found support in the work of Mohmoud and Hafes., (2010), El-Sayed Hameda et al., (2011), and Rizk et al (2013). Contrary results were reported by Suh et al., (2014) that soil application of humic acid had no effect on potato physical or chemical properties.

Table 4. Effect of humic acid and fertilization level on tuber physical properties.

Treatments	Tuber length (cm)		Tuber diameter (cm)		Average of tuber weight (g)	
	First	Second	First	Second	First	Second
T0	7.66 c	8.62 b	4.3 c	5.4 b	130 c	169 b
T1	9.66 ab	9.37 b	6.0 ab	6.1 a	161 ab	190 ab
T2	9.83 ab	8.92 b	6.0 ab	5.7 ab	167 ab	196 ab
T3	10.50 a	10.27 a	6.0 ab	6.2 a	184 a	237 a
T4	10.16 a	9.55 ab	5.8 b	5.8 ab	149 bc	224 a
T5	8.83 b	9.32 b	7.3 a	6.1 a	159 ab	206 ab

Means of same letter/s in a column don't differ significantly at $p= 0.05$ (Duncan's multiple range test).

4. Conclusion

The local farmers have accustomed to use excess of chemical fertilizers which increased nitrate level in wells water for domestic use. Humic acid was applied on soil grown with potato cv. Spunta at 15 or 20 kg ha^{-1} to decrease level of fertilization program of Ministry of Agriculture. The two humic acid doses under the different fertilizers applications could improve vegetative growth, yield and tuber properties. The most economic and effective results were obtained by application of 83-N+40-P$_2$O$_5$+ 40-k$_2$O kg ha^{-1} + 15M^3 cattle manure ha^{-1} + 20kg humic acid ha^{-1}.

References

[1] Abdel Mawgoud, A., El Greadly, M.R.N., Helmy, Y.I. and Singer, S.M. 2007. Responses of tomato plants to different rates of humic based fertilizer and NPK fertilization. Journal of Applied Sciences Research 3, 169-174.

[2] Biondi, F.A., Figholia, A., Indiati, R., and Izza, C. 1994. Effect of fertilization with humic acidson soil and plant metabolism: a multidisciplinary approach. Note III: phosphorus dynamics and behavior of some plant enzymatic activities. In Humic Substances in the Global Environment and Implications on Human Health, ed. Senesi N & Miano TM. Elsevier, New York, pp. 239-244.

[3] Chen, Y. and Aviad, T. (1990): Effect of humic substances on plant growth. In: MacCarthy P., Clapp C.E., Mal- 28 colm R.L., Bloom P.R. (eds): Humic Substances in Soil and Crop Sciences: Selected Reading. Soil Science Society of America, Madison, 161–187.

[4] Cimrin, K.M. and Yilmaz, I. 2005. Humic acid applications to lettuce do not improve yield but do improve phosphorus availability. Acta Agriculturae Scandinavica, Section B - Soil & Plant Science, 55 (1): 58-63.

[5] David, P.P., Nelson, P.V. and Sanders, D.C. 1994. A humic acid improves growth of tomato seedling solution culture. Journal of Plant Nutrition, 17 (1): 173-184.

[6] Dursun, A.I Guvenç I. and Turan, M. 2002. Effects of different levels of humic acid on seedling growth sand macro and micronutrient contents of tomato and eggplant. Acta Agrobotanica, 56: 81-88.

[7] EL-Desuki, M. 2004. Response of onion plants to humic acid and mineral fertilizers application. Annals of Agricultural Science, Moshtohor, 42 (4): 1995-1964.

[8] El-Hefn, EM. 2010. Effect of saline irrigation water and humic acid application on growth and productivity of two cultivars of cowpea *Vigina unguiculata* L. Australian Journal of Basic and Applied Science, 4 (12): 6154- 6168.

[9] El-Sayed Hameda, E.A., Saif El Dean, A., Ezzat, S., and El Morsy, A.H.A. 2011. Responses of productivity and quality of sweet potato to phosphorus fertilizer rates and application methods of the humic acid. International Research Journal of Agriculture Science and Soil Science, 1 (9): 383-393.

[10] Erik, B., Feibert G., Shock C.C., Saundres, L.D. 2000. Evaluation of humic acid and other non conventional fertilizer additives for onion productivity. Malheur Experiment Station, Oregon State University Ontario.

[11] FAO. 2011. FAO Statistical Database. Production Crops. Rome, Italy, http://faostat.fao.org/ (verified April 2013).

[12] Gracia, M.C.V., Estrella F.S., Lopez M.J. and Moreeno, J. 2008. Influence of composite amendment on soil biological properties and plants. Dynamic soil, Dynamic Plant, 1: 1-9.

[13] Hafez. M.M. 2003. Effect of some sources of nitrogen fertilizer and concentration of humic acid on the productivity of squash plant. Egyptian Journal of Applied Science, 19 (10): 293-309.

[14] Hartwigson, J.A. and Evans, M.R. 2000. Humic acid seed and substrate treatments promote seedling root development. HortScience, 35 (7): 1231-1233.

[15] Kashif, S.R., Yaseen, M., Arshad, M. and Abbas, M. 2007. Evaluation of response of calcium carbide as a soil amendment to improve nitrogen economy of soil and yield of okra. Soil and Environment, 26 (1): 69-74.

[16] Mahmoud. A.R, Hafez M.M. 2010. Increasing productivity of potato plants (*Solanum tuberosum, L*) by using potassium fertilizer and humic acid application. International Journal of Academic Research, 2 (2):83-88.

[17] Mehdi, K., Tobeh, A., Gholipoor, A., Jahanbakhsh, S., Hassanpanah, D. and Sofalian, O. 2011. Effects of different N fertilizer rate on starch percentage, soluble sugar, dry matter, yield and yield components of potato cultivars. Australian Journal of Basic and Applied Sciences, 5 (9): 1846-1851.

[18] Ministry of Agriculture. 2010. Annual report on vegetables production in Gaza strip. MOA, Gaza, Palestine.

[19] Ohta, K., Morishitai S., Sudai, K., Kobayashii, N. and Hosoki, T. 2004. Effects of chitosan soil mixture treatment in the seedling stage on the growth and flowering of several ornamental plants. Journal of Japanese Society for Horticultural Science, 73: 66-68.

[20] Rajpar, I., Bhatti, M., Hassan, Z. and Shah, A. 2011. Humic acid improves growth, yield and oil content of *Brassica compestris* L. Parkisan Journal Agriculture Engineering and Veterinary sciences. 27 (2): 125-133.

[21] Rizk, F.A., Shaheen, A.M., Singer, S.M. and Sawan, O.A. (2013) The Productivity of potato plants affected by urea fertilizer as foliar Spraying and humic acid added with irrigation water. Middle East Journal of Agricultural Research, 2 (2): 76-83.

[22] Russo, R., O. and Berlyn, G.P. 1990. The use organic biostimulants to help low input sustainable agriculture. Journal of Sustainable Agriculture, 1(2): 19-42.

[23] Selim, E.M., Mosa, A.A. and El-Ghamry, A.M. 2009. Evaluation of humic acid fertigation through surface and subsurface drip irrigation systems on potato grown under Egyptian sandy soil conditions. Agricultural Water Management, 96: 1218-1222.

[24] Serenella, N., Pizzeghelloa, D., Muscolob, A. and Vianello, A. 2002. Physiological effects of humic substances on higher plants. Soil Biology & Biochemistry, 34: 1527-1536.

[25] Steel, R.G.D. and Torrie, J.H. 1980. Principals and procedures of statistics. Approach Second Edition, pp. 633.

[26] Suganya, S. and Sivasamy, R. 2006. Moisture retention and cation exchange capacity of sandy soil as influenced by soil additives. Journal of Applied Sciences Research, 2 (11): 949-951.

[27] Suh H.Y., Yoo K.S. and Suh S.G. 2014. Tuber growth of potato (*Solanum tuberosum* L.) as affected by foliar or soil application of flavic and humic acid. Horticulture, Environment, and Biotechnology. 55 (3): 183-189.

[28] Sure, S., Arooie H., Sharifzade K., Dalirimoghadam, R. 2012. Responses of productivity and quality of cucumber to application of the two bio-fertilizers(humic acid and nitroxin) in fall planting. Agriculture Journal. 7 (6): 401-404.

[29] Taha, Z. Sarhan, 2011. Effect of humic acid and seaweed extracts on growth and yield of potato plant (*Solomun tubersum* L.). Desirce cv. Mesopotamia Journal of Agriculture, (31): 2.

10

Utilization of Mangrove Forest Plant: Nipa Palm (*Nypa fruticans* Wurmb.)

Md. Farid Hossain[1], Md. Anwarul Islam[2]

[1]School of Agriculture and Rural Development, Bangladesh Open University, Gazipur, Bangladesh
[2]School of Education, Bangladesh Open University, Gazipur, Bangladesh

Email address:
faridhossain04@yahoo.com (Md. F. Hossain), anwarul2003islam@yahoo.com (Md. A. Islam)

Abstract: This review paper discusses the production, uses and importance of Nipa palm (*Nypa fruticans* Wurmb., Arecaceae). It is a mangrove palm that grows well in 'Sundarbans' mangrove forest of Bangladesh. Nipa palm is locally called '*Golpata*' used for multipurpose such as roof thatching, partitioning, foods, medicinal purposes and as a source of fuel wood. The sugary sap from the inflorescence stalk is used as a source of treacle (molasses), amorphous sugar, vinegar and alcohol. Newly developed shoots are to be used as a vermicide. Ash from Nipa palm is used as an analgesic against tooth pain and headache. Dry leaves, petiole, stem wood, fruit residues etc. are used as fuel. In fishing, rhizomes of Nipa palm are extensively used, facilitating the fishing net to float over the water surface. This palm helps stabilizing soils, protecting against erosion, reducing the forces of cyclones and high sea waves in the coastal zones. The demand of Nipa palm products is increasing day by day in the different countries of the world including Bangladesh.

Keywords: Nipa Palm, Production, Uses, Importance, Mangrove Forest, Bangladesh

1. Introduction

Nipa palm (*Nypa fruticans* Wurmb., Arecaceae) is an important component of the East Asian mangrove vegetation. It is one of the oldest living palms [1, 2]. In Bangladesh, the natural distribution of Nipa palm is restricted to the 'Sundarbans', the largest single continuous tract of mangrove forest in the world [3, 4]. It grows along coastlines and estuarine habitats in the Indian and Pacific Ocean. It is a stem-less palm with tall erect fronds and underground rhizomatous stem [5] possessing an extensive root system, well suited to resist swift running water [6]. This familiar palm, growing naturally in the coastal areas and the demand of products is increasing day by day in Bangladesh [7]. Total nipa palm (*Golpata*) production was 320016 mounds in reserve forest of Bangladesh in 2011-2012 [8]. It is used mainly for housing, food, fuel, fence-making, medicine, cigarette wrapping, molasses, wine, fishing etc. The kernels of immature fruits are used as food. Another important product '*Gur*' (molasses) to be sold, which generally followed a short marketing chain; this product is sold directly to the end-users in villages or in local markets [7]. The role of nipa palm is invaluable for both the rural and urban livelihood economies of Bangladesh. People living around the 'Sundarbans' depend on Nipa palm and about 80% of houses in the area are made of Nipa palm [9]. Its sap may be a prospective source for production of sugar, vinegar and alcohol. Nipa palm serves as the first line of defense against the impacts of tsunami, hurricanes, and cyclones that reduce the damages in the coastal zones. The paper is an overview of nipa palm considering origin, distribution, habitat, uses and importance aspect. The information out lined in this article have been collected from different national and international agricultural, food and forest journals, different reports of FAO and visited of useful websites etc.

2. Origin and Distribution of Nipa Palm

It was postulated that this species had an original distribution from Asia extending to Europe, Africa, and America. Its current range is now confined within the tropical Indo-West Pacific region, from Sri Lanka through Asia to Northern Australia and the Western Pacific islands, suggesting changing climatic conditions and/or the loss of versatile genotypes that can tolerate wider environmental conditions [10, 11].

3. Nipa Palm in Mangrove Forest

'Mangroves' is an ecological term referring to a taxonomically diverse assemblage of trees and shrubs that form the dominant plant communities in tidal and saline wetlands along sheltered tropical and subtropical coasts. Economically, mangroves are a great source of timber, poles, thatch and fuel and the bark is used for tanning materials; some species have food or medicinal value [12]. *Nypa fruticans* is a species best adapted to grow in mangrove coastal areas with moderate only salt load, and circumscribing quite well the actual areas of occurrence of this palm in the gradient from seawater habitats to inland sites [13]. It is a mangrove palm that grows well in calm estuaries and coastal zones. The species can prevail in a simple channel or complex tributaries, bays, tidal flats and creeks, as long as there is a tide and a freshwater outflow action [14]. Nipa palm usually thrives well in the sediments deposited by an accreting process by the sea, creating a clayish type of soil, with brackish water that promotes an anaerobic system [15]. It grows in soft mud, usually where the water is calmer, but where there is regular inflow of freshwater and nutritious silt. They can be found inland, as far as the tide can deposit the Palm's floating seeds. It can tolerate infrequent inundation, so long as the soil does not dry out for too long. Its horizontal creeping stem stabilizes river banks preventing soil erosion. New fronds emerge quickly after damage and so quickly protect the land after storms and also continuously produce useful products for the locals [16].

4. The Coastal Zone of Bangladesh

The coastal zone covers 19 out of 64 districts/zilas facing or in proximity to the Bay of Bengal, encompassing 153 upazilas [17]. The zone constitutes 32 percent of the area and 28 percent of the population of Bangladesh [18].

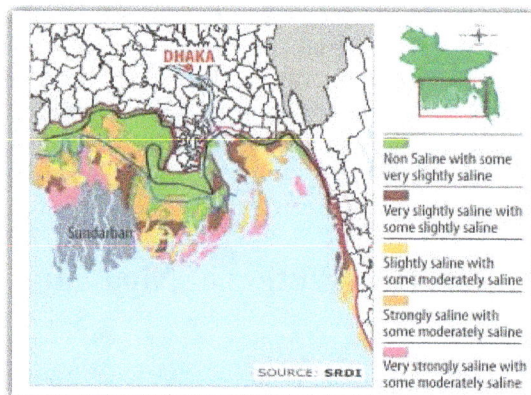

Figure 1. The coastal zone of Bangladesh [19].

In 12 of these districts, 51 upazilas face a combination of cyclone risk, salinity and tidal water movement above critical levels and are designated as "exposed coast" (Figure 1, green areas). The coastal zone covers an area from the shore of 37 to 195 kilometers, whereas the exposed coast is limited to a distance of 37 to 57 kilometers [20].

5. Morphology of Nipa Palm

Nipa is a monocious and pleonanthic palm; it also exhibits viviparous germination [21] as in many other mangrove species. The leaves of nipa palm can grow up to 10 m, and arise from a dichotomously branched underground rhizome that grows to about 50 cm in length [1, 22]. The species lacks a visible upright trunk, and the leaves appear from the ground. The younger leaves appear from the middle of the crown and push the older leaves aside before they dry and fade away, leaving bulbous leaf bases or scars behind. The diameter of the cluster could be up to 75 cm and a single leaf may attain a height of 8 m. The mature crown may contain 6 to 8 living leaves and 12 to 15 bulbous leaf bases at a time [15].

Figure 2. Nipa palm plant [23].

The collection of leaves normally started from the 6-7 year old plants within the plantation. A slanting cut, maintaining a 45° angle is used. The cutting height above the ground depends on the planting density. If the density is high, cutting is undertaken at 7 or 8 cm above the ground and in case of low density, at 5 or 6 cm. The season of harvesting was January to February as new shoot development begins in March [7].

6. Uses of Nipa Palm

Nipa palm is utilized by humans for several purposes, such as roof thatching, wall partitioning, making of sun hats and mats, foods like edible young seed, aromatic tea from leaf blade, sugar from xylem sap, medicinal purposes, bio-ethanol production, and remediation of heavy metal from polluted sites [24-26]. Newly developed shoots are to be used as a vermicide. Ash from Nipa palm is used as an analgesic against tooth pain and headache. Dry leaves, petiole, stem wood, fruit residues etc. are used as fuel. In fishing rhizomes of Nipa palm are extensively used, facilitating the fishing net to float over the water surface. Farmers also report that Nipa palm in the river or sea attracts deep-water fish. The juice is used for making molasses and alcohol [7]. The tapping of the palm for beverages such as wine or toddy and identify this as an ancient and traditional practice in Pan-Pacific and South and Southeast Asian countries [12, 27]. Nipa palm has a great potential for commercial use in housing and medicine in Bangladesh [28]. It is using for housing and other important purposes [29]. The long, pinnate leaves (fronds) provide material for thatching houses. In the Philippines, Malaysia, Indonesia and Thailand the fabrication of thatching panels, called locally 'shingles', 'pawid' or 'atap', is a significant local source of income. Leaflets and midribs are used for manufacturing of brooms, baskets, mats and sunhats. The white endosperm of immature seeds is sweet and jelly-like and is consumed as a snack. The cuticle of young, unfurled leaves has locally been used as cigarette wrapping. Various parts of nipa palm are a source of traditional medicines such as juice from young shoots is used against herpes, ash of burned nipa palm material against toothache and headaches. Nipa palm material also use for salt extraction. The use of the hard shell (mesocarp) in the making of buttons, necklaces and other fashion apparels is successful in Nigeria. Nipa fronds are commonly used as sails by local fishermen [26]. The leaves have traditionally been used for roof thatching and parts are used for making umbrellas, raincoats, hats, mats, brooms, baskets, cigarette wrappers, ropes, and as a source of fuel wood. The sugary sap from the inflorescence stalk is used to make vinegar, and like those of other palms such as the coconut, its sap is also used to make a popular alcoholic beverage better known as "toddy" in Malaysia, India, and Bangladesh. The gelatinous endosperm from the young seeds is edible and can be eaten raw or preserved in 'heavy syrup' while the hardened ones from the ripened fruits are used as vegetable ivory and buttons. Parts of the palm like young shoots, decayed wood, and the burned roots and leaves are also used as traditional medicinal remedies for the treatment of headaches, toothaches, and herpes [30]. Before the inflorescence blooms, it is tapped to collect a sweet sap. Young Nipa Palm shoots can be eaten. The petals of the flower can be brewed to make an aromatic tea. The immature fruits are white translucent and hard jelly-like. Called attap chee, they are a common ingredient in local desserts [16]. In South-East Asia, there is a long tradition of using palm sap obtained by tapping the inflorescence stalks (peduncle) as a source of treacle (molasses), amorphous sugar ('gula malacca'), alcohol or vinegar. The slightly fermented sap called 'toddy' ('nera' in Indonesia and Malaysia; 'tuba' in the Philippines) is sold and consumed as local beer. In Papua New Guinea, there is no tradition of using the sap [26].

7. Molasses (Gur) Production from Nipa Palm Sap

Sap is a product of photosynthesis; frond (leaf) biomass is likely related to sap production. *Nypa fruticans* frond biomass is the crucial factor for sap/sugar production. The growth properties of the nipa fruit stalk, as well as the water

status, were shown to affect sap production [31].

Figure 3. *Sap collection from Nipa palm plant [32].*

Nipa palm is also much valued for the sweet sap tapped from the stalk of the inflorescence. It was reported that tapping normally commences from the Nipa palm shoots after four years and continues up to 15 years or more. The shoots of 9–12 year old stems are reported to be the highest yielding, providing up to 1500–1900 ml of sap per stem per season. The stems of 15 years or more were reported to yield a reduced sap production. The farmers also reported that palms from canals, ditches or riversides gave the highest amount of sap. Tapping began from the first week of December when the fruit begins to mature and turns tan and this was continued up to the next mid-March. The stalk of inflorescence was then pulled down and every morning and evening, a thin slice of stalk was removed and sap collected in a container. It was also reported that 1 kg molasses was produced from 7–8 liters of sap. Most of the farmers were reluctant to tap the stems because of the decline in the quality of leaf production. However, it was observed that 10% of the landless farmers tapped the stems for wine [7].

8. Factors Affecting Growth and Sap of Nipa Palm

Nipa palm is a tropical plant. The average minimum temperature in its growing areas is 20°C and the maximum 32-35°C. Its optimum climate is sub-humid to humid with more than 100 mm rainfall per month throughout the year. Nipa palm thrives only in a brackish water environment. It is rarely seen directly on the seashore. Optimum conditions are when the base and the rhizome of the palm are regularly inundated by brackish water. For this reason, nipa palm occupies estuarine tidal floodplains of rivers. The optimum salt concentration is 1-9 per mil. Nipa palm swamp soils are muddy and rich in alluvial silt, clay and humus; they have a high content of various inorganic salts, calcium, and sulphides of iron and manganese, contributing to the typical odour and dark color. The pH is around 5; oxygen content is low with the exception of the topmost layers. Typically, nipa palm forms pure stands, but in some areas it grows mixed with other mangrove trees [26]. In general, mangrove forests are composed of salt-tolerant plant species that have special adaptations to NaCl stress, including salt-filtering roots and salt-excreting glands on leaves [33]. Nipa palm is distributed at mangrove coasts preferably at the hypo saline end of the sea water-inland gradient and extends more upstream than the dicot mangrove trees at the banks of river mouths [10]. In salt-affected glycophytic palm species, Na^+ is accumulated in the cells, and plants show toxicity symptoms such as chlorophyll degradation, diminished chlorophyll fluorescence, reduced net photosynthesis, and overall growth inhibition [34, 35]. Free proline is an osmolyte that plays a role in osmo-regulatory defense mechanisms in higher plants exposed to salt or drought stress. Free proline accumulation in salt-stressed seedlings of glycophytic palm species has already been investigated [36, 33]. The overall growth performance and physiology of 6-month-old seedlings of Nipa palm was unaffected by mild salt stress (8.9–16.6 dS m^{-1}), whereas seedlings grown under severe salt stress (EC = 57.2 dS m^{-1}) had lower chlorophyll content and fluorescence, reduced net photosynthesis and transpiration, which resulted in reduced growth of the plants. Na^+ contents in leaf, petiole, and root tissues increased considerably under salt stress, depending upon the NaCl levels in the soil solution. Under salt-stress K^+ content declined, whereas Ca^{2+} content increased somewhat, in parallel to Na^+. Free proline accumulated in plants growing under high salt stress (EC = 57.2 dS m^{-1}). In contrast, soluble sugars were enriched under inter-mediate levels of salt stress (EC = 16.6 dS m^{-1}). *Nypa fruticans* is a species best adapted to grow in mangrove coastal areas with moderate only salt load, and circumscribing quite well the actual areas of occurrence of this palm in the gradient from seawater habitats to inland sites [13]. Sap production would be controlled by the nipa growth and the amount of Na^+ in soil. Both the frond biomass surrounding fruit stalk and above-ground biomass are related to sap production. Frond biomass explained the variance of sap production by 48% and above-ground biomass explained by 57%. Sap production will increase by an increase of soil organic matter content and when the influence of brackish water decreases. Sap could be produced more when the farm is managed such that the influence of brackish water is reduced [31]. From recent studies of mangrove mortality at several locations including Guiana, Gambia, Côte d'Ivoire, Kenya, India and Bangladesh, it appears that these coastal ecosystems are so specialized that any minor variation in their hydrological or tidal regimes causes noticeable mortality. Each species of mangrove (but particularly those belonging to the genera *Rhizophora, Bruguiera, Sonneratia, Heritiera* and *Nypa*) occurs in ecological conditions that approach its limit of tolerance with regard to salinity of the water and soil, as well as the inundation regime [37]. Sap sugar concentration of *Nypa fruticans* vary due to collection time. Maximum sugar content (10%) was in November 1996 whereas, minimum sugar content (8%) was found in April 1997 under an experiment at the Chakaria Sundarbans area of Bangladesh [38].

9. Conclusion

Nipa Palm is growing naturally in the coastal areas with a brackish water environment. It sap may be a prospective source for production of sugar, vinegar and alcohol. The role of Nipa palm is important in the coastal areas people's livelihood of Bangladesh. Multipurpose uses and export potentiality of Nipa palm products can increase the worldwide demand day by day.

Acknowledgements

The author thanks to Dr. N. M. Talukder, former Professor, Department of Agricultural Chemistry, Bangladesh Agricultural University, Mymensingh, Bangladesh for his valuable suggestions and cordial cooperation.

References

[1]   Badve, R. M. and Sakurkar, C.V. 2003. On the disappearance of palm genus Nypa from the west coast with its present status in the Indian subcontinent. Curr. Sci. 85:1407–1409.

[2]   Gee, C.T. 1989. On the fossil occurrences of the mangrove palm Nypa. Paper presented at the symposium; Paleofloristic and paleoclimatic changes in the Cretaceous and Tertiary. Prague.

[3]   Chaudhuri, A.B. and Naithani, H.B.1985. A comprehensive Survey of Tropical Mangrove Forests of Sundarbans and Andaman's. Part I. International Book Distributors, Dehra Dun, India. 41 pp.

[4]   Akhtaruzzaman, A.F.M. 2000. Mangrove Forestry Research in Bangladesh. In: Asia-Pacific Cooperation on Research for Conservation of Mangroves, Proceedings of an International Workshop, 26–30 March, 2000, Okinawa, Japan. pp. 139–146.

[5]   Shahidullah, M. 2001. Nursery and Plantation Techniques of Golpata (Nypa fruticans). In: Siddiqui, N.A. and M.W. Baksha (eds.), Proceedings of the National Workshop on Mangrove Research and Development at Bangladesh Forest Research Institute, Chittagong, Bangladesh 15-16 May, 2001.pp.33–38.

[6]   Percival, M. and Womersley, J.S. 1975. Floristics and Ecology of the Mangrove Vegetation of Papua New Guinea. Botany Bulletin No. 8, Department of Forests, Papua New Guinea. 94pp.

[7]   Miah, M.D.; Ahmed, R. and Islam, S.J. 2003. Indigenous Management Practices of Golpata (Nypa fruticans) in Local Plantations in Southern Bangladesh. PALMS. 47(4):185-190.

[8]   BBS. 2013. Statistical Year Pocket Book. Bangladesh Bureau of Statistics. Ministry of Planning. Peoples Republic of Bangladesh.

[9]   Faizuddin, M.; Rahman, M.M.; Shahidullah, M.; Helalsiddiqui, A.S.M.; Hasnin, M. and Rashid, M.H. 2000. Golpata (Nypa fruticans) important forests produce of the Sundarbans. Mangrove silviculture division, Bangladesh Forest Research Institute (BFRI), Khulna, Bangladesh.

[10]   Duke, N. 2006. Australia's Mangroves. The Authoritative Guide to Australia's Mangrove Plants. University of Queensland, Queensland. 200 pp.

[11] Dransfield, J.; Uhl, N.W.; Asmussen,C.B.; Baker,W.J.; Harley, M.M. and Lewis,C.E. 2008. Genera Palmarum: The Evolution and Classification of Palms. Royal Botanic Gardens, Kew. 732 pp.

[12] Hamilton, L. S. and Murphy, D.H.1988. Use and management of Nipa Palm (*Nypa fruticans*, Arecaceae): A review. Economic Botany. 42(2): 206–213.

[13] Theerawitaya, C.; Samphumphaung,T.; Cha-um, S.; Nana Yamada, N. and Takabe, T. 2014. Responses of Nipa palm (*Nypa fruticans*) seedlings, a mangrove species, to salt stress in pot culture. Flora - Morphology, Distribution and Functional Ecology of Plants 209 (10):597-603.

[14] Fong, F.W. 1982. Nypa swamps in Peninsular Malaysia. Proceedings of the First International Wetlands Conference. Wetlands: Ecol. Manage., pp. 31-38.

[15] Rozainah, M. Z and Aslezaeim, N. 2010. A demographic study of a mangrove palm, *Nypa fruticans*. Scientific Research and Essays.5 (24):3896-3902.

[16] http://www.naturia.per.sg/buloh/plants/palm_nipah.htm

[17] MoWR. 2006. Coastal Development Strategy. Ministry of Water Resources. Government of the People's Republic of Bangladesh.

[18] Islam, M.R. (Ed.). 2004. Where land meets the sea: a profile of the coastal zone of Bangladesh. Dhaka, the University Press Limited. 317 pp.

[19] http://www.thedailystar.net/beta2/wp-content/uploads/2013/07/fr0151.jpg

[20] Islam, M.R.; Ahmad, M.; Huq, H. and Osman, M.S. 2006. State of the coast 2006. Dhaka, Program Development Office for Integrated Coastal Zone Management Plan Project, Water Resources Planning Organization.

[21] Tomlinson, P. S. 1986. The Botany of Mangroves. Cambridge University Press, New York. 413 pp.

[22] Teo, S.; Ang, W.F.; Lok, A.F.S.L.; Kurukulasuriya, B.R. and Tan, H.T.W. 2010. The status and distribution of the nipah, *Nypa fruticans* Wurmb (Arecaceae), in Singapore. Nat. Singapore. 3: 45–52.

[23] http://www.vietnam-pictures.com/tour_to_vietnam/vietnam_photos_big/DSC_2929.jpg

[24] Okugbo, O.T.; Usunobun, U.; Esan, A.; Adegbegi, J.A., Oyedeji, J.O. and Okiemien, C.O. 2012. A review of Nipa palm as a renewable energy source in Nigeria. Res. J. Appl. Sci. Eng. Technol. 4, 2367–2371.

[25] Tsuji, K.; Ghazalli, M.N.F.; Ariffin, Z.; Nordin, M.S.; Khaidizar, M.I.; Dulloo, M.E. and Sebas-tian, L.S. 2011. Biological and ethnobotanical characteristics of nipa palm

(*Nypa fruticans* Wurmb.): a review. Sians Malaysia 40: 1407–1412.

[26] Wankasi, D.; Horsfall Jr., M. and Spiff, A.I., 2006. Sorption kinetics of Pb2+ and Cu2+ ions from aqueous solution by Nipah palm (*Nypa fruticans* Wurmb.) Shoot biomass. Elect. J. Biotechnol. 9:587–592(www.cabi.org/isc/.../36772).

[27] FAO. 1995. Integrated Resource Development of the Sunderbans Reserved Forest, Bangladesh. Draft final report. FAO: DP/BGD/84/056.

[28] Shiva, M.P. 1994. Report on Mangrove Non-wood Forest Products. FAO/UNDP project BGD/84/056, Khulna, Bangladesh. 147 pp.

[29] Killmann, W.; Wong,W.C. and Shaari, K.1989. Utilization of Palm stems and leaves: an annotated bibliography. Forest Research Institute, Kuala Lumpur, Malaysia.

[30] Burkill, I. H.1966. A Dictionary of the Economic Products of the Malay Peninsula. Volume II. Crown Agents for the Colonies, London. 2444 pp.

[31] Matsui, N.; Okimori,Y.;, Fumio Takahashi,F.; Matsumura, K. and Bamroongrugsa, N. 2014. Nipa (*Nypa fruticans* Wurmb) Sap Collection in Southern Thailand II. Biomass and Soil Properties. Environment and Natural Resources Research.4 (4).

[32] http://img1.photographersdirect.com/img/26650/wm/pd2650443.jpg

[33] Ame, R.B.; Ame, E.C. and Ayson, J.P., 2011. Management of the Nypa mangrove as a mitigating measure against resource over-utilization in Pamplona. Cagayan. Kuroshio Sci. 5, 77–85.

[34] Cha-um, S.; Takabe, T. and Kirdmanee, C. 2010. Ion contents, relative electrolyte leakage, proline accumulation, photosynthetic abilities and growth characters of oil palm seedlings in response to salt stress. Pak. J. Bot. 42, 2020–2191.

[35] Youssef, T. and Awad, M.A. 2008. Mechanisms of enhancing photosynthetic gas exchange in date palm seedlings (*Phoenix dactyifera* L.) under salinity stress bya 5-aminolevunic acid-based fertilizer. J. Plant Growth Regul. 27, 1–9.

[36] Al-Khayri, J.M. 2002. Growth, proline accumulation, and ion content in sodium chloride-stressed callus of date palm. In Vitro Cell. Dev. Biol. Plant. 38:79–82.

[37] Blasco, F.; Saenger, P. and Janodet, E. 1996. 'Mangroves as indicators of coastal change', Catena. 27(3-4):167-178. (Available at http://dx.doi.org/10.1016/0341-8162 (96)00013-6).

[38] Das, S.C.; Akhter, S.; Wazihullah A.K.M and Rahman, M.S. 2000. Yield of Vinegar, Alcohol and Sugar from the Sap of *Nypa fruticans*. Bangladesh J. Forest Sc. 29: 92–96.

11

Soil Fertility Status of Rice Field in Paundi Watershed, Lamjung District, Nepal

Ram Kumar Shrestha

Tribhuvan University, Institute of Agriculture and Animal Science, Lamjung Campus, Lamjung, Nepal

Email address:

Shresthark_2004@yahoo.com

Abstract: From May to November 2014, a research was carried out to study the soil fertility status of lowland paddy field differed in the cropping system in Paundi watershed, Nepal. A total of 20 soil samples were collected and analyzed, and a household survey was carried out to collect the information regarding soil fertility management practices being adopted along with crops yield. Average annual inputs of the organic manure, urea, Diammonium phosphate (DAP) and Muriate of potash (MoP) were 21 t ha^{-1}, 143 kg ha^{-1}, 116 kg ha^{-1} and 16 kg ha^{-1} respectively. Maize field received significantly higher amount of the organic manure, whereas the rice crop received the higher amount of the urea and DAP. Terrace riser slicing and the legume integrations were the other soil fertility management strategies being adopted by farmers. Soils were silt clay loam and were acidic. The soil organic matter in paddy field was low though the level was significantly higher in rice-rice cropping system than that of in rice-maize system. Most of the soils were low in the soil total nitrogen and available phosphorus. Potassium appeared to be low in the study area. Available zinc was found to be adequate in both types of the paddy field. The yield of the wet season rice, dry season rice and maize crop were 3.75, 2.0 and 2.6 t ha^{-1} respectively. Appropriate soil fertility management practices should be adopted to improve the soil fertility level in the study area.

Keywords: Cropping System, Paddy Field, Paundi Watershed, Soil Fertility

1. Introduction

Rice-based cropping systems are prevalent in lowland terai and the foot hills of Nepal covering about 765,000 hectares [1]. The agricultural environment of the mid-hills of Nepal is degrading at a high rate [2] & the issues are becoming critical for the productivity of midhills. Nepalese Soils across the country are low in organic matter (OM), mostly acidic in reaction, deficient in nitrogen (N) and phosphorus (P); exchangeable potassium (K), zinc (Zn), boron (B), cupper (Cu), manganese (Mn) and molybdenum [3][4][5][6]. Negative nutrient balances have resulted significant depletion of soil nutrients in irrigated rice areas of tropical Asia [7]. This problem is severe in the hilly regions due to the loss of the soil and soil nutrients by soil erosion. As high as ten tons of the soils per hectare are annually lost even from the well-managed paddy terraces [8] causing soil fertility degradation. These all have been threatening the sustainability of agriculture [9]. Soil fertility status is highly determined by both natural factors such as parent materials, climate, and soil age as well as the socioeconomic conditions

such as inputs of manure and fertilizers [10, 11] which are highly specific to the local conditions. Limited studied have been carried out to assess the soil fertility status in different part of the country and study specific to this location too is yet to be done. Assessment of soil fertility status provides a basis not only for evaluating the effectiveness and efficiency of the existing soil fertility management strategies but also provides the basis for site specific fertilizer recommendation for the optimum crop yield. The main objective of this study was to evaluate the soil fertility status in relation with soil fertility management practices and crops yield of the low land paddy field in Paundi Watershed.

2. Material and Methods

This study was carried out at paddy production sites in Paundi Watershed (Lamjung and Tanahun district), Nepal from May to November, 2014. A total of twenty farm households (HHs)-10 each for rice-rice system and rice-maize system- were purposively selected. A household survey using pretested semi-structured interview schedule was conducted to assess the soil fertility management practices and crop yield. A total of twenty composite samples were collected from 0-20 cm soil depth at the vegetative growth stage (15-20 days after transplanting) during July 2014 and analyzed for important physical and chemical properties that influences soil fertility in the Soil Science Laboratory of Soil Management

Directorate, Hariharbhawan, Lalitpur by using standard methods as described in table 1.

For the interpretation of the soil test value for different soil chemical parameters, this study used the ranking system as described by Pradhan [12]. Data collected from the field survey and from soil analysis were entered into SPSS Window version 17.0. Treatment mean separation were done using Tukey Test and figures were created using MS Excel 2008.

Table 1. Methods of the soil analysis.

Soil Characteristics	Standard Method
Soil Texture	Hydrometer method
Soil pH	Electronic pH meter (1:1 soil water suspension method)
Soil organic matter (%)	Wakley-Black method
Soil Total Nitrogen (%)	Microjeldahl method
Soil available Phosphorus (kg ha^{-1})	Modified Olsen's bicarbonate method
Soil available Potassium (kg ha^{-1})	Neutral normal ammonium acetate extraction by flame photometer Method
Soil available Zinc (kg ha^{-1})	DTPA extraction by Atomic Absorption Spectrometer

3. Results and Discussion

3.1. Soil Fertility Management Practices

Result shows that farmers were using mainly FYM as the sources of organic manure- other major sources being goat manure and poultry manure. Organic manures were applied mainly for the maize and dry season rice (Table 2). Rice-maize system received significantly higher amount of the organic manure annually (31 t ha^{-1}) than that of the rice-rice system (12 t ha^{-1}).

Table 2. Inputs of organic and inorganic sources (mean±SE) of the plant nutrients in rice-based system.

Manure/Fertilizer	Rice-Rice system		Rice-Maize system	
	Dry season rice	Wet season rice	Maize	Rice
FYM t ha^{-1}	6.6±2.27	1.86±0.3	19.4±0.83	2.58±1.72
Poultry manure t ha^{-1}	0.81±0.61	0.83±0.61	2.67±1.29	3.28±1.37
Goat manure t ha^{-1}	0.95±0.53	0.48±0.32	1.18±0.89	1.71±0.89
Urea (kg ha^{-1})	80.13±7.37	82.0±2.0	42.3±6.75	81.6±3.38
DAP (kg ha^{-1})	86.96±17.49	88.5±10.85	7.5±6.02	49.2±6.71
MOP (kg ha^{-1})	5.0±3.41	9.0±6.40	5.0±3.41	12.2±4.15

Applications of the urea and diammonium phosphate (DAP) were the customary practices for the crop production. Amount of their application, however, varied with the crop and cropping system. Rice crop- irrespective of the season- received significantly higher amount of urea (>80 kg ha^{-1}). Farmer's applied >80 Kg DAP per hectare when double rice system was being adopted. Rice-Maize cropping system received significantly lower amount of DAP. Few farmers had been applying muriate of potash (MOP). None of the farmers were using micronutrient in both types of cropping systems. Almost all the farmers were adopting the practices of terrace riser slicing and legume crop inclusion as a soil fertility management strategy in rice field- bean and soybean being the dominant leguminous crops intercropped in maize

and wet season rice field respectively.

It was observed that soil testing based fertilizer application was almost zero in the study site. Application of twenty tons per hectare of the FYM for maize production was very high than the recommended dose of 6 ton ha^{-1} [13]. Similar kind of the practice has been observed by Regmi and Zoebisch [14].

3.2. Soil Fertility Status

3.2.1. Soil Texture, Soil pH, Organic Matter and Nitrogen

The area was mainly dominated with silt loam and silty clay loam soil. The result shows that the soils were mostly acidic in reaction in rice-rice field (pH 5.28) and was slightly acidic in rice-maize field (pH 6.11). On average, the medium level of the soil OM content in the rice-rice cropping system

(2.27%) did not significantly differed with that of the rice-maize cropping system (2.08%). Sixty percent of the soil samples from rice-rice system showed low level of OM and this reached to 80% for that of the rice-maize cropping system. Paddy field with both types of the cropping systems showed low level of total N (<0.1%). However, the rice-maize system showed significantly higher level of total N than rice-rice system (Table 3). About 80% soils were low in N in rice-rice system and this figure was 60% for rice-maize system.

The soil acidity is the basic characteristics of the soil in mid hills of Nepal. Acid parent materials and the unbalance use of the chemical fertilizers as well as depletion of the soil OM are reported to be the major causes of the soil acidity throughout the country [5]. Relatively higher amount of the OM in the rice-rice system may be due to flooding of the field for longer duration than that of the rice-maize system. Low level of the soil total N in the study area seems due to the low level of the N application (Table 2). This may be also attributed to the higher N loss from the double rice system.

3.2.2. Available Phosphorus and Potassium

Rice-rice system, on average, contained low level of the available P (28 kg ha^{-1}) which was significantly lower than medium level of P (64 kg ha^{-1}) in the rice-maize system. It was found that 60% of the soils in rice-rice system were low in available P (<31 kg ha^{-1}). Higher P in rice-maize system can be attributed to the higher amount of the organic manures addition. Though rice-rice system often received higher amount of the DAP, the added P might be immobilized as phosphate of aluminum or iron in acidic soil.

Though, soils of both types of the cropping systems were low in available K (<110 kg ha^{-1}), this was significantly higher for rice-maize system (<84 kg ha^{-1}) than that of the rice-rice system (64 kg ha^{-1}). Unbalanced use of the fertilizers-excluding K fertilizers- might be the causes of the depletion of the available K in the soil.

3.2.3. Available Zinc

The result showed that the supply of the Zn is not limiting in both types of cropping systems (>0.5 mg Kg^{-1} soil) and the concentration is significantly higher in rice-maize system (1.05 mg Kg^{-1} soil) than rice-rice system (0.77 mg Kg^{-1} soil) (Table 3). Though the farmers were not applying Zn fertilizer to the field, under the prolong flooding conditions, higher level of the carbonates might be responsible for the Zn immobilization in rice field [7].

3.3. Crop Yield

Irrespective of the cropping system, the average yield of the wet season rice did not differ. The yield of the wet season rice, however, was significantly higher than that of the dry season rice (Fig 1). The yield of the dry season rice was found as 2.3 t ha^{-1}. And the average maize yield was found to be 2.6 t ha^{-1}.

The yield of the wet season rice appeared to be above the national average yield of 3.21 t ha^{-1} in 2012/013. However, the yield of the maize was lower as compared to the national average yield of 4.44 t ha^{-1} for the same year [15]. Such higher yield in spite of the lower fertility status suggests that factors other than the plant nutrients- such as use of hybrid varieties- under the study highly influence the crop yield.

4. Conclusion

Soil fertility level is affected by both natural factors and management practices, and its assessment is needed to develop appropriate fertility management strategies for better crop production. From the above mentioned results, it can be concluded that paddy fields in the study area- affected by cropping system- vary in terms of the different soil fertility parameters, and are low in OM, total N, available P, available K, and available Zn. Reduced use of the organic manures and increasing dependency on the chemical fertilizers may further decrease the availability of various plant nutrients and eventually the crop yield. Integrated approach of the plant nutrient management that relies on the nutrient balance is needed to maintain the optimum level of plant nutrients for the production of crops in sustainable manner.

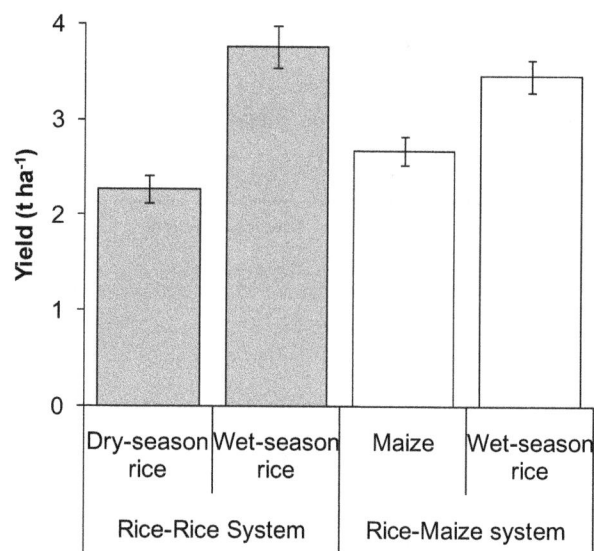

Figure 1. Yield of the crops under different cropping system in the paddy field.

Acknowledgements

Funding from the University Grant Commissions, Tribhuvan University, Nepal to conduct this study is highly acknowledged. The author is especially thankful to the farmers in the study area, who helped to explore the information required for this study.

Table 3. Analysis of soil samples from Paundi Watershed (0-20 cm depth).

Soil Sample I. D.	Soil pH	OM (%)	Total N (%)	Available P (Kg ha^{-1})	Available K (Kg ha^{-1})	Available Zn (Kg ha^{-1})
Rice-Rice System						
1.	4.90	2.0	0.06	17.0	75.0	0.65
2.	5.20	2.3	0.07	26.0	50.0	0.77
3.	4.80	1.9	0.07	21.0	40.0	0.56
4.	5.20	2.5	0.11	41.0	90.0	0.71
5.	5.40	2.3	0.06	32.0	75.0	0.96
6.	5.20	2.6	0.04	31.0	99.0	1.04
7.	5.54	1.8	0.11	22.0	46.0	0.77
8.	5.30	2.2	0.06	19.0	57.0	0.64
9.	5.10	2.6	0.07	30.0	47.0	0.58
10.	6.20	2.5	0.05	41.0	69.0	1.06
Mean	5.28	2.3	0.07	28.0	64.8	0.77
Standard Deviation	0.39	0.3	0.02	8.6	20.0	0.19
Rice-Maize System						
11.	5.65	2.1	0.08	67.0	86.0	0.94
12.	6.32	2.4	0.05	76.0	86.0	0.99
13.	6.50	2.2	0.16	34.0	97.0	1.27
14.	6.49	2.7	0.15	84.0	105.0	0.87
15.	6.55	2.0	0.07	35.0	78.0	1.09
16.	6.47	2.4	0.07	42.0	94.0	1.26
17.	5.37	1.7	0.13	101.0	106.0	0.94
18.	6.16	2.6	0.06	51.0	99.0	1.33
19.	5.43	2.2	0.09	64.0	47.0	0.75
20.	6.19	2.0	0.12	90.0	48.0	1.01
Mean	6.11	2.2	0.09	64.4	84.6	1.05
Standard Deviation	0.46	0.3	0.04	23.5	21.4	0.19

References

[1] IRRI, "Rice in Nepal. International Rice Research Institute, Phillippines, 2013.

[2] P. P. Regmi, Agricultural development through eco-restructuring in different ecological zones across Nepal. PhD Dissertation AIT AC99-2. Asian Institution of Technology, Bangkok, Thailand. Pp. 102-145, 1999.

[3] T. B. Khatri-Chhetri, and E. E. Schulte, Response of maize to the application of secondary and micronutrients in the soils of Chitawan Valley, Nepal II: Result of Multilocation Trials". J. Inst. Agric. Anim. Sci., 6, 59-75, 1985.

[4] S. N. Jaisy, K. K. Shrestha, and K. P. Baral. Effect of zinc on grain yield of rice". Paper presented at the 14th Summer Crop Workshop, January 1989, Parwanipur, Nepal, 1989.

[5] B. Carson, The Land, The Farmer & The Future: A Soil Fertility Management Strategy for Nepal. ICIMOD Occasional Paper No. 21. International Centre for Mountain Development, Kathmandu Nepal, 1992.

[6] B. P. Tripathi, Soil Fertility Status in the Farmers' Fields of the Western Hills of Nepal. Lumle Seminar Paper No. 99/4. Kaski (Nepal): Lumle Agricultural Research Centre, 1999.

[7] A. Dobermann, and T. Fairhurst, Rice: Nutrient Disorders and Nutrient Management". Potash and Phosphate Institute (PPI), Potash and Phosphate Institute of Canada (PPIC) and International Rice Research Institute, 2000.

[8] MoEST, Rural Energy Policy. Ministry of Environment Science and Technology, Kathmandu, 2006.

[9] G. B. Thapa, Land Use Management and Environment in a Subsistence Mountain Economy in Nepal. Agriculture Ecosystems & Environment, 57, 57-71, 1996.

[10] David Boansi. Yield Response of Rice in Nigeria: A Co-Integration Analysis. *American Journal of Agriculture and Forestry.* Vol. 2, No. 2, 2014, pp. 15-24. doi: 10.11648/j.ajaf.20140202.11

[11] Roland Nuhu Issaka, Moro Mohammed Buri, Satoshi Nakamura, Satoshi Tobita. Comparison of Different Fertilizer Management Practices on Rice Growth and Yield in the Ashanti Region of Ghana. *Agriculture, Forestry and Fisheries.* Vol. 3, No. 5, 2014, pp. 374-379.doi: 10.11648/j.aff.20140305.17

[12] S. B. Pradhan, Soil and Plant Tissue Analysis. Nepal Agricultural Research Council, Nepal, 2006.

[13] K. H. Maskey, Soil Sampling Technique. Soil Management Directorate, Lalitpur, Nepal, 1998.

[14] B. D. Regmi, and M. A. Zoebisch, Soil Fertility Status of Bari and Khet Land in a Small Watershed of Middle Hill Region of Nepal. Nepal Agric. Res. J., 5, 38-44, 2004.

[15] MoAD. Statistical information Nepalese Agriculture. Agri-Business, Promotion and Statistics Division, Ministry of Agriculture Development, Gov. of Nepal, 2013.

12

Supply Chain to Industrialization of Pig Excrete by Biotransformation to Increase Corn Performance

José Antonio Valles Romero[1], Emilio Raymundo Morales Maldonado[2]

[1]Industrial Engineering Department, National Technological Institute, ITESHU, Mexico
[2]Sustainable Agronomy Innovation Department, National Technological Institute, ITESHU, Mexico

Email address:

avallesdoc@gmail.com (J. A. V. Romero)

Abstract: The objective of this work is evaluate manure of Mexican Hairless Pig (MHP) from the three energy levels: high 3.5, medium 2.5 and under 1.5 of metabolizable energy for maintenance (MEM) kg body weight^{-1} processed as fertilizer in shape in dry, mature and fresh for corn yield. In this work was established a design completely randomized with factorial arrangement 3 x 3 with 3 replications. The excreta of pigs fed with medium and low levels promoted ($p \leq 0.05$ y $p < 0.01$) higher content of forage in dry matter (DM) and excreta from high levels increased ($p \leq 0.05$ y $p < 0.01$) the content of nitrogen retained in the plant. The manure dry and mature increased 12.5% diameter and 15% leaf area in corn plants, with respect to fresh excreta. The dry manure promoted 6.1% the number of leafs with respect to mature and fresh excreta. The manure dry and mature increased 21.2% content DM in plant, 18.8% cob, and 26% grain with respect to fresh excreta. The dry excrete high level, excelled in most of the variables evaluated, followed by excrete dry from the medium and excrete mature from the level medium and lower.

Keywords: Organic Fertilizers, Sustainability, Corn Production, Corn Supply Chain

1. Introduction

The processing methods to produce fertilizer from excreta are diverse, to be main the composting, microbial process that converts organic waste materials into a product of agricultural interest [14], another method is to dry gradually excreta to remove moisture facilitating handling this [3] and the implementation of activities of grazing pigs, being an appropriate and sustainable form of land use [1]. Such methods provide a management system for optimizing animal waste and crop production contributing to an integrated management of recycling nutrients to the soil.

However, during processing the excreted lost nutrients due among other things to handling food animal that affect the characteristics excreta; It is known about pigs fed with forage in their diet have changes in the characteristics and nitrogen partition in excreta [7] and [13].

In this sense the Mexican hairless pig (MHP) show a lower nitrogen retention when feed with low protein and low energy densities in the food, which requires complete feeding forage [18], which It ensures a diet that generate less manure nitrogen losses during processing as fertilizer and greater contribution to the requirements of the plant.

The aim of this study was to evaluate manure of Mexican hairless Pig (MHP) from three energy levels: high 3.5, medium 2.5 and under 1.5 metabolizable energy for maintenance (MEM) kilogram weight live^{-1} processed as manure in form dry (without maturation) mature (subject to a composting process) and fresh (defecated during the stay of pigs on pasture) on corn yield.

2. Materials and Methods

2.1. Experimental Area

This work was done from June 30 to October 27, 2011 (120 days) under reed conditions on the campus of Biological and Agricultural Sciences located in the downtown area of Yucatán at 21°06 'N latitude and 89°27'W longitude, to an altitude of 8 msnm. The climate is tropical humid with summer rains, according to the Köppen classification, modified by [12]. The annual average temperature is 25.8°C with relative humidity of 75-80% and average rainfall of 984.4 mm annually accumulating 81% during the months of May to October [3]. See Figure 1.

Figure 1. Place where the experiment was conducted.

2.2. Origin of Organic Material

The excreted dry and mature was obtained from Mexican hairless Pigs (MHP) stabled in individual pens and divided into 3 energy levels: high 3.5, medium 2.5 and under 1.5 metabolizable energy for maintenance (MEM) kg live weight^{-1}; while fresh excreted was incorporate by nine pigs in Luvisol floor, divided into 3 plots with a different level of energy (high, medium and low), each pig stay in a corral. The experiment occupied an area of 2,160 m^2. See Figure 2.

Figure 2. Mexican hairless pig.

2.3. Excreted Dry (Without Maturation)

Figure 3. Dry excreted.

Excreta were collected individually by animal and empty in aluminum trays that stay under shadow 3-5 days to remove moisture, then they were stored in paper bags labeled with the number of repetition and belonging MEM level. See Figure 3.

2.4. Mature Excreted (Compost)

A portion of the dry excreta collected from each pig matured through composting, in individual concrete containers 1.50 m long x 0.90 m wide and 0.5 m high covered with wood top. Excreta were removed twice weekly until the temperature showed no change (seven weeks). Finally they were kept in paper bags labeled with the number of repetition and belonging MEm level. See Figure 4.

Figure 4. Composting.

2.5. Fresh Excreta (Defecate for Pigs While in Grazing)

9 pigs were used, 33 kilos. Food distributed in three levels: high 3.5, medium 2.5 and under 1.5 kg weight live^{-1} MEM present in the parcels, with 3 pigs per level. Each individually Remained pig in a pen of 6 x 10 m made of welded mesh of 5 mm, a trough of metal and a shelter for 40 days. See Figure 5.

Figure 5. Mexican hairless pigs grazing.

2.6. Transplant and Plant Fertilization

In each plot were transplant 180 maize plants with a spacing of 0.6 m between rows and 0.5 m between plants (30,000 plants ha^{-1}). During the transplant were done a hole 20cm in the ground, where it was deposited and rolled a ratio of 100 g of excreta (1 g N plant^{-1}) equivalent to 30 kg N ha^{-1} immediately placing a seed corn, the same was made dose of nitrogen was applied as fresh excreted by pigs being defecated while in the plots.

2.7. Variables Assessed

One day after transplant were randomly selected and labeled 15 corn plants. At the end of the crop (120 days) the following variables were measured: plant height (cm), stem diameter (cm) The number of leaves and leaf area of the leaf (cm^2) was estimated by the equation: (length leaf) × AH (sheet width) x 0.75 (correction factor) [2] and [4] plant performance, fodder, corn and grain (Ton MS ha^{-1}) and total nitrogen content retained in plant (Kg N plant^{-1}).

2.8. Experimental Design

Corn plants are distributed in a completely randomized design 3 × 3 factorial arrangement with 3 replications. Factor 1, corresponded to three energy levels: high 3.5, medium 2.5 and under 1.5 (MEM) kg body weight live^{-1} and factor 2, the processing of excreta: dry, mature and fresh. The data obtained were analyzed using the SAS statistical software version 9.0 and comparison of means was performed using the Tukey test (p <0.05 y p <0.01).

3. Results and Discussion

3.1. Corn Plants

The excreta of pigs fed with high levels 3.5 and 2.5 medium of metabolizable energy for maintenance (MEM) kgweight live^{-1} increased (p ≤ 0.05 y p ≤ 0.01) the height of corn plants, respect to the level under 1.5 EMM kg^{-1} live weight with 7.72% difference compared to high and similar to medium level. Dry and mature increased 12.5% diameter and 15% leaf area respect to fresh excreta. The dry excreta increased 6.1%the number of leaf respect to mature and fresh excreta. As for the interaction of levels MEM kg body weight live^{-1} and the excreted processing, the excrete dry of high level excelled (p ≤ 0.05 y p ≤0.01) in all variables measured during maize growth, results they are seen in table 1.

The disposition of excreta nutrients used as manure during growing maize [9] and [8] has been successful, which it was checked using dry excrete of high level, reporting a plant height similar to found by [16] applying 5 to 10 g N plant^{-1} dry cattle manure in forage corn with an average height of 262.5 cm. Similar results were obtained with the mature excreted high level and fresh excreted from the low level to that reported by [11] to apply for 3 years (2001, 2002 and 2003) the average amount of 74.3 t ha^{-1} (220 kg N ha^{-1}) dry cow dung to 15,200 plants ha^{-1} in forage maize (hybrid SB-302) with height of 228 cm on the ground. apply for three years (2009, 2010 and 2011) consecutive 0.7 g N plant^{-1} (ammonium nitrate) equivalent to 70 kg N ha^{-1} corn (variety Anjou 387) to 95,000 plants ha^{-1} with heights of 214.5 cm and 263 cm^2 leaf area apply for three years (2009, 2010 and 2011) consecutive 0.7 g N plant^{-1} (ammonium nitrate) equivalent to 70 kg N ha^{-1} corn (variety Anjou 387) to 95,000 plants ha^{-1} with heights of 214.5 cm and 263 cm^2 leaf area. apply for three years (2009, 2010 and 2011) consecutive 0.7 g N plant^{-1} (ammonium nitrate) equivalent to 70 kg N ha^{-1} corn (variety Anjou 387).

In this paper were obtained results higher height and leaf area as reported by [17] by applying 5 to 10 g N plant^{-1} dry cattle manure in forage maize with 210 cm height and leaf area of 76.02 cm2 respectively [4].

Table 1. *Variables measured in corn used as manure excreta processed manure dry, mature and fresh.*

Treatment	Height (cm)	Diameter (cm)	Leaf area (cm^2)	Number of leaves
Excrete	ns	**	**	**
Dry	252	1.45^a	500^a	13.9^a
Mature	251	1.43^a	503^a	12.9^b
Fresh	241	1.26^b	426^b	13.2^b
EEM	6.99	0.02	12.1	0.17
Level	**	Ns	Ns	ns
High	259^a	1.39	481	13.5
Medium	247^{ab}	1.38	478	13.4
Low	239^b	1.36	470	13.1
Interaction (N x E)	**	**	*	ns

Means within a row followed by different letters are significantly different (Tukey test) ns: not significant * p <0.05; ** p <0.01.

3.2. Dry Matter and Yield to Plant, Corn, Grain and Forage

The excreta of pigs fed with levels medium and low (2.5 and 1.5 kg weight live^{-1} (MEM) increased (p ≤ 0.05 y p ≤ 0.01) the content of forage with difference of 21% compared to the high level. The excrete dry and mature (p ≤ 0.05) increased 21.2% in (DM) plan, 18.8% cob and 26% grain respect to fresh excreta. In interaction of levels MEM kg body weight^{-1} and the processing excrete, the mature excreted from the level middle and mature excreted from the low level, increased the DM content, followed by dry excreted from the high level dry excreted from the middle level; in contrast, the lowest results were obtained with the use of dried excreta of low level, high level excreted mature and fresh excreta of three energy levels. The results are shown in Table 2.

Table 2. *Performance of plant, fodder, corn and grain corn used as manure excrete processed dry, mature and fresh.*

Treatment	Plant	Fodder	Cob	Grain
	(Ton DM ha^{-1})			
Excreta	**	**	**	**
Dry	5.11^a	0.72^b	3.34^a	1.03^a
Mature	5.25^a	0.82^a	3.45^a	0.97^a
Fresh	4.08^b	0.58^c	2.75^b	0.74^b
EEM	0.19	0.02	0.16	0.06
Level	ns	**	ns	**
High	4.59	0.60^b	3.11	0.87^{ab}
Medium	4.98	0.72^a	3.21	1.04^a
Low	4.86	0.80^a	3.23	0.83^b
Interaction (N x E)	**	**	**	**

Means within a row followed by different letters are significantly different (Tukey test) ns: not significant * p <0.05; ** p <0.01.

A high number of plants ha^{-1} does not mean a higher content of DM however, the DM can change its distribution in plant organs as reported [10] who indicate that population density has a signification effect on the distribution of DM in the vegetative and reproductive areas of the plant. [6] to study three doses of nitrogen fertilization (125, 250 and 375 kg N ha^{-1}) and 5 densities on forage maize they found that more than 9 plants / m2 densities did not significantly increase the performance of DM in plant. In this paper superior results were obtained in MS floor, while on the cob and forage low production yield, in contrast to the findings by [4] to apply were reported 0.7 g N plant^{-1} (70 kg N ha^{-1}) in corn (variety Anjou 387) for three consecutive years, to 95,000 plants ha-1 reporting a DM content in plant, ear and forage 3.1, 4.7 and 2.6 ton DM ha^{-1}, respectively; while, [15] to determine the production and quality of forage with 180 and 240 kg N ha^{-1} and three densities: 60,000, 80,000 and 100,000 plants ha^{-1} hybrid corn (H-376 and Lobo) concluded that the best dose was 240-90-00 (N-P$_2$O$_5$-K$_2$O) increasing from 1.3 to 3.3 ton DM ha^{-1} fodder by increasing planting density 60 000 to 100 000 plants ha^{-1} with the H-376 variety. Different results forage yield were obtained as reported by [20] to apply on average over three years the amount of 74.3 t ha^{-1} (220 kg N ha^{-1}) of dry cattle manure to 15,200 plants ha-1 corn (hybrid SB-302) obtained 18.5 ton DM ha^{-1} fodder. For his part [11] evaluated the effect of the application of organic fertilizers (Biocompost and vermicompost) in forage production of a hybrid yellow corn determined that the best performance corresponded to vermicompost con13 Ton DM ha^{-1}. In this paper 100 g apply excreted plant-1 (30 kg N ha^{-1}) in corn (variety Nal-xoy) where similar results were obtained on the cob with the use of dried excreta of intermediate and mature excreted level average (3.20 and 3.73 t DM ha^{-1}); It is different in forage and grain as reported by [15] using 360 g of excreta and veal (7.8 g N plant^{-1}) equivalent to 5 t ha^{-1} (109.5 kg N) in native maize race Oloton temporarily to 13,889 plants ha-1 with forage yield, cob and grain: 4.56, 3.29 and 2.62 t DM ha^{-1} respectively. [1] to apply 600 g of compost (9 g N plant^{-1}) equivalent to 6 t ha^{-1} (90 kg N ha^{-1}) in corn to 10,000 plants ha-1 recorded a yield of 2.7 t ha^{-1} grain and [14] to apply 20, 30 and 40 ton ha-1 of dry cattle manure in maize genotype San Lorenzo obtained yields of 2.63, 5.60 and 4.81 respectively Ton MS grain.

3.3. Total Nitrogen Retained in the Plant

Table 3. Content of total N retained at the end of the crop plant used as manure excreted processed dry, mature and fresh.

Treatment	Total nitrogen retained (Kg N plant^{-1})
Excrete	**
Dry	0.18^b
Mature	0.17^b
Fresh	0.25^a
EEM	0.01
Level	**
High	0.23^a
Medium	0.19^b
Low	0.18^b
Iteraction (N x E)	**

Means within a row followed by different letters are significantly different (Tukey test) ns: not significant * p <0.05; ** p <0.01.

The excreta of pigs fed high levels EMM 3.5 kg live^{-1} increased (p ≤ 0.05) 19.5% N content in ground held regarding

pig excreta of lower and middle level MEm kg live^{-1}. Using fresh excreta as fertilizer increased (p≤0.05) 30% N plant, relative to dry and mature excreted. As for the interaction of levels MEM and excreted processing, dry manure favored the high level (p ≤ 0.05 y p ≤ 0.01) total N content in plant^{-1}, being similar to using fresh excreted from the middle level, fresh excreta the low level and high level of dry excreta; in contrast with the lower results obtained with the use of dry excreta low level. See Table 3.

It is assumed that the fresh excreta, incorporated into the soil while in pigs grazing, initially presented a temporary immobilization of N as pointed out by [17] with the pattern of mineralization found in pig feces to start a phase of immobilization and after 6 weeks net mineralization. However, during maize growth increased availability to the plant is presumed, consistent with those reported by [19] in a study of forage sorghum (Sorghum vulgare) with different amounts composting considering that one of the risks of adding amendments is to cause temporary immobilization of N depending on organic content, thus limiting the availability of this nutrient for growing and restricting the initial growth in the plant.

Higher than that reported by [15] to apply for 3 years an average of 74.3 ton values were ha^{-1} (220 kg N ha^{-1}) of cattle manure to 15,200 plants ha^{-1} corn (hybrid SB-302) obtained 0.012 kg N retained in plant^{-1} at the end of cultivation and [5] using different amounts of zeolite and fertilizer (200 kg N ha^{-1}) in corn determining that the end of culture was retained 0.03 kg N plant^{-1}.

4. Conclusions

The dried manure pigs fed with high levels of maintenance metabolizable energy per kilogram live weight (3.5 MEM kg live weight^{-1}) as fertilizer favored the diameter, leaf area and number of leaves on corn; followed by mature manure the average level (2.5 kg MEM weigh live t^{-1}) and the low level of fresh manure (1.5 kg MEM weight live $^{-1}$). Likewise, the average level of mature manure increases dry matter content in plant, ear and grain; followed by dry excreta high level and low level fresh excreta. The application of fresh manure promoted a high level of higher nitrogen content in maize plants, with respect to dried excreta of low and medium level of mature manure.

References

[1] Álvarez R. C., Torres D.M. Chamorro E.R. Ambrosio M.A. 2009. Descompactación de suelos franco limosos en siembra directa: efectos sobre las propiedades edáficas y los cultivos. Facultad de agronomía de launiversidad de buenos aires argentina. CI. Suelo (Argentina) 27(2): 159-169.

[2] Avendaño A.C., Molina, J.D., Trejo C.C., López C. y Cadena. I. J. 2008. Respuesta a altos niveles de estrés hídrico en maíz. Agronomía Mesoamericana 19(1): 27-37.

[3] Bautista F., Palma L. D. y Huchin M. W. 2005. Actualización de la clasificación de los suelos del estado de Yucatán. 105- 122.

[4]   Bolaños A. E. y Jean C. E. 2012. Distancia entre surcos en el rendimiento y calidad de la materia seca de maíz. Revista Mexicana de Ciencias Pecuarias. 2(3):299-312.

[5]   Civeira, G. y Rodríguez. M. B. 2011. Nitrógeno residual y lixiviado del fertilizante en el sistema suelo-planta-zeolitas. CI. Suelo (Argentina) 29(2): 285-294.

[6]   Cueto W.J., Reta D.S.G., Barrientos R.J.L. Gonzales C.G. y Salazar. S.E. 2006. Rendimiento de maíz forrajero en respuesta a la fertilización nitrogenada y densidades de población. Revista Fitotecnia Mexicana. 29 (97) 101.

[7]   Delve R.J., Cadish G., Tanner J.C., Thorpe W., Thorne P.J. y Giller. K.E. 2000. Implications of livestock feeding managements on soil fertility in the smallholder farming systems of sub-Sahara Africa. Agricultural Ecosistem Environment 84: 227-243.

[8]   Del pino A.C., Repetto C.A. y Mori C.P. 2008. Patrones de descomposición de estiércoles en el suelo. Terra latinoamericana 26: 43-52.

[9]   Eghball B. J., Wienhold J. E. Gilley J. E. y Eigenberg R.A. 2002. Mineralization of manure nutrients. J. Soil Water Conserv. 57: 470-473.

[10] Fallah S. y Tadayyon. A. 2010. Absorción y eficiencia del nitrógeno en maíz forrajero: efecto del nitrógeno y la densidad de población. Agrociencia. 44: 549-560.

[11] Fortis H. M., Leos R.J.A., Preciado R.A., Orona C.I. García, S.J.A., García H.J.L. y Orozco V.J.A. 2009. Aplicación de abonos orgánicos en la producción de maíz forrajero con riego por goteo. Terra latinoamericana. 27 (4): 329-336.

[12] García E. 1988. Modificaciones al sistema de clasificación de Köppen. México, D.F.

[13] Gerard L., Velthof J., Nelemans O. y Kuikman. P. 2005. Gaseous nitrogen and carbon losses from pig manure derived from different diets. Waste Management.Technical reports. Journal Environment Quality. 34: 698-706.

[14] López M.J.D., Diáz A.E. Martínez E. R. y Valdez C. 2001. Abonos orgánicos y su efecto en propiedades físicas y químicas del suelo y rendimiento en maíz. Universidad Autónoma de Chapingo. Terra Latinoamericana. 19(4):293-299.

[15] Peña R.A., González C.F. y Robles. E.F. 2010. Manejo agronómico para incrementar el rendimiento de grano y forraje en híbridos tardíos de maíz. Revista Mexicana de Ciencias Agrícolas 1 (1).

[16] Salazar S. E., Trejo, E. H., Vázquez, I. V.C. López M.J. Fortis H.M., Zuñiga T.R. y Amado A.J. 2009. Distribución de nitrógeno disponible en suelo abonado con estiércol bovino en maíz forrajero,Terra Latinoamericana 27 (4): 373-382.

[17] Salazar S.E., Trejo E.H., Lopez M.J., Vazquez V.C., Serrato C.S., Orona C.I. y Flores M.J.P.2010. Efecto residual de estiércol bovino sobre el rendimiento de maíz forrajero y propiedades del suelo. Terra Latinoamericana. Vol.28, (4):381-390.

[18] Trejo L.W. 2005. Strategies to improve the use of limited nutrient resources in pig production in the tropics, supplement 85. unikasselversitat. Journal of agriculture and rural developmen in the tropics and subtropics. 37- 38.

[19] Wing C.R., Rojas A. y Quan. A. 2005. Nitrógeno orgánico y químico en sorgo negro con cobertura permanente de maní forrajero. Características nutritivas y de reproducción. Agronomía Costarricense. 29(1): 29-39.

13

Estimates of Tree Biomass, and Its Uncertainties Through Mean-of-Ratios, Ratio-of-Means, and Regression Estimators in Double Sampling: A Comparative Study of Mecrusse Woodlands

Tarquinio Mateus Magalhães[1, 2, *], **Thomas Seifert**[2]

[1]Department of Forest Engineering, Eduardo Mondlane University, Main Campus, Maputo, Mozambique
[2]Department of Forest and Wood Science, University of Stellenbosch, Stellenbosch, South Africa

Email address:

tarqmag@yahoo.com.br (T. M. Magalhães), seifert@sun.ac.za (T. Seifert)

Abstract: Frequently, biomass expansion factors (BEFs), the respective biomass densities, and their uncertainties are computed without taking into account the appropriate estimators. The objective of this study was to compare the estimates of BEF, BEF-based biomass densities, and their uncertainties using different estimators (mean-of-ratios, ratio-of-means, and regression estimators) in double sampling. Our results demonstrated that increased uncertainty is associated with regression-based biomass densities, and that the computation of BEF using merchantable timber volume should utilize regression estimators, not the usual ratio estimators, which preferably, should be avoided altogether, as they are found to be subjective and more susceptible to errors and personal judgment.

Keywords: Tree Component, Additivity, Belowground Biomass, Aboveground Biomass, Biomass Expansion Factor (BEF), *Androstachys Johnsonii* Prain

1. Introduction

In double sampling, the variables of interest (e.g., biomass) are determined through auxiliary variables, either by ratio estimators or regression estimators [1–6]. The ratio estimators, in turn, are divided into two classes: mean-of-ratios and ratio-of-means estimators [1, 3, 4]. When the regression line of the variable of interest (e.g., component biomass) and the auxiliar variable (e.g., stem volume) does not pass through the origin, the regression estimators are preferable [1–4, 6]. Ratio estimators are appropriate when the regression takes the form of a straight line passing through the origin [1–6].

An example of ratio estimators in forestry is the estimate of biomass density with the aid of biomass expansion factor (BEF), where BEF is the ratio. Vegetation BEFs can be estimated either by mean-of-ratios [7–9] or by ratio-of-means [10–14].

BEF values are generally calculated from the ratio of tree component or total tree biomass to stem volume [7–14] or merchantable timber volume [15–17]. When BEF values are calculated using the first option (using stem volume), if stem volume is zero, then concurrently, tree component biomass is zero; therefore, the ratio estimators are deemed appropriate [1, 3–5]. In the second option, however, merchantable timber volume can be zero when a tree component biomass is nonzero; e.g., merchantable timber volume, defined as the volume of the stem excluding the portion with diameter < 7cm [11, 18, 19], and trees with DBH < 7cm [20] can be zero when the stem biomass or other component biomass is nonzero; therefore, regression estimators are preferable [1–6] over ratio estimators. Nevertheless, in the second case, the BEFs and respective biomasses are computed using ratio estimators [15–17], which may lead to biased estimates.

The computation of uncertainties in double sampling depends on the estimators used, otherwise the uncertainties may be under- or overestimated. In BEF-based biomass estimation, those uncertainties are mainly attributed to BEF estimates [16], and thus represent a major gap in biomass and carbon accounting at regional and national levels [19].

Therefore, this study was aimed at comparing estimates of BEF and BEF-based biomass densities, as well as their uncertainties, using different estimators (mean-of-ratios, ratio-of-means, and regression estimators) in double sampling.

2. Material and Methods

2.1. Study Area

Mecrusse is a forest type characterized by the dominant canopy species *Androstachys johnsonii* Prain, the relative cover of which varies from 80% to 100% [21]. In Mozambique, mecrusse woodlands are mainly found in Inhambane and Gaza provinces in the Massangena, Chicualacuala, Mabalane, Chigubo, Guijá, Mabote, Funhalouro, Panda, Mandlakaze, and Chibuto districts. The easternmost mecrusse forest patches, located in Mabote, Funhalouro, Panda, Mandlakaze, and Chibuto districts, were defined as the study area and encompassed 4,502,828 ha [22], of which 226,013 ha (5%) were mecrusse woodlands. The climate throughout the study region is dry tropical, with the exception of humid tropical areas in western Panda and southwestern Mandlakaze districts [22–27]; their warm or rainy season occurs from October to March, and their cool or dry season occurs from March to September [23–27]. More description on *Androstachys johnsonii* Prain and mecrusse woodlands can be foud in Magalhães [28] and Magalhães and Seifert [29–32].

The mean annual temperature generally exceeds 24 °C, and mean annual precipitation varies from 400 to 950 mm [22–27]. According to the United States Food and Agriculture Organization (FAO) classification [33], soils are mainly Ferralic Arenosols across more than 70% of the study region [22]. Arenosols, Umbric Fluvisols, and Stagnic soils are predominant in the northernmost part of the study region [22]. There is a shortage of water resources and precipitation throughout the study region; only Chibuto and Mandlakaze districts have water resources [22–27], either from precipitation or from lakes and rivers.

2.2. Data Collection

We used a two-phase sampling design to determine stem volume and biomass. In the first phase, we measured diameter at breast height (DBH) and stem height of 3574 trees (m_1) within 23 randomly located circular plots (20 m radius) for estimation of stem volume; only trees with DBH $\geq$ 5 cm were considered. In the second phase, 93 trees (m_2) with DBHs varying from 5 to 32 cm were randomly selected from those analysed during the first phase for destructive measurement of biomass and stem volume. The felled trees were divided into the following components: (1) taproot; (2) lateral roots; (3) root system (1 + 2); (4) stem wood; (5) stem bark; (6) stem (4 + 5); (7) branches; (8) foliage; (9) crown (6 + 7); (10) shoot system (6 + 9); and (11) whole tree (3 + 10). Tree components were sampled and the dry weights estimated as follows.

2.2.1. Root System

The stump height was predefined as being 20 cm for all trees and considered as part of the taproot, as recommended by Parresol [34] and because in larger *A. johnsonii* trees this height (20 cm) is affected by root buttress; therefore, the root collar was also considered part of the taproot. The root system was divided into 3 sub-components: fine lateral roots, coarse lateral roots, and taproot. Lateral roots with a diameter at insertion point on the taproot < 5 cm were considered fine roots, and those with diameters ≥ 5 cm were considered coarse roots.

First, the root system was partially excavated to the first node, using hoes, shovels, and picks, to expose the primary lateral roots (Figure 1a, b). The primary lateral roots were numbered and separated from the taproot with a chainsaw (Figure 1a, b), then removed from the soil, one by one. This procedure was repeated in the subsequent nodes until all primary roots were removed from the taproot and the soil. Finally, the taproot was excavated and removed (Figure 1c, d). The complete removal of the root system was relatively easy because 90% of the lateral roots of *A. johnsonii* are located in the first node, which is located close to ground level (Figure 1a, b); the lateral roots grow parallel to the ground level (they do not grow downwards); and because the taproots had, at most, only 4 nodes and at least 1 node (at ground level).

Fresh weight was obtained for the taproot, each coarse lateral root, and all fine lateral roots. A sample was obtained from each sub-component, fresh weighed, marked, packed in a bag, and brought to the laboratory for oven drying. For the taproot, the samples consisted of two discs: one taken immediately beneath ground level, and another from the middle of the taproot. For the coarse lateral roots, two discs were also taken, one from the insertion point on the taproot, and another from its middle. For fine roots, the sample was 5 to 10% of the fresh weight of all fine lateral roots. Oven drying of all samples was performed at 105°C to constant weight; hereafter, referred to as dry weight.

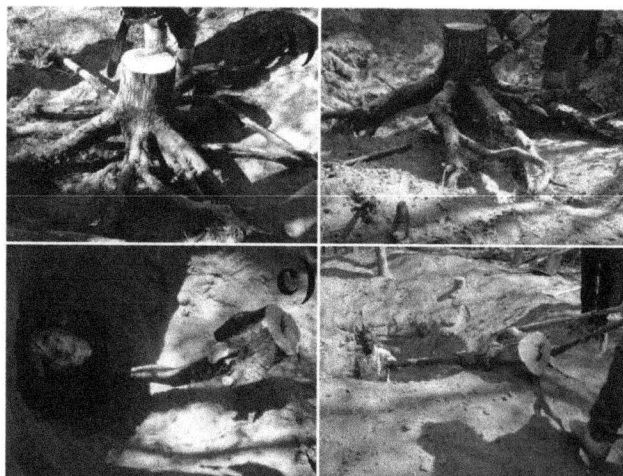

Figure 1. *Separation of lateral roots from the taproot (a, b), and removal of the taproot including the root collar and the stump (c, d).*

2.2.2. Stem Wood and Stem Bark

Felled trees were scaled up to a 2.5 cm top diameter. The stem was defined as the length of trunk from the stump to the height that corresponded to 2.5 cm diameter. The remainder (from the height corresponding to 2.5 cm diameter to the tip of the tree) was considered a fine branch. The stem was divided into sections: the first with 1.1 m length, the second with 1.7 m, and the remaining with 3 m, except the last, the remainder, whose length depended on the length of the stem. Discs were removed on the bottom and top of the first section, and on the top of the remaining sections; i.e., discs were removed at heights of 0.2 m (stump height), 1.3 m (breast height), and 3 m; with successive discs removed at intervals of 3 m to the top of the stem. Their fresh weights were measured using a digital scale.

Diameters over and under bark were taken from the discs in the North-South direction (previously marked on the standing tree) with the help of a ruler. The volumes over and under the bark of the stem were obtained by tallying the volumes of each section, calculated using Smalian's formula [6]. Bark volume was obtained from the difference between volume over bark and volume under bark.

The discs were dipped in drums filled with water, until constant weight (3 to 4 months), for their saturation and subsequent determination of the saturated volume and basic density. Saturated volume of the discs was obtained based on the water displacement method [35], using Archimedes' principle. This procedure was conducted twice: before and after debarking. Hence, we obtained saturated volume under and over bark.

Wood discs and respective barks were oven dried at 105°C to constant weight. Basic density was obtained by dividing oven dry weight of the discs (with and without bark) by the relevant saturated wood volume [36, 37]. Consequently, two distinct basic densities were calculated: (1) basic density of the discs with bark and (2) basic density of the discs without bark.

We estimated the basic density at the point of geometric centroid of each section, using the regression function of density over height [38]. This density value was established as representative of each section [38].

2.2.3. Crown

The crown was divided into two sub-components: branches and foliage. Primary branches, originating from the stem, were classified in two categories: large branches, or primary branches with diameter at insertion point on the stem ≥ 2.5 cm, and fine branches, or those with diameter < 2.5 cm. Large branches were sampled similarly to coarse roots, and fine branches and foliage were sampled similarly to fine roots.

2.2.4. Tree Component Dry Weights

We determined dry weight of the taproot, lateral roots, branches, and foliage by multiplying the ratio of oven-dry- to fresh-weight of each sample by the total fresh weight of the relevant component. Dry weights of the root system and crown were obtained by tallying the relevant sub-components' dry weights. Dry weights of each stem section (with and without bark) were obtained by multiplying respective densities by relevant stem section volumes. Stem (wood + bark) and stem wood dry weights were obtained by tallying each section's dry weight with and without bark, respectively. The dry weight of stem bark was determined from the difference between the dry weights of the stem and stem wood. We determined the dry weight of major components (root system, shoot system, and crown) and the whole tree by tallying the dry weights of their constituent components.

2.3. Data Processing and Analysis

We divided the stem of each felled tree into 10 segments of equal length, and we measured the diameter of each segment at the midpoint, starting from the bottom of the stem. Stem volume was computed using Hohenadl's method (Eq. 1) [39]:

$$v_{i2} = \frac{\Pi L}{40}\left(d_{.05}^2 + d_{.15}^2 + d_{.25}^2 + d_{.35}^2 + d_{.45}^2 + d_{.55}^2 + d_{.65}^2 + d_{.75}^2 + d_{.85}^2 + d_{.95}^2\right) \ [m^3] \tag{1}$$

where v_{i2} is the stem volume of the ith tree from the second sampling phase, L is the stem length, and $d_{.i}$ is the diameter measured at the proportional distance along the stem of the ith tree.

Additionally, we determined merchantable stem volume of the trees of the second phase (v_{mi2}) (Eq. 2), defined as the volume of the stem excluding the portion with diameter < 7cm [11, 18, 19] and trees with DBH < 7cm [20]:

$$v_{mi2} = \frac{\Pi l}{4}\sum_{i=1}^{k} d_m^2 \ [m^3] \tag{2}$$

where d_m is the diameter at midpoint of each segment, l is the

length of each segment, and k is the number of equal segments to the height that corresponded to 7 cm diameter.

Individual stem volume of the ith tree of the jth plot from the first sampling phase (v_{ij1}) was calculated using Eq. 3 as follows:

$$v_{ij1} = \frac{\Pi}{4}DBH^2 \times H \times f_h \ [m^3] \tag{3}$$

where H is stem height and f_h is the average Hohenadl's form factor obtained from the trees of the second sampling phase. Individual f_h was computed using Eq. 4 as:

$$f_h = 0.1\left(1 + \frac{d_{.15}^2}{d_{.05}^2} + \frac{d_{.25}^2}{d_{.05}^2} + \frac{d_{.35}^2}{d_{.05}^2} + \frac{d_{.45}^2}{d_{.05}^2} + \frac{d_{.55}^2}{d_{.05}^2} + \frac{d_{.65}^2}{d_{.05}^2} + \frac{d_{.75}^2}{d_{.05}^2} + \frac{d_{.85}^2}{d_{.05}^2} + \frac{d_{.95}^2}{d_{.05}^2}\right) \ [dimensionless] \tag{4}$$

Individual merchantable timber volume of the ith tree of the jth plot from the first sampling phase (v_{mij1}) was calculated using Eq. 5:

$$v_{mij1} = v_{ij1} \times F_{timber} \ [m^3] \qquad (5)$$

where

$$F_{timber} = \frac{\overline{v}_{m2}}{\overline{v}_2} = \frac{\sum\limits_{i=1}^{m_2} v_{mi2}}{\sum\limits_{i=1}^{m_2} v_{i2}} \ [dimensionless] \qquad (6)$$

is the fraction of the total stem that is merchantable timber, and $\overline{v}_{m2}$ and $\overline{v}_2$ are the average merchantable timber volume (per tree) and average stem volume (per tree) of the trees of the second phase, respectively.

The main auxiliary variables (the first-phase variables) consist of the average number of trees per hectare (N_1), and the stand-level stem volume (m^3 ha^{-1}), the last estimated from Eq. 7:

$$V_1 = \frac{\sum\limits_{j=1}^{n} \sum\limits_{i=1}^{m_j} v_{ij1}}{n \times a} = \overline{v}_1 \times N_1 \ [m^3 \ ha^{-1}] \qquad (7)$$

where m_j is the number of trees in the jth plot, n is the number of plots, a is the plot area (ha), $\overline{v}_1$ is the average stem volume of the trees of the first phase (m^3), and N_1 is the average number of trees per hectare estimated from the first sampling phase. Stem height of trees from the first phase was obtained by subtracting predefined stump height from the whole-tree height (TH) to standardize the definitions of stem height and stem length (for phase-1 trees).

We determined tree component BEF-based biomass density or the ratio estimate of biomass density (W_h) using Eq.8:

$$W_h = BEF_h \times V_1 = W_{hi} \times N_1 \ [Mg \ ha^{-1}] \qquad (8)$$

where

$$W_{hi} = BEF_h \times \overline{v}_1 \ [Mg] \qquad (9)$$

is the estimated average component biomass per tree, which yields W_h when multiplied by the number of trees per hectare (N_1); BEF_h is the vegetation BEF of the hth tree component (taproot+stump, lateral roots, root sytem, stem wood, stem bark, stem, branches, foliage, crown, shoot system or whole tree).

We calculated the sample BEF_h using mean-of-ratios [7–9] (Eq.10) and ratio-of-means [10–14] (Eq. 11)

$$BEF_h^I = \frac{\sum\limits_{i=1}^{m_2} \dfrac{w_{hi2}}{v_{i2}}}{m_2} = \frac{\sum\limits_{i=1}^{m_2} BEF_{hi}}{m_2} \ [Mg \ m^{-3}] \qquad (10)$$

$$BEF_h^{II} = \frac{\overline{w}_{h2}}{\overline{v}_2} = \frac{\sum\limits_{i=1}^{m_2} w_{hi2}}{\sum\limits_{i=1}^{m_2} v_{i2}} \ [Mg \ m^{-3}] \qquad (11)$$

where the superscripts I and II denote mean-of-ratios and ratio-of-means, respectively; therefore BEF_h^I and BEF_h^{II} correspond to BEF_h computed using mean-of-ratios and ratio-of-means, respectively; BEF_{hi} is the BEF of the hth component of the ith tree; w_{hi2} is the biomass of the hth component of the ith tree measured during the second phase; v_{i2} is the stem volume of the ith tree measured during the second phase; m_2 represents the number of trees of the second sampling phase, and $\overline{w}_{h2}$ is the average tree component biomass of the trees from the second sampling phase.

For regression-based tree component biomass density, we computed W_{hi} and W_h unsing Eq. 12 [1–6] and Eq. 13, respectively:

$$W_{hi} = \overline{w}_{h2} + b\left(\overline{v}_{m1} - \overline{v}_{m2}\right) \ [Mg] \qquad (12)$$

$$W_h = N_1 \times W_{hi} \ [Mg \ ha^{-1}] \qquad (13)$$

where

$$b = BEF_h^{III} = \frac{\sum\limits_{i=1}^{m_2} \left(w_{hi2} - \overline{w}_{h2}\right)\left(v_{mi2} - \overline{v}_{m2}\right)}{\sum\limits_{i=1}^{m_2} \left(v_{mi2} - \overline{v}_{m2}\right)^2} \ [Mg \ m^{-3}] \qquad (14)$$

is the regression slope.

Here the superscript III denotes regression estimators, so BEF_h^{III} is BEF_h computed using regression estimators, and $\overline{v}_{m1}$ and $\overline{v}_{m2}$ are the average merchantable timber volume per tree of the first and second sampling phase, respectively.

The regression slope b is an estimate of the change in tree component biomass (w_{h2}) when merchantable timber volume (v_{m2}) is increased by unity [2]. The same definition holds for BEF_h with regard to stem volume; since in ratio estimators, the ratio R (e.g. BEF_h) is the regression slope when the regression line passes through the origin [40]. Therefore the regression slope b was, in this study, considered and treated as BEF_h computed with aid to merchantable timber volume (Eq. 14).

Using mean-of-ratios estimators (I), the variance of the estimated W_{hi} (Eq. 9) was computed according to Freese [1, 3] as follows:

$$VAR_{W_{hi}}^I = \overline{v}_1^2 \left(\frac{S_{BEF_h}^2}{m_2}\right)\left(1 - \frac{m_2}{m_1}\right) + \frac{S_{w_{h2}}^2}{m_1}\left(1 - \frac{m_1}{M}\right) \ [Mg^2] \qquad (15)$$

Rearranging Eq. 9 as $BEF_h = \dfrac{W_{hi}}{\overline{v}_1}$, the variance of the estimated BEF_h becomes [4]:

$$VAR_{BEF_h}^I = \frac{VAR_{W_{hi}}^I}{\overline{v}_1^2} \; [Mg^2 \; m^{-6}] \qquad (16)$$

$$VAR_{W_h}^I = N_1^2 \times VAR_{W_{hi}}^I \; [Mg^2 \; ha^{-2}] \qquad (17)$$

Similarly, the variance of the estimated W_h is:

For ratio-of-means (II) the variance of the estimated W_{hi} (Eq. 9) was computed using Eq. 18 [1, 3]:

$$VAR_{W_{hi}}^{II} = \left(1 - \frac{m_2}{m_1}\right)\left(\frac{\overline{v}_1}{\overline{v}_2}\right)^2 \left(\frac{S_{w_{h2}}^2 + BEF_h^2 \times S_{v_2}^2 - 2 \times BEF_h \times S_{yx}}{m_2}\right) + \frac{S_{w_{h2}}^2}{m_1}\left(1 - \frac{m_1}{M}\right) \; [Mg^2] \qquad (18)$$

Analogously, the variances of the estimated BEF_h (Eq.11) and Wh are computed as in Eqs. 19 and 20

Finally, using regression estimators (III), we calculated the variances of the estimated W_{hi} (Eq. 12) and W_h (Eq. 13) also according to Freese [1, 3] (Eqs. 21 and 22) and the variance of the estimated regression slope b according to Jayaraman [5] (Eq. 23):

$$VAR_{BEF_h}^{II} = \frac{VAR_{W_{hi}}^{II}}{\overline{v}_1^2} \; [Mg^2 \; m^{-6}] \qquad (19)$$

$$VAR_{W_h}^{II} = N_1^2 \times VAR_{W_{hi}}^{II} \; [Mg^2 \; ha^{-2}] \qquad (20)$$

$$VAR_{W_{hi}}^{III} = S_r^2 \left(\frac{1}{m_2} + \frac{(\overline{v}_{m1} - \overline{v}_{m2})^2}{SS_{v_2}}\right)\left(1 - \frac{m_2}{m_1}\right) + \frac{S_{w_{h2}}^2}{m_1}\left(1 - \frac{m_1}{M}\right) \; [Mg^2 \; m^{-6}] \qquad (21)$$

$$VAR_{W_h}^{III} = N_1^2 \times VAR_{W_{hi}}^{III} \; [Mg^2 \; ha^{-2}] \qquad (22)$$

$$VAR_b = VAR_{BEF_h}^{III} = \frac{S_r^2}{SS_{v_2}} \; [Mg^2 \; m^{-6}] \qquad (23)$$

where $S_{BEF_h}^2 = \dfrac{\sum BEF_{hi}^2 - \dfrac{\left(\sum BEF_{hi}\right)^2}{m_2}}{m_2 - 1}$ is the variance of

BEF_h for the second phase; $S_{w_{h2}}^2 = \dfrac{\sum w_{hi2}^2 - \dfrac{\left(\sum w_{hi2}\right)^2}{m_2}}{m_2 - 1}$ is

the variance of w_{h2}; w_{h2} is the component biomass for the

second phase; $S_{v_2}^2 = \dfrac{\sum v_{i2}^2 - \dfrac{\left(\sum v_{i2}\right)^2}{m_2}}{m_2 - 1}$ is the variance of

stem volume of the trees of the second phase (v_2); SS_{v_2} is the sum of squares of v_2; S_{yx} is the covariance of w_{h2} and v_2;

$S_r^2 = \dfrac{SS_{w_{h2}} - \dfrac{\left(SP_{xy}\right)^2}{SS_{v2}}}{m_2 - 2}$ is the squared standard deviation from

regression; $SS_{w_{h2}}$ is the sum of squares of w_{h2}; and SP_{xy} is the sum of products of w_{h2} and v_2. The finite population correction factor $\left(1 - \dfrac{m_1}{M}\right)$ was eliminated in all formulae because m_1 was very small relative to M, which was

unknown.

The square root of Eqs. 16, 19, 23 is the absolute standard error of the estimated BEF_h; and the square root of Eqs. 17, 20, 22 is the absolute standard error of the estimated W_h. Dividing these values by BEF_h and W_h, respectively, then multiplying them by 100, provides the respective percent standard error. The absolute and 95% confidence limits (CI) were computed by multiplying the absolute and percent standard error by the student's t-value.

3. Results

3.1. Data Presentation

The average number of trees per ha and average stem volume per ha were, approximately, 1237 ha^{-1} (min = 551, max = 2220, SD = 477, CV = 38.56%) and 115 m^3 ha^{-1} (min = 66.90, max = 170.47, SD = 25.44, CV = 22.09%), respectively. The average stem volume of the trees of the first and second sampling phase were, respectively, 0.0933 m^3 (min = 0.0020, max = 1.6463, SD = 0.1153, CV = 123.69%) and 0.1890 m^3 (min = 0.083, max = 0.5806, SD = 0.1512, CV = 79.98%), with an average Hohenadl's form factor of 0.4460 (min = 0.3002, max = 0.6128, SD = 0.0592, CV = 13.27%). The merchantable timber volume of the trees of the first and second sampling phase were, respectively, 0.0899 m^3 (min = 0.0019, max = 1.5866, SD = 0.1112, CV = 123.62%) and 0.1821 m^3 (min = 0.0000, max = 0.5784, SD = 0.1552, CV = 26.83%), with an F_{timber} of 0.9638. The average component dry weights per tree varied considerably (Table 1).

Table 1. *Average component dry weight (kg) per tree from the trees in the second phase, their standard deviation (SD), and coefficient of variation (CV).*

Tree component	Average	Minimum	Maximum	SD	CV (%)
Taproot	23.7	1.4735	71.9260	18.9255	80.0187
Lateral roots	24.1	0.7455	100.8152	23.9455	99.4281
Root system	**47.7**	**2.5450**	**162.1045**	**41.2099**	**86.3314**
Stem wood	124.1	4.9469	357.3484	99.4971	80.1955
Stem bark	14.2	0.6774	55.8045	12.3722	87.1382
Stem	**138.3**	**5.6355**	**413.1529**	**110.5770**	**79.9738**
Branches	55.6	2.5827	211.3196	57.3549	103.1827
Foliage	2.8	0.3333	15.1000	2.4929	88.8182
Crown	**58.4**	**3.0377**	**216.6946**	**59.0769**	**101.1720**
Shoot system	**196.7**	**9.8230**	**590.8628**	**163.7135**	**83.2473**
Total tree	**244.4**	**12.4844**	**752.5709**	**204.3297**	**83.6068**

The major components and their values are indicated in bold font.

3.2. BEF and Biomass Estimates

The BEF_h values and the BEF-based component biomass densities, and their uncertainties computed using the different estimators, are shown in Tables 2 and 3, respectively. The total tree and the shoot system BEFs obtained using mean-of-ratios estimators were > 100% of the stem volume. The total tree biomass density was approximately 150 Mg ha^{-1}, of which 80, 24, 56 and 20% were from the shoot system, crown, stem, and roots system, respectively. The uncertainties of the estimated BEF_h and W_h, as measured by percent standard error SE (%), were < 5% for 7 of 11 tree components and < 10% for 10 of the 11 tree components; denoting high levels of precision.

Using ratio-of-means estimators the BEF_h varied from 0.0149 for foliage to 1.2932 for total tree. As in the mean-of-ratios, the total tree and the shoot system BEFs were > 100% of the stem volume and the total tree biomass density was also approximately 150 Mg ha^{-1}. The uncertainties (SE (%)) of the estimated BEF_h and W_h were < 5 % for 7 of the 11 components and < 8 % for all tree components.

The BEF_h values computed using regression estimators were similar to those obtained using mean-of-ratios and ratio-of-means estimators. The estimated total tree and shoot system biomass densities were approximately, 157 and 127 Mg ha^{-1}. The uncertainties, SE (%), of the estimated BEF_h were < 5 % for 7 of the 11 components and < 7 % for 10 of the 11 components. On the other hand, the uncertainties, SE (%), of the estimated W_h were < 10 % for 7 of the 11 components and < 15 % for all tree components.

Table 2. *Component biomass expansion factors (BEF_h), their variances (VAR_{BEF}), standard errors (SE), and 95% confidence intervals (CI), computed using mean-of-ratios, ratio-of-means and regression estimators.*

#	Tree component	BEF_h (Mg m^{-3})	VAR_{BEF} (Mg^2 m^{-6})	SE (Mg m^{-3})	SE (%)	95% CI (Mg m^{-3})	95% CI (%)
Mean-of-ratios estimates							
1	Taproot + stump	0.1407	3.6E-05	0.0060	4.2382	± 0.0119	± 8.4764
2	Lateral roots	0.1162	4.4E-05	0.0067	5.7232	± 0.0133	± 11.4465
3	**Root system (1 + 2)**	**0.2569**	**1.0E-04**	**0.0100**	**3.8930**	**± 0.0200**	**± 7.7860**
4	Stem wood	0.6569	3.6E-04	0.0191	2.9046	± 0.0382	± 5.8092
5	Stem bark	0.0765	1.3E-05	0.0036	4.7534	± 0.0073	± 9.5068
6	**Stem (4 + 5)**	**0.7334**	**4.4E-04**	**0.0210**	**2.8615**	**± 0.0420**	**± 5.7230**
7	Branches	0.2928	3.1E-04	0.0177	6.0590	± 0.0355	± 12.1180
8	Foliage	0.0242	6.6E-06	0.0026	10.6242	± 0.0051	± 21.2483
9	**Crown (7 + 8)**	**0.3170**	**3.6E-04**	**0.0190**	**5.9973**	**± 0.0380**	**± 11.9946**
10	**Shoot system (6 + 9)**	**1.0504**	**1.2E-03**	**0.0340**	**3.2345**	**± 0.0679**	**± 6.4690**
11	**Total tree (3 + 10)**	**1.3072**	**1.8E-03**	**0.0428**	**3.2736**	**± 0.0856**	**± 6.5472**
Ratio-of-means estimates							
1	Taproot + stump	0.1251	2.9E-05	0.0054	4.3388	± 0.0109	± 8.6775
2	Lateral roots	0.1274	4.6E-05	0.0068	5.3245	± 0.0136	± 10.6490
3	**Root system (1 + 2)**	**0.2526**	**9.8E-05**	**0.0099**	**3.9259**	**± 0.0198**	**± 7.8517**
4	Stem wood	0.6565	4.3E-04	0.0208	3.1691	± 0.0416	± 6.3382
5	Stem bark	0.0751	1.4E-05	0.0037	4.9522	± 0.0074	± 9.9043
6	**Stem (4 + 5)**	**0.7316**	**5.2E-04**	**0.0228**	**3.1195**	**± 0.0456**	**± 6.2389**
7	Branches	0.2941	3.9E-04	0.0197	6.6939	± 0.0394	± 13.3878
8	Foliage	0.0149	1.4E-06	0.0012	7.8790	± 0.0023	± 15.7579
9	**Crown (7 + 8)**	**0.3090**	**4.0E-04**	**0.0201**	**6.5118**	**± 0.0402**	**± 13.0236**
10	**Shoot system (6 + 9)**	**1.0406**	**1.3E-03**	**0.0366**	**3.5201**	**± 0.0733**	**± 7.0402**
11	**Total tree (3 + 10)**	**1.2932**	**2.1E-03**	**0.0457**	**3.5343**	**± 0.0914**	**± 7.0685**
Regression estimates							
1	Taproot + stump	0.1113	2.7E-05	0.0052	4.6931	± 0.0104	± 9.3861
2	Lateral roots	0.1418	4.1E-05	0.0064	4.4969	± 0.0128	± 8.9939
3	**Root system (1 + 2)**	**0.2531**	**7.1E-05**	**0.0084**	**3.3266**	**± 0.0168**	**± 6.6532**
4	Stem wood	0.6262	2.1E-04	0.0144	2.3006	± 0.0288	± 4.6011

#	Tree component	BEF$_h$ (Mg m^{-3})	VAR$_{BEF}$ (Mg^2 m^{-6})	SE (Mg m^{-3})	SE (%)	95% CI (Mg m^{-3})	95% CI (%)
5	Stem bark	0.0703	1.6E-05	0.0039	5.6200	± 0.0079	± 11.2400
6	**Stem (4 + 5)**	**0.6965**	**2.5E-04**	**0.0158**	**2.2624**	**± 0.0315**	**± 4.5248**
7	Branches	0.3136	4.2E-04	0.0205	6.5404	± 0.0410	± 13.0807
8	Foliage	0.0105	1.6E-06	0.0013	12.1947	± 0.0026	± 24.3894
9	**Crown (7 + 8)**	**0.3240**	**4.4E-04**	**0.0209**	**6.4639**	**± 0.0419**	**± 12.9277**
10	**Shoot system (6 + 9)**	**1.0205**	**7.8E-04**	**0.0280**	**2.7449**	**± 0.0560**	**± 5.4898**
11	**Total tree (3 + 10)**	**1.2736**	**1.2E-03**	**0.0350**	**2.7475**	**± 0.0700**	**± 5.4950**

The major components and their values are indicated in bold font.

Table 3. *Component biomass density (W_h), their variances (VAR_{Wh}), standard errors (SE), and 95% confidence intervals (CI), computed using mean-of-ratios, ratio-of-means and regression estimators.*

#	Tree component	W$_h$ (Mg ha^{-1})	VAR$_{Wh}$ (Mg^2 ha^{-2})	SE (Mg ha^{-1})	SE (%)	95% CI (Mg ha^{-1})	95% CI (%)
Mean-of-ratios estimates							
1	Taproot + stump	16.2192	0.4725	0.6874	4.2382	± 1.3748	± 8.4764
2	Lateral roots	13.4005	0.5882	0.7669	5.7232	± 1.5339	± 11.4465
3	**Root system (1 + 2)**	**29.6197**	**1.3296**	**1.1531**	**3.8930**	**± 2.3062**	**± 7.7860**
4	Stem wood	75.7526	4.8413	2.2003	2.9046	± 4.4006	± 5.8092
5	Stem bark	8.8182	0.1757	0.4192	4.7534	± 0.8383	± 9.5068
6	**Stem (4 + 5)**	**84.5708**	**5.8565**	**2.4200**	**2.8615**	**± 4.8400**	**± 5.7230**
7	Branches	33.7612	4.1845	2.0456	6.0590	± 4.0912	± 12.1180
8	Foliage	2.7923	0.0880	0.2967	10.6242	± 0.5933	± 21.2483
9	**Crown (7 + 8)**	**36.5535**	**4.8058**	**2.1922**	**5.9973**	**± 4.3844**	**± 11.9946**
10	**Shoot system (6 + 9)**	**121.1243**	**15.3491**	**3.9178**	**3.2345**	**± 7.8356**	**± 6.4690**
11	**Total tree (3 + 10)**	**150.7440**	**24.3521**	**4.9348**	**3.2736**	**± 9.8696**	**± 6.5472**
Ratio-of-means estimates							
1	Taproot + stump	14.4316	3.9E-01	0.6262	4.3388	± 1.2523	± 8.6775
2	Lateral roots	14.6951	6.1E-01	0.7824	5.3245	± 1.5649	± 10.6490
3	**Root system (1 + 2)**	**29.1267**	**1.3E+00**	**1.1435**	**3.9259**	**± 2.2869**	**± 7.8517**
4	Stem wood	75.7039	5.8E+00	2.3991	3.1691	± 4.7983	± 6.3382
5	Stem bark	8.6635	1.8E-01	0.4290	4.9522	± 0.8581	± 9.9043
6	**Stem (4 + 5)**	**84.3674**	**6.9E+00**	**2.6318**	**3.1195**	**± 5.2636**	**± 6.2389**
7	Branches	33.9173	5.2E+00	2.2704	6.6939	± 4.5408	± 13.3878
8	Foliage	1.7126	1.8E-02	0.1349	7.8790	± 0.2699	± 15.7579
9	**Crown (7 + 8)**	**35.6300**	**5.4E+00**	**2.3202**	**6.5118**	**± 4.6403**	**± 13.0236**
10	**Shoot system (6 + 9)**	**119.9974**	**1.8E+01**	**4.2240**	**3.5201**	**± 8.4480**	**± 7.0402**
11	**Total tree (3 + 10)**	**149.1240**	**2.8E+01**	**5.2704**	**3.5343**	**± 10.5409**	**± 7.0685**
Regression estimates							
1	Taproot + stump	16.5475	1.5E+00	1.2190	7.3670	± 2.4381	± 14.7339
2	Lateral roots	13.6022	2.2E+00	1.4938	10.9822	± 2.9877	± 21.9645
3	**Root system (1 + 2)**	**30.1497**	**4.2E+00**	**2.0468**	**6.7888**	**± 4.0936**	**± 13.5777**
4	Stem wood	81.9714	1.4E+01	3.7913	4.6251	± 7.5826	± 9.2503
5	Stem bark	9.5409	8.3E-01	0.9095	9.5323	± 1.8189	± 19.0647
6	**Stem (4 + 5)**	**91.5122**	**1.7E+01**	**4.1665**	**4.5529**	**± 8.3330**	**± 9.1059**
7	Branches	32.9616	2.2E+01	4.6851	14.2139	± 9.3703	± 28.4278
8	Foliage	2.2760	8.2E-02	0.2869	12.6051	± 0.5738	± 25.2101
9	**Crown (7 + 8)**	**35.2376**	**2.3E+01**	**4.7876**	**13.5866**	**± 9.5752**	**± 27.1731**
10	**Shoot system (6 + 9)**	**126.7499**	**5.0E+01**	**7.0566**	**5.5674**	**± 14.1132**	**± 11.1347**
11	**Total tree (3 + 10)**	**156.8996**	**7.8E+01**	**8.8133**	**5.6172**	**± 17.6266**	**± 11.2343**

The major components and their values are indicated in bold font.

4. Discussion

4.1. BEF and Biomass Density

The BEF values computed from different estimators are consistent with one another. The reasons the ratio-based BEFs (either mean-of-ratios or ratio-of-means) were similar to regression-based BEFs involve the fact that F_{timber} is very close to 1 [16], which causes stem volume values to be close to merchantable timber values; as well as the fact that in phase two only 7 trees were not considered in calculation of merchantable timber volumes because their DBH was < 7 cm.

However, it should be mentioned that, in most tropical tree species, and specially in broadleaf species (as opposed to conifers), procuring a minimum top diameter of 7 cm to define merchantable tree height, and thus, merchantable timber volume, is somewhat impractical because the merchantable height is limited by branching, irregular form or defects, which can cause the top diameter to be substantially larger than 7 cm (and inconsistent in each tree), and thus, F_{timber} much smaller than 1. This will lead to larger values of BEF and overestimation of biomass densities. On the other hand, BEF values computed using merchantable timber volume as defined in this study, disregard younger trees (DBH < 7 cm); which are found to be very important in the United Nations Framework Convention on Climate

Change (UNFCCC) reporting process [41]. Moreover, merchantable tree height (i.e., to 7 cm top diameter) measurement/estimation (and thus, merchantable timber volume) in standing trees is subjective and more susceptible to measurement error than total tree height, since the 7 cm top diameter on the stem is more difficult to identify than the tip of the tree.

The biomass densities computed based on mean-of-ratios are slightly larger than those based on ratio-of-means, except for root system and branches. On the other hand, considerable discrepancies were found between ratio-based (either mean-of-ratios or ratio-of-means) and regression-based biomass densities, with regression-based biomass densities calculated as larger than ratio-based ones, except for 3 components. However, despite those discrepancies, the component biomass densities derived from the 3 estimators lie in any estimator's 95% CI.

Although the BEF values obtained from the different estimators are similar, the regression-based BEFs (i.e., computed based on merchantable timber volume) exclude the portion of the stem with diameter < 7 cm and the trees with DBH < 7 cm; and therefore, might not be suitable for estimation of C storage in forested ecosystems, as claimed by Black et al. [20], especially in cases where F_{timber} is found to be much smaller than 1.

The estimated total tree and aboveground BEF values by any of the estimators are consistent with those obtained by Ducta et al. [8], Marková and Pokorný [9], Lehtonen et al. [11], Cháidez [13], Segura and Kanninen [42], Kamelarczyk [43], and Sanquetta et al. [44].

The aboveground biomass (AGB) densities calculated by any estimator are in agreement with those estimated for Mozambique by Brown [45] for dense forests growing in moist-dry season (120 Mg ha^{-1}) and in moist-short dry season (130 Mg ha^{-1}), but are higher compared to dense forests growing in dry seasons (70 Mg ha^{-1}). However, the AGB density estimates based on mean-of-ratios and ratios-of-means estimators (121 and 120 Mg ha^{-1}, respectively) are much closer to those of dense forests growing in moist-dry season; and the regression-based AGB densities (127 Mg ha^{-1}) are close to those of dense forests growing in moist-short dry season. Yet, mecrusse woodlands are typically from dry season regions [22–27], implying that the biomass productivity of mecrusse woodlands are, approximately, twice as larger than the average productivity of dense forests growing in dry seasons in Mozambique.

Our estimates of AGB densities are much larger than the estimates of miombo woodlands, the primary woodlands in Mozambique [46]. Mate et al. [47] found that the AGB density in miombo woodlands in Inhambane and Sofala provinces were 27.3 Mg ha^{-1}; and Ribeiro et al. [48] found that the tree biomass density in miombo woodlands in Niassa National Reserve were approximately 59 Mg ha^{-1}. Low stem density (380–400 ha^{-1}) and basal area (7–19 m^2 ha^{-1}) in miombo woodlands [49] can explain the lower estimates for biomass densities compared to our estimates within mecrusse woodlands (1237 ha^{-1} and 22 m^2 ha^{-1}, respectively).

For the different estimators used to estimate biomass densities, the property of additivity was achieved automatically for the major tree components (root system, shoot system, stem, and crown) and for total tree biomass; i.e., the biomass estimate of the relevant minor tree components' sum to the estimate of relevant major component biomasses and the total tree biomass, which is a desired and logical feature. This is so because stem volume (for ratio estimators) or merchantable timber volume (for regression estimators) is the single auxiliary variable for all tree components.

4.2. Uncertainty

The percent standard errors in the regression-based BEFs are smaller (more precise) than those of the ratio-based ones, in 8 tree components for mean-of-ratios and 6 tree components for ratio-of-means. The BEFs computed based on mean-of-ratios are more precise in 9 components than are those computed based on ratio-of-means estimators. On the other hand, the regression-based tree component biomass densities are less precise (more uncertain) than the ratio-based ones (both mean-of-ratios and ratio-of-means).

The estimated uncertainty (SE (%)) in our ratio-based BEF values (2.9%–10.6% for mean-of-ratios; and 3.1%–7.9% for ratio-of-means) and regression-based BEF values (2.7%–12.1%) were lower than those of Lehtonen et al. [11, 19] (3%–21%), and Jalkanen et al. [7] (4%–13%). The component biomass and stem volume values used here to calculate BEF were obtained directly using destructive sampling, whereas those by Lehtonen et al. [11, 19] and Jalkanen et al. [7] were based on values obtained indirectly using regression models. These different approaches might explain the differences among BEF estimates and the higher uncertainty reported by those authors, because they also incorporate uncertainty from the regression models.

The total uncertainty of the estimate of AGB biomass density, as measured by SE (%) and 95% CI, using ratio estimators (either mean-of-ratios or ratio-of-means) and regression estimators, were approximately four-fold and two-fold smaller, respectively, than those obtained by Chave et al. [50] (SE = 24%) and Brown et al. [51] (95% CI = ± 20%).This denotes that our estimates of ABG biomass are two to four times more precise and accurate than those obtained by Chave et al. [50] and Brown et al. [51].

The low level of uncertainty in our BEF and biomass estimates is, presumably, attributed to the homogeneity of mecrusse woodlands and site characteristics, as *A. johnsonii* is the only canopy species in mecrusse woodlands, and because our phase two sample size (n = 93) is large, as defined by Freese [1, 3], Husch et al. [6], Stellingwerf [52], and Stauffer [53], in which the sample size should be >30 to be considered large. Stellingwerf [52] suggested that the 95% CI should not exceed ± 20% of the mean. Our 95% CI for estimates of component BEFs fell well within these accepted limits (<20%); except for foliage BEF, in which it was 21.25% and 24.39% for ratio-of-means-based and regression-based BEF, respectively.

5. Conclusion

The computation of BEF using merchantable timber volume should utilize regression estimators, not ratio estimators as usually done. Tree component biomass densities computed with the aid of regression-based BEFs were found to be more uncertain than those computed with the aid of ratio-based BEFs. BEFs computed using merchantable timber volume (regression-based BEFs) are subjective and more susceptible to errors, as the definition of merchantable stem height is subjective, and susceptible to errors and personal judgement, especially in standing trees; and the use of a fixed top diameter to define merchantable height is limited by branching, irregular form, or defects in most tropical tree species and particularly in broadleaf tree species. Furthermore, BEFs computed using merchantable timber volume exclude trees that have not achieved a predefined minimum merchantable DBH and the portion of the tree without minimum top diameter, and that therefore, might not be suitable for estimation of C storage in forested ecosystems.

Acknowledgments

This study was funded by the Swedish International Development Cooperation Agency (SIDA).

References

[1]　F. Freese, *Elementary Forest Sampling*, United States Department of Agriculture, Washington DC; 1962.

[2]　W. G. Cochran, *Sampling Techniques*, John Wiley & Sons, New York, 3rd edition, 1977.

[3]　F. Freese, *Statistics for Land Managers*, Paeony Press, Edinburgh, 1984.

[4]　P. G. de Vries, *Sampling Theory for Forest Inventory*, Springer-Verlag, New York, 1986.

[5]　K. Jayaraman, *A Statistical Manual for Forestry Research*, FORSPA-FAO, Bangkok, 2000.

[6]　B. Husch, T. W. Beers, and J. A. Kershaw Jr., *Forest Mensuration*, John Wiley & Sons, New York, USA, 4th edition, 2003.

[7]　A. Jalkanen, R. Mäkipää, G. Stahl, A. Lehtonen, and H. Petersson, "Estimation of the biomass stock of trees in Sweden: comparison of biomass equations and age-dependent biomass expansion factors", *Ann. For. Sci.*, vol. 62, pp. 845–851, 2005.

[8]　I. Dutca, I. V. Abrudan, P. T. Stancioiu, V. Blujdea, "Biomass conversion and expansion factors for young Norway spruce (*Picea abies* (L.) Karst.) trees planted on non-forest lands in Eastern Carpathians", *Not Bot Hort Agrobot Cluj*, vol. 38, no. 3, pp. 286–292, 2010.

[9]　I. Marková, and R. Pokorný, "Allometric relationships for the estimation of dry mass of aboveground organs in young highland Norway spruce stand", *Acta Univ Agric Silvic Mendel Brun*, vol. 59, no. 6, pp. 217–224, 2011.

[10]　E. M. Nogueira, P. M. Fearnside, B. W. Nelson, R. I. Barbosa, and E. W. H. Keizer, "Estimates of forest biomass in the Brazilian Amazon: New allometric equations and adjustments to biomass from wood-volume inventories", *Forest Ecology and Management*, vol. 256, pp. 1853–1867, 2008.

[11]　A. Lehtonen, R. Mäkipää, J. Heikkinen, R. Sievänen, and J. Lisk, "Biomass expansion factors (BEFs) for Scots pine, Norway spruce and birch according to stand age for boreal forests", . *Forest Ecology and Management*, vol. 188, pp. 211–224, 2004.

[12]　P. Soares, and M. Tome, "Biomass expansion factores for *Eucalyptus globulus* stands in Portugal", *Forest Systems*, vol. 21, no. 1, pp. 141-152, 2012.

[13]　J. J. N. Cháidez, "Allometric equations and expansion factors for tropical dry forest trees of Eastern Sinaloa, Mexico", *Trop Subtrop Agroecosys*, vol. 10, pp. 45–52, 2009.

[14]　F. M. Silva-Arredondo, and J. J. N. Návar-Cháidez, "Factores de expansión de biomasa en comunidades florestales templadas del Norte de Durango, México", *Rev Mex Cien For*, vol. 1, no. 1, pp. 55–62, 2010.

[15]　S. Brown, "Measuring carbon in forests: current and future challenges", *Environ Pollut*, vol. 116, pp. 363–372, 2002.

[16]　P. E. Levy, S.E. Hale, and B. C. Nicoll, "Biomass expansion factors and root: shoot ratios for coniferous tree species in Great Britain", *Forestry*, vol. 77, no. 5, pp. 421–430, 2004.

[17]　Z. Somogyi, E. Cienciala, R. Mäkipää, P. Muukkonen, A. Lehtonen, and P. Weiss, "Indirect methods of large-scale forest biomass estimation", *Eur J Forest Res*, vol. 126, pp. 197–207, 2007.

[18]　P. N. Edwards, and J. M. Christie, *Yield models for forest management*, HMSO, London, 1981.

[19]　A. Lehtonen, E. Cienciala, F. Tatarinov, R. Mäkipää, "Uncertainty estimation of biomass expansion factors for Norway spruce in the Czech Republic", *Ann For Sci*, vol. 64, pp. 133–140, 2007.

[20]　K. Black, B. Tobin, G. Siaz, K. A. Byrne, and B. Osborne, "Improved estimates of biomass expansion factors for Sitka spruce", *Irish Forestry*, pp. 50–65, n.d.

[21]　J. Mantilla, and R. Timane, *Orientação para maneio de mecrusse*, SymfoDesign Lda, Maputo, Mozambique, p. 1, 2005.

[22]　Dinageca, *Mapa Digital de Uso e Cobertura de Terra*, Cenacarta, Maputo, Moçambique, 1990.

[23]　Mae, *Perfil do Distrito de Chibuto, Província de Gaza*, Mae, Maputo, Moçambique, 2005.

[24]　Mae, *Perfil do Distrito de Mandhlakaze, Província de Gaza*, Mae, Maputo, Moçambique, 2005.

[25]　Mae, *Perfil do Distrito de Panda, Província de Inhambane*, Mae, Maputo, Moçambique, 2005.

[26]　Mae, *Perfil do Distrito de Funhalouro, Província de Inhambane*, Mae, Maputo, Moçambique, 2005.

[27]　Mae, *Perfil do Distrito de Mabote, Província de Inhambane*, Mae, Maputo, Moçambique, 2005.

[28] T.M. Magalhães, "Allometric equations for estimating belowground biomass of *Androstachys johnsonii* Prain", Carbon Balance and Management, 10:16, 2015.

[29] T.M. Magalhães, T. Seifert, "Estimation of tree biomass, carbon stocks, and error propagation in mecrusse woodlands", Open Journal of Forestry vol. 5, pp. 471–488, 2015.

[30] T.M. Magalhães, T. Seifert, "Biomass modelling of *Androstachys johnsonii* Prain – a comparison of three methods to enforce additivity", International Journal of Forestry Research vol. 2015, pp.1–17, 2015.

[31] T.M. Magalhães, T. Seifert, "Tree component biomass expansion factors and root-to-shoot ratio of Lebombo ironwood: measurement uncertainty", Carbon Balance and Management 10: 9, 2015.

[32] T.M. Magalhães, T. Seifert, "Below- and aboveground architecture of *Androstachys johnsonii* Prain: Topological analysis of the root and shoot systems", Plant and Soil, 2015. DOI 10.1007/s11104-015-2527-0.

[33] FAO, *FAO Map of World Soil Resources,* Italy, Rome, 2003.

[34] B. R. Parresol, "Additivity of nonlinear biomass equations", Can. J. For. Res., vol. 31, pp. 865–878, 2001.

[35] M. A. M. Brasil, R. A. A. Veiga, and J. L. Timoni, "Erros na determinação da densidade básica da madeira", CERNE, vol. 1, no. 1, pp. 55–57, 1994.

[36] I. A. Gier, *Forest mensuration (fundamentals)*; International Institute for Aerospace Survey and Earth Sciences (ITC), The Netherlands, p. 20, 21, 1992.

[37] J. Bunster, *Commercial timbers of Mozambique,* Technological catalogue, Traforest Lda, Maputo, Mozambique, 2006.

[38] T. Seifert, and S. Seifert, "Modelling and simulation of tree biomass", in *Bioenergy from Wood: Sustainable Production in the Tropics*; T. Seifert, Ed., pp. 42–65, Springer, Managing Forest Ecosystems, vol. 26, 2014.

[39] S. A. Machado, A. Figueiredo Filho, *Dendrometria*, Unicentro, Paraná, Brazil, 2005.

[40] E. W. Johnson, *Forest Sampling Desk Reference,* CRC Press LLC, Florida, USA, 2000.

[41] K. Black, B. Tobin, G. Siaz, K. A. Byrne, and B. Osborne, "Allometric regressions for an improved estimate of biomass expansion factors for Ireland based on a Sitka spruce chronosequence", *Irish Forestry*, Vol. 61, no. 1, pp. 50–65, 2004.

[42] M. Segura, M. Kanninen, "Allometric models for tree volume and total aboveground biomass in a tropical humid forest in Costa Rica", *Biotropica*, vol. 37, no. 1, pp. 2–8, 2005.

[43] K. B. F. Kamelarczyk, "Carbon stock assessment and modelling in Zambia: a UN-REDD programme study", United Nations–Reducing Emissions from Deforestation and Forest Degradation, Zambia, 2009.

[44] C. R. Sanquetta, A. P. D. Corte, F. Silva, "Biomass expansion factors and root-to-shoot ratio for Pinus in Brazil", *Carbon Balance and Management,* vol. 6, pp. 1–8, 2011.

[45] S. Brown, "Estimating biomass and biomass change of tropical forests: a primer," FAO Forest Paper 134, 1997.

[46] A. A. Sitoe, and N. S. Ribeiro, *Miombo Book Project (Case Study of Mozambique),* Universidade Eduardo Mondlane (UEM). Maputo, Mozambique, 1995.

[47] R. Mate, T. Johansson, and A. Sitoe, "Biomass equations for Tropical Forest tree species in Mozambique", *Forests*, vol. 5, pp. 535-556, 2014.

[48] N. S. Ribeiro, C. N. Matos, I. R. Moura, R. A. Washington-Allen, and A. I. Ribeiro (2013) "Monitoring vegetation dynamics and carbon stock density in miombo woodlands", *Carbon Balance and Management*, vol. 8, no. 1, pp. 1–9, 2014.

[49] N. Ribeiro, A. A. Sitoe, B. S. Guedes, and C. Staiss, *Manual de Silvicultura Tropical*, Food and Agriculture Organisation of the United Nations, Maputo, Mozambique, 2002.

[50] J. Chave, R. Condic, S. Aguilar, A. Hernandez, S. Lao, and R. Perez, "Error propagation and scaling for tropical forest biomass estimates", *Phil. Trans. R. Soc. Lond. B*, vol. 309, pp. 409–420, 2004.

[51] I. F. Brown, L. A. Martinelli, W. W. Thomas, M. Z. Moreira, C. A. C. Ferreira, R. A. Victoria, "Uncertainty in the biomass of Amazonian forests: An example from Rondônia, Brazil", *Forest Ecology and Management*, vol. 75, pp. 175–189, 1995.

[52] D. A. Stellingwerf, *Forest inventory and remote sensing,* International Training Centre for Aerial Survey (ITC), Enschede, 1994.

[53] H. B. Stauffer, *Some sample size tables for forest sampling,* Ministry of Forests, British Columbia, Canada, 1983.

14

Comparative Study on the Effect of *Citrillus lanatus* and *Cucumis sativus* on the Growth Performance of *Archachatina marginata*

Ufele Angela Nwogor

Zoology Department, Nnamdi Azikiwe University, Awka, Nigeria

Email address:

ufeleangel@yahoo.com

Abstract: Feeding accounts for a reasonable percentage of the cost of livestock production and a major factor that determines the viability and profitability of livestock farming ventures. Also scarcity and high cost of meat for human consumption has necessitated the need for intensive rearing of some non-conventional livestock such as the snail hitherto hunted from the wild, in view of this, this study evaluated the effects of two fruits, Cucumber (*Cucumis sativus*) and Watermelon (*Citrillus lanatus*) on the growth performance of *Archachatina marginata*. The study was conducted using One hundred and thirty five (135) snails, 15 snails per treatment and each treatment was replicated three times and the experiment lasted for a period of eight weeks. In terms of growth, results showed that the snails fed with watermelon and cucumber performed generally better than others but statistically, no significant difference (P>0.05) existed between the snails (in terms of weight, length, circumference of the snail) fed with the two fruits and the combination of both. From the result, it was observed that the mean weight gain of snails fed with *Citrillus lanatus* (Cage A) was 5.43g, those fed with *Cucumis sativus* (Cage B) had mean weight gain of 1.22g , while those fed with both fruits (Cage C) had 7.53g. The mean shell length increase in snails fed with *Citrillus lanatus* was 10.74mm, those of *Cucumis sativus* was 10.52mm, while those fed with both had 12.24mm. The mean shell circumference of snails fed with *Citrilus lanatus* was 14.47mm, those fed with *Cucumis sativus* had 14.15mm, while those fed with both had 15.16mm. From the results, snails fed with both *Citrillus lanatus* and *Cucumis sativus* performed best while snails fed with *Citrillus lanatus* performed better than those fed with *Cucumis sativus* in all parameters measured. Therefore, snail farmers are advised to use the combination of cucumber and watermelon for a better yield.

Keywords: Snails, Growth Performance, Cucumber (*Cucumis sativus*), Watermelon (*Citrillus Lanatus*)

1. Introduction

The scarcity and high cost of meat for human consumption has necessitated the need for intensive rearing of some non-conventional livestock such as the snail hitherto hunted from the wild (Alikwe *et al*., 2014). The low capital and simple management practices involved have also drawn the attention of many farmers to snail farming (Mogbo *et al*., 2013). Snail meat is often regarded as a form of bush meat or game meat to be eaten occasionally instead of being a nutritious meat to be relished on a daily basis just like the meat of other conventional livestock (Malik and Dikko, 2009). Snail meat often referred to as Congo meat is a high quality food rich in protein, low in fats and source of iron (Orisawuyi, 1989) calcium, magnesium and zinc (Ademolu *et al*., 2004). Imevbore and Ademosun (1988) accessed the nutritional

value of snail and observed that it has a protein content of 88.37% which compares favourably with conventional animal protein sources. Adeyeye (1996) noted that snails contain almost all the amino acids required by man. Its tenderness and fine texture makes it the most suitable for all ages (Okonta, 2012). The low content of fat (1.3%) and low cholesterol level make snail a good antidote for vascular diseases such as heart attack, cardiac arrest, hypertension, stroke, high blood pressure and other fat related ailments (Akannusi, 2002). Other curable ailments by snails in Nigeria include whooping cough, anaemia, ulcer, asthma, age problems, hypertension and rheumatism (Abere and Lameed, 2008). The meat content of snails has been reported to cause reduction in the labour pain and loss of blood during labour, restoration of virility and fertility in human beings (Agbogidi *et al*., 2008). Imevbore and Ademosun, (1988) also maintained that the serotonin secreted in the snail's body is effective in the maintenance of normal behavior after mental depression. Not only is the flesh of snail a valued delicacy, but the shells and offal have also gained considerable value in the manufacturing of feed for animals of different types (Ayodele and Ashimolowo, 1999). According to Cobbinah *et al.*,(2008), crushed snail shells may be applied in chicken feed or liming to improve the quality of acidic (fish pond) soil.

In view of the high quality of protein obtained from snails, they have secured high demand in many cuisines both locally and internationally (Ngenwi *et al.*, 2010). The popularity of giant land snails in the world is increasingly reduced by indiscriminate hunting and deforestation which destroys the snail habitat, therefore rearing of the giant land snails as a domestic animal would therefore help in some measure to satisfy the demand for the meat and to ensure the survival of the species (Ademolu *et al.*, 2004). Usually, snails become scarce during the dry season hence expensive at this period (Amusan, 2002) their domestication could make them more readily available all year round as well as reduce their prices to a reasonable extent (Okonta, 2012). Awah (1992) said that *Archachatina marginata* feed on a wide variety of feed including both fruits and leaves of plants.

Cucumis sativus (cucumber) and *Citrillus lanatus* (watermelon) are fruits with high water content. They belong to the family Cucurbitaceae. They are readily available in Nigeria especially during the wet season. These fruits were used for this experiment in accordance to Amata,(2014), who stated that moisture content of snails feed is usually very high because of the method of feeding by snails which prefer feed in fluid.

Snail farming is one of the most lucrative and prolific farming in recent times (Ufele *et al.*, 2013). This has raised the interest of improving the culturing and rearing of snails within Nigeria to increase protein intake through eating snail meat (Ufele *et al.*, 2013). Snail farming is also a tool for poverty alleviation (Moyin-jesu and Ajao 2008).

2. Materials and Methods

2.1. Procurement of Experimental Animal

One hundred and thirty five adult snails of the specie *Archachatina marginata* of average weight 102.5g were used for the experiment. The snails were allowed to acclimatize in their new environment for one week before the commencement of the experiment.

2.2. Experimental Treatments

One hundred and thirty five adult snails were used for the study and the experiment lasted for a period of eight weeks. The snails were randomly grouped into three of fifteen snails per group and were assigned to three dietary treatments. Treatment one was fed with Watermelon (*Citrillus lanatus*), Treatment two was fed with Cucumber (*Cucumis sativus*), while Treatment three was fed with both cucumber and watermelon. Each treatment was replicated three times. The snails were measured on weekly basis and parameters measured were length, shell circumference and weight. The snails were measured individually.

2.3 Data Analysis

The length, shell circumference and weight of the snails were taken weekly using a sensitive weighing balance and caliper. The result of the experiment was analyzed using Analysis of variance (ANOVA). The comparison of mean was separated using a post Hoc test (Least Significant Difference), (William and George, 2008).

3. Results

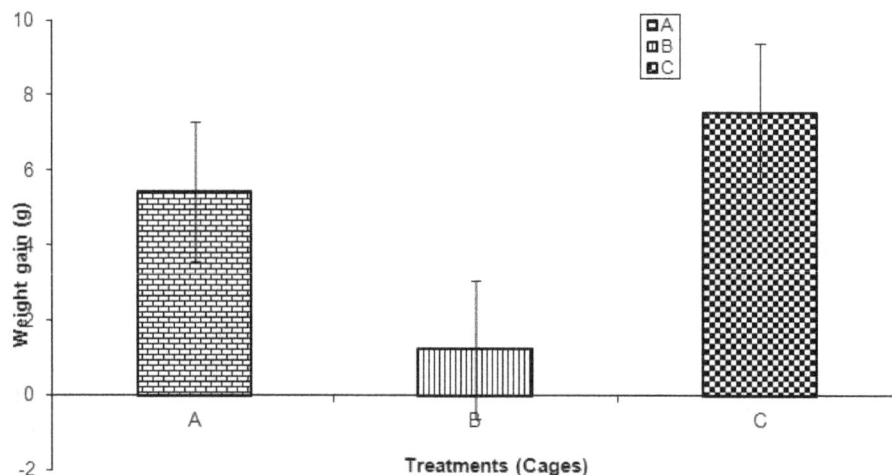

Figure 1. Mean Weight gain of snails.

Figure 1 shows the mean weight of snails fed with different treatments. From the result it was observed that snails fed with combination of Cucumber (*Cucumis sativus*) and Watermelon (*Citrillus lanatus*) (Cage C) had the highest mean weight (7.53g). Followed by those fed with Watermelon (*Citrillus lanatus*) (Cage A) (5.43g) and those fed with Cucumber (*Cucumis sativus*) (Cage B) had the lowest mean weight gain (1.22g).

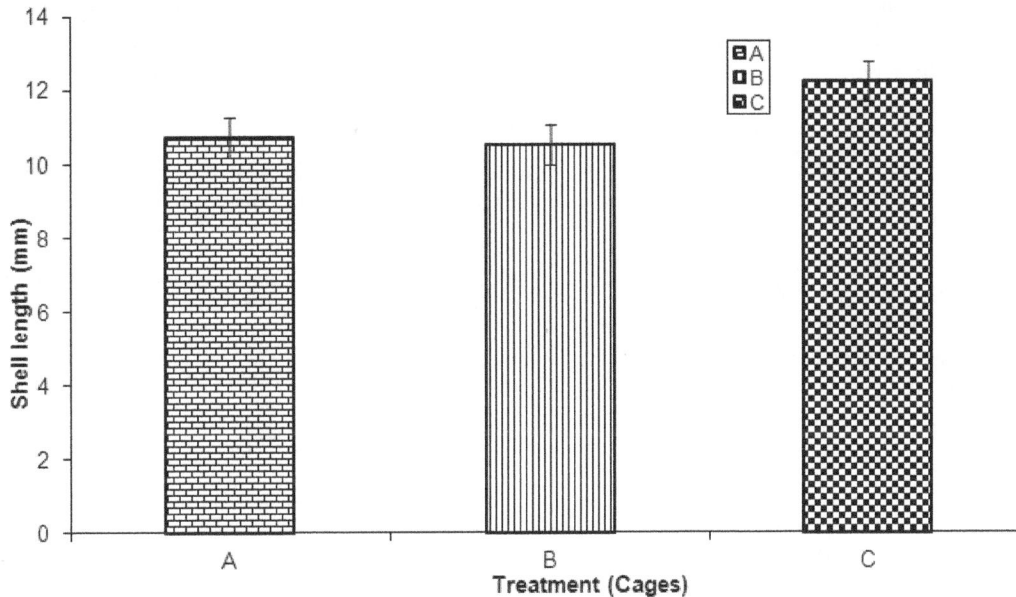

Figure 2. Shell length of snails.

Figure 2 shows the shell length increase of the snails fed with different treatments. From the figure it was observed that snail fed with combination of Cucumber (*Cucumis sativus*) and Watermelon (*Citrillus lanatus*) (Cage C) had the highest increase in the shell length(12.24mm). Followed by those fed with Watermelon (*Citrillus lanatus*) (Cage A) (10.74mm) and those fed with Cucumber (*Cucumis sativus*) (Cage B) had the lowest increase in the shell length (10.52mm).

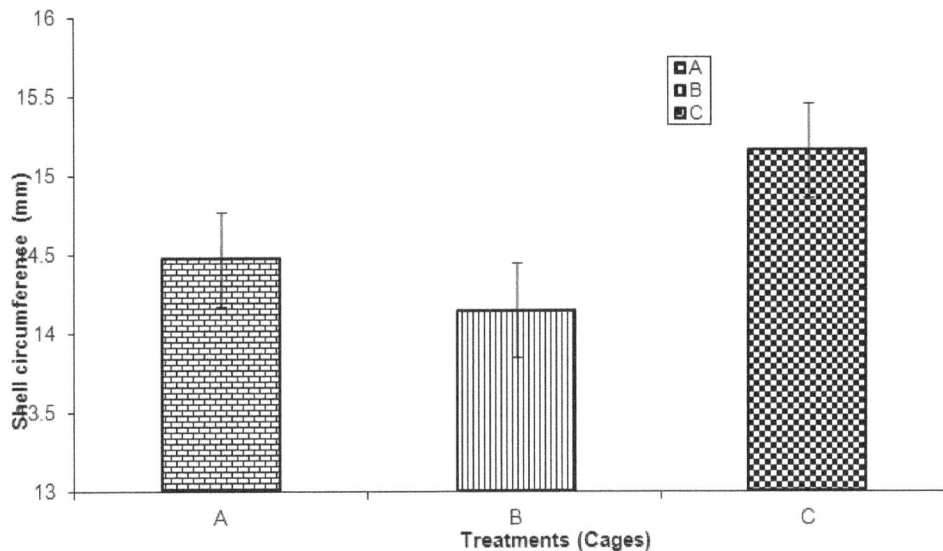

Figure 3. Shell Circumference of snails.

Figure 3 shows the shell circumference of snails fed with different treatments. From the result of Figure 3, it was observed that snails fed with combination of Cucumber (*Cucumis sativus*) and Watermelon (*Citrillus lanatus*) (Cage C) had the highest increase in the shell circumference (15.16mm). Followed by those fed with Watermelon (*Citrillus lanatus*) (Cage A) (14.47mm) and those fed with Cucumber (*Cucumis sativus*) (Cage B) had the lowest increase in the shell circumference (14.15mm).

4. Discussion

From the results above, it is observed that both feeds have positive impact on the growth performance of the experimental animal. It was observed that snails fed with combination of both fruits (Cage C) performed best, followed by those fed with watermelon (Cage A). While those fed with cucumber had the least performance, though statistically, there was no significant difference in the growth performance of the snails in all the treatments ($p < 0.05$). This result is in accordance to Alexander (1997) who stated that food attractiveness is important in snail's nutrition; if the food is appetizing the snails will eat a lot and grow quickly. There was no significant difference in the growth performance of the snails fed with the three treatments, this might be as a result that the two fruits belong to the same family and have almost same nutritive components. Also, there was increase in both length and circumference of the snail. As the body weight increased, there seemed to be increase in size probably due to expansion of the shell. This observation is consistent with the finding of Omole et al., (1999) who noted an increase in the length and circumference of land snail fed on fruits. From the result, the snails in all the treatments increased in both weigth, shell length and shell circumference, this may be attributed to the findings of Ajayi, et al., (1978) who stated that snails are generally heterotrophic animals and feed on a wide variety of young dicotyledonous plants and cultivated plants. It was also stated that snails are vegetarians and will accept many types of food. Snails feed on leaves which may include cocoyam, kola, pawpaw, cassava, okra, egg plant, cabbage, lettuce. They also feed on fruits like pawpaw, mango, banana, egg plant, pear, oil palm, fig, tomato, cucumber which are rich in vitamins and minerals (Akinnusi, 1998). In another research it was also discovered that the younger snail appear to prefer leaves to fruits while older and larger snails would go for fruits when offered a choice between leaves and fruits (Ejidike, 2001). This is in relation to this research.

5. Conclusion

In conclusion, the result of this research shows that the snail *Archachatina marginata* can be fed with cucumber (*Cucumis sativus*) and watermelon (*Citrillus lanatus*) since the snails enjoy it and it had positive effect on their growth performance, it is better to combine the two fruits for a better yield. Snail farmers are advised to feed their snails with cucumber and watermelon.

References

[1] Abere, S.A. and Lameed, G.A. (2008). The Medicinal Utilization of Snails in Some Selected States in Nigeria. *Proceeding of the First National Conference of the Forest and Forest Products Society (FFPs).* pp. 233-237.

[2] Ademolu, K. O., Idowu, A. B., Matiana, C. F. and Osinowo, O. A. (2004). Performance proximate and mineral analysis of African giant land snail *(Archachatina marginata)* fed different nitrogenous sources. *African Journal of Biotechnology* 3 (8): 412 – 417.

[3] Adeyeye, E.I. (1996). Waste yield, Proximate and mineral Composition of three different types of land snail found in Nigeria. *International Journal of Food Science and Nutrition* 42(2): 111-116.

[4] Agbogidi, O.M. Okonta, B.C. and Ezeani, E.L. (2008) Effects of two edible fruits on the growth performance of African giant land snail, *(Archachatina marginata)*. *Journal of Agricultural and Biological Sciences* 3 (3): 26– 29.

[5] Ajayi, S. S, Tewe, O. O., Moriarty, C. and Awesu, M. O. (1978): Observation on the Biology and Nutritive Value of African Giant Snail *(Archachatina marginata)*. *East African Wildlife Journal.* 16:85 – 95.

[6] Akinnusi, O. (1998). A Practical Approach to Back Yard Snail Farming. *Nigerian Journal of Animal Production.* 25:193–197.

[7] Akinnusi, O. (2002). *Introduction to snail and snail farming.* Triolas Exquisite ventures. Abeokuta. 25pp.

[8] Alexander, B.O. (1997). *Snails and Snail Farming* (Nigeria Edible Land Snails). University Publishers Ibadan. Pp 1-65.

[9] Alikwe, P.C.N, Yeigba, J., Akinnusi, B., Oyenike, F.A, Ohimain, E.I (2014). Performance and carcass characteristics of Giant African Land Snails fed *Alchornea cordifolia* leaf meal in replacement for soybean meal. *International Journal of Research in Agriculture and Food Sciences.* 1(6):1-4.

[10] Amata, I.A. (2014). Effects of Microhabitat on the Egg Hatchability and Hatchling Growth Performance of the Giant African Land Snails *(Archachatina marginata)*. *Global Journal of Biology, Agriculture and Health Sciences.* 3(2):27-31.

[11] Amusan, O.M. (2002). *The techniques of snail farming as a viable and profitable venture.* Oak Ventures Publishers, Lagos. 23pp

[12] Awah, A.A. (1992). Snail Farming in mature Rubber Plantation: Studies on the aspect of specialized Production Techniques of farming *Archachatina marginata*: *Snail Farming Research* 4: 33-39

[13] Ayodele, I.A, and Asimalowo, A.A (1999). *Essentials of snail farming.* Agape Prints, U.I, Ibadan, p. 51.

[14] Cobbinah, J.R, Vink, A., Onwuka, B. (2008). *Snail Farming: Production, processing and marketing.* Agromisia Foundation, Wageningen. First Edition 78pp.

[15] Ejidike. B. N. (2001) Comparative effect of supplemental and complete diets on the performance of African giant land snail (*A. marginata*). *Proceedings of the 26th Annual Conference of the Nigerian Society for Animal Production.* 26: 151 - 153.

[16] Imevbore, E. and Ademosun, A. A. (1988). The nutritive value of African Giant land snail (*Archachatina marginata*). *Nigerian Journal of Animal Production.* 15: 109 – 112.

[17] Malik, A.A. and Dikko, A.H. (2009). Heliculture in Nigeria, the potentialities, opportunities and challenges (a review). *Proceedings of the 34th Annual Conference of Nigerian Society for Animal Production.* Pp 120-1124.

[18] Mogbo, T.O, Okeke, J.J, Ufele, A.N, Nwosu, M.C, Ibemenuga, K.N (2013). Preliminary investigation on the influence of housing types on reproductive characteristics of snail *(achatina achatina)*. *American journal of bioscience.* 1(4):54-58.

[19] Moyin-Jesu, E.I, and Ajao, K. (2008). Raising of giant snails *(Archachatina marginata)* in urban cities using soil amendments and feeding materials for food security. *African Journal of Science and Technology (AJST) Science and Engineering series.* 9(1):118-124.

[20] Ngenwi, A.A, Mafeni, J.M, Etchu, K.A, Oben, F.T (2010). Characteristics of snail farmers and constraints to increased production in West and Central Africa. *African Journal of Environmental Science and Technology.* 4(5):274-278.

[21] Okonta, B.O. (2012). Performance of Giant African Land Snails *Archachatina marginata* fed with selected diets. *Global Journal of Bioscience and Biotechnology.* 1(2):182-185.

[22] Omole, A.J. Oluokun, J.A. Oredein, A.O. Tiomiyu, A.K. Afolabi, A.O. Adetoro, F.O. and Adejuyibe, A.P. (1999) Snail production potential for increasing animal protein intake in West Africa. *Proceedings of the 26th annual conference of the Nigeria Society of Animal Production.* Pp. 393–401.

[23] Orisawuyi, Y. A. (1989). *Practical guides to snail rearing.* Gratitude Enterprises, Lagos. 27pp.

[24] Ufele, A.N, Nnajidenma, U.P., Ebenebe, C.I., Mogbo, T.C., Aziagba, B.O. and Akunne, C.E. (2013). The effect of *Azardirachta indica* (neem) leaf extract on longevity of Snails (*Achatina achatina*). *International Research Journal of Biological Sciences.* 2(1): 61-63.

[25] William, A.C., and George, W.S. (2008). Statistical Methods, 6th Ed., The Iowa State University Press. Ames, Iowa, USA. Pp. 167-263.

15

Analysis of Genetic Diversity Using Simple Sequence Repeat (SSR) Markers and Growth Regulator Response in Biofield Treated Cotton (*Gossypium hirsutum* L.)

Mahendra Kumar Trivedi[1], **Alice Branton**[1], **Dahryn Trivedi**[1], **Gopal Nayak**[1], **Mayank Gangwar**[2], **Snehasis Jana**[2, *]

[1]Trivedi Global Inc., Henderson, NV, USA
[2]Trivedi Science Research Laboratory Pvt. Ltd., Bhopal, Madhya Pradesh, India

Email address:

publication@trivedisrl.com (S. Jana)

Abstract: Cotton is the most important crop for the production of fiber that plays a key role in economic and social affairs. The aim of the study was to evaluate the impact of biofield energy treatment on cotton seeds regarding its growth, germination of seedling, glutathione (GSH) concentration, indole acetic acid (IAA) content and DNA fingerprinting using simple sequence repeat (SSR) markers for polymorphism analysis. The seeds of cotton cv. Stoneville-2 (*Gossypium hirsutum* L.) was obtained from DNA Land Marks Inc., Canada and divided into two groups. One group was remained as untreated, while the other was subjected to Mr. Trivedi biofield energy and referred as treated sample. The growth-germination of cotton seedling data showed higher germination (82%) in biofield treated seeds as compared to the control (68%). The alterations in length of shoot and root of cotton seedling was reported in the treated sample with respect to untreated seeds. However, the endogenous level of GSH in the leaves of treated cotton was increased by 27.68% as compared to the untreated sample, which may suggest an improved immunity of cotton plant. Further, the plant growth regulatory constituent *i.e.* IAA concentration was increased by 7.39%, as compared with the control. Besides, the DNA fingerprinting data, showed polymorphism (4%) between treated and untreated samples of cotton. The overall results suggest that the biofield energy treatment on cotton seeds, results in improved overall growth of plant, increase germination rate, GSH and IAA concentration were increased. The study assumed that biofield energy treatment on cotton seeds would be more useful for the production of cotton fiber.

Keywords: Biofield Energy, DNA Fingerprinting, Polymorphism, Cotton cv. Stoneville-2, Glutathione

1. Introduction

The cotton genus has more than 50 species reported worldwide in arid, semi-arid regions. It is indigenous to the tropic and subtropics regions [1]. Cotton is regarded as a vital source of seed oil and protein meal and is the major cash crop in the World. Cotton (*Gossypium* spp.), belongs to family *Malvaceae*, and is among the most important non-food crops, which occupies a significant position from both agricultural and manufacturing sectors points of view. It is the major source of one of the basic human need *i.e.* clothing apart from the other fiber sources *viz.* jute, silk and synthetics. Hence, it is one among the most cultivated and traded commodities in the World. Countries such as USA, China, Sudan, Egypt, Australia and

India are the major producers of cotton. Cotton industry throughout the world has wide economic market of about $500 billion per year [2]. Apart from its economic importance, cotton has been regarded as the standard experimental model system to study polyploidization, cell elongation, cellulose, and cell wall biosynthesis [3, 4]. However, it is the only common plant, which yields single-celled fibers [3].

To promote the growth, development, and yield of agricultural crops, plant hormones plays a major role [5]. Indole-3-acetic acid (an auxin), is an endogenous phytohormone, mainly produced in the meristemic tissues of root apices, stem, and young developing leaves [6], and plays an important role in growth and development of root [7]. Multiple roles of glutathione in plant metabolism have been reported such as signaling of sulfur status, heavy metal tolerance, pathogen response, resistance to xenobiotics, and antioxidative defense and redox control [8]. The level of growth hormones, seed variety, use of pesticides, environmental factors, etc. plays a major role in final yield of the cotton. Recent reports from textile industry states the deficiencies in the quality of cotton, which include poor seed inputs, poor fiber attributes, rapid deterioration of fiber quality, wide range of contaminants, poor soil and rain-fed situations, etc. These all factors contributes to poor yield and increases the cost of cultivation [9]. Despite of several advances in agricultural sciences, some safe and natural approach is still required to improve the agricultural crops yield. In search of some cost effective and safe approach, authors studied the impact of biofield energy treatment on cotton seeds in terms of overall growth and yield of cotton.

Biofield energy, the electromagnetic field/energy that permeates and surrounds the living organisms is reported to have the capacity to improve the germination rate, enhance the biochemical markers and improved other agrochemical parameters [10, 11]. However, the energy can exists in various form such as kinetic, potential, electrical, magnetic, and nuclear, and human nervous system consist of chemical information in the form of electromagnetic signals. Biofield involves regulation of electromagnetic information, which regulating hemodynamics. Energy medicine is one of the complementary and alternate medicine (CAM). According to National Health Interview Survey (NHIS), conducted by the Centers for Disease Control and Prevention's (CDC) and National Center for Health Statistics (NCHS), energy therapy was reported to be very common among adults [12]. Biofield treatment on agricultural crops will be new and upcoming approach worldwide to improve the agricultural productivity. Mr. Mahendra Kumar Trivedi possesses unique biofield energy, which has been reported to alter the growth characteristics in the field of agricultural science research [13], and plant biotechnology [14]. Mr. Trivedi's unique biofield treatment is also termed as The Trivedi Effect®.

After considering the significant outcomes of biofield energy treatment, the study was designed to evaluate the impact of The Trivedi Effect® on cotton with respect to growth, yield, and genetic variability parameters (DNA fingerprinting) using standard molecular method.

2. Materials and Methods

Cotton cv. Stoneville-2 (*Gossypium hirsutum* L.) was obtained from DNA LandMarks, Montreal, Canada. The seeds of cotton were divided into two parts, one part was considered as control, no treatment was given. The other part was coded as treated and subjected to Mr. Trivedi's biofield treatment. The DNA fingerprinting analysis was performed using SSR markers.

2.1. Biofield Treatment Strategy

The treated group sample of cotton seeds was subjected to Mr. Trivedi's biofield treatment under laboratory conditions. Mr. Trivedi provided the biofield treatment through his unique energy transmission process to the treated group without any

touch for few minutes. The treated sample was assessed for growth germination of seedlings, glutathione (GSH) level and indole acetic acid (IAA) content in roots and shoots of cotton plant [15].

2.2. Impact of Biofield Energy on Juvenile Growth of Cotton Plant

Control and treated cotton seeds were soaked for 6 hours in distilled water. The water soaked seeds were wrapped with moist tissue paper and kept in dark condition for germination. Juvenile growth was measured by recording the percent of germinated seeds, time taken for emergence of both radicle and plumule, and the length of shoot and root of young seedlings [15].

2.3. Estimation of Glutathione in Cotton Leaves

For the extraction of GSH approximate 5 gm of cotton leaves were crushed and mixed with 5 mL of 80% cold methanol (as a solvent). Further, the extract was sonicated for 10 minutes, and 1 mL of 5% tricholoroacetic acid (TCA) was added to the extract. This sample was used for the analysis of GSH content. The GSH levels were estimated as per Moron *et al.* and TCA was taken as blank [16].

2.4. Estimation of Indole Acetic Acid (IAA) Content in Cotton Seedlings

For the extraction of IAA approximate 200 mg plant tissue was grinded with 5 mL of 80% chilled methanol. The extract was filtered through Whatmann filter paper (No. 1). After filtration the final volume of extract was made up to 10 mL using 80% ice-cold methanol. Then optical density was measured after 30 minutes at 530 nm using ultra-violet visible spectrophotometer. IAA was analyzed using Tang and Bonner's method. Freshly prepared Salkowski's reagent was used for the detection of IAA content in cotton seedlings [17].

2.5. DNA Fingerprinting

2.5.1. Plant Material and Primers

Two series of seed aliquots (series A and B) were prepared for cotton cv. Stoneville-2. The seeds of the untreated sample (series "B") and seeds on the treated sample (series "A") were sowed. When plants reached the appropriate stage, leaf discs were harvested from each plant. DNA extraction was performed according to DNA LandMarks standard protocols. Purified DNA was then diluted to a concentration of approximately 25 ng/µL. The marker of cotton was selected for PCR reactions and were amplified following DNA LandMarks. These markers were based on probes selected from a cDNA library, as per standard protocol for simple sequence repeat (SSR) on cotton samples [18, 19].

2.5.2. Data Collection and Scoring

Detection of amplified fragments was performed on ABI 3700 DNA sequencer (Applied Biosystems®, Massachusetts, USA) and integrated systems for sequencing. Fragment size was generated by GeneScan software (Applied Biosystems,

Massachusetts, USA) based on an internal size standard (GS-500) loaded with each cotton sample. Relative size of each detected fragment was then binned into categories to associate an allele size to this specific fragment using Genotypes software (Applied Biosystems, Massachusetts, USA). Each bin was defined as being ± 0.7 base pair apart from any given allele size for a specific marker. Any data falling outside this range was reanalyzed and binned manually or declared "failed" by the scorer.

2.5.3. Level of Polymorphism and Similarity Analysis

The level of polymorphism was evaluated by calculating the ratio between the numbers of markers giving a true allelic variation type polymorphism by the total number of markers amplifying on the samples. Class 2 markers were removed from the calculation because of the uncertainty that such polymorphism can represent. The fingerprint of each sample was compared by similarity analysis using software "NTSYSpc V2.10" (NTSYS-PC 2.10, Applied Biostatistics, Setauket, NY, USA) with the Jaccard coefficient. A similarity coefficient of 1 indicates identity of the sample. The smaller the coefficient, the more diverse are the lines [20].

2.6. Statistical Analysis

Data from growth germination of seedling and indole acetic acid (IAA) were expressed as mean ± S.E.M. between control and treated cotton seeds at the end of the experiment.

3. Results and Discussion

3.1. Effect of Biofield on the Growth of Young Cotton Seedlings

Plant population, growth, and canopy of cotton plant are the major yield contributing parameters, that overall contributes the final yield of cotton crop. Proper irrigation system and the management of various physical parameters plays a vital role in final yield of cotton. To increase the yield of crop, use of sophisticated equipment for crop growth, use of chemicals, pesticides, etc. were practiced. All these methods have their own merits or limitations. Growth of plant seedlings are also depends on spacing between the newly grown crops, which results in growth modifications that affect the final yield [21]. The plant height and number of monopodial branches per plant are some of the important vegetative factor which will effect and direct the cotton yield. The number of nodes and sympodia of cotton plant was also direct with plant height [22]. The rate of germination of cotton seedling data, and length of control and treated samples are shown in Table 1. It was observed that biofield energy treatment on cotton seeds results in improved percentage of germination, and length of the plant root and shoot. Based on the results, the control seeds of cotton showed 68% germination, while the biofield treated seeds showed 82% germination. Some seeds of control group failed to germinate as compared with the treated seeds, might be due to less supply of oxygen, which causes metabolic

abnormalities. Contrarily, biofield treatment might provide the sufficient energy to the seeds through biofield energy, which resist the oxygen deficit conditions and results in improved germination rate. Biofield treated cotton seeds resulted in better germination, without any associated infections in leaves and steam, and improved plant height (Figure 1). After germination, plants shoot and roots were measured, which showed slight improved length of shoot, while decreased length of roots as compared with the control group. However, various environmental factors such as light, temperature, oxygen, etc. play an important role for final yield as they affects the seed germination to emergence during the first 10 days [23]. Biofield treatment, could be a new and alternative approach to improve the overall growth, and yield of cotton crops.

Table 1. *Effect of biofield energy treatment on growth-germination rate of cotton seedlings on 10 day old plant.*

| Group | Germination (%) | Length (cm) Mean ± S.E.M. | |
		Shoot	Root
Control	68	7.59 ± 0.014	6.80 ± 0.060
Treated	82	7.60 ± 0.020	5.78 ± 0.047

n = 50; S.E.M.: Standard error of mean

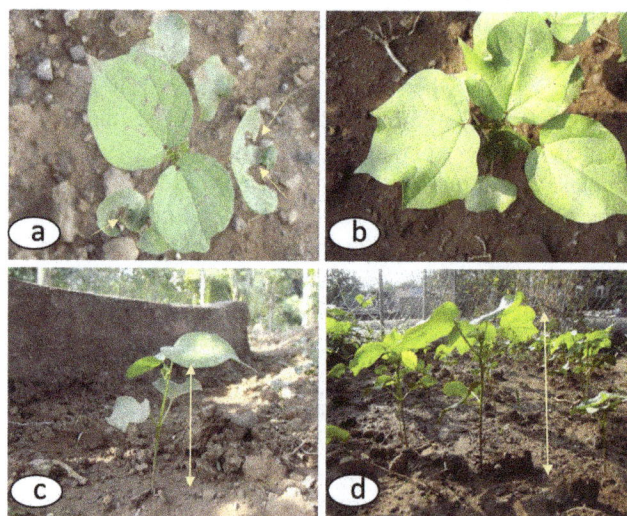

Figure 1. *Effect of biofield energy treatment on cotton plant (a) control cotton crops marked with high infection in most of the leaves (b) biofield treatment showed leaves were bright green and glossy, with more pods per plant and free from any kind of disease or pest attack, (c) control cotton plant showed less plant height and with slow growth, (d) biofield treated cotton showed plants with more length as compared with control with more secondary and tertiary branches.*

3.2. Estimation of Glutathione Level in Cotton Leaves

GSH is one of the important chemical entities in plants involved in the detoxification of reactive oxygen species (ROS). Studied in support of the role of GSH in oxidative stress have been reported. Stress can be induced due to exposure to chemical oxidants [24], gaseous pollutants [25], high air temperatures, and water stress [26]. Effect of Mr. Trivedi's biofield energy treatment on cotton seeds showed

0.176 mM GSH concentration in control group, while 0.225 mM in treated group. Overall, 27.68% increased level of GSH was reported after biofield treatment as compared with the control (Table 2). The plant glutathione system is regarded as the best stress marker in plant ecophysiology in drought tolerance conditions, by maintaining its redox status [27]. It is related to the sequestration of xenobiotics and heavy metals, while is an essential component of the cellular anti-oxidative defense system, which will regulate and control the ROS [28]. As redox reaction and antioxidantive defense plays a central role, it made glutathione system as stress marker in the plants.

During infection by fungal pathogens, plant cells respond by expressing a battery of disease response genes, which can result in the production of various toxic plant products, including active oxygen species and phytoalexins. In addition, an invading fungus or microbe, may produce stress-inducing chemicals, such as phytotoxins, resulting in significant stress and damage to the host cells. One main noticeable response of plants are the increased expression of glutathione S-transferase (GST) genes following infection by pathogens [29]. In the present work glutathione level in plant cell, as a biochemical marker for immunity was increased, that might be due to increased immunity after biofield treatment. Biofield energy might redefines the ionic strength and may provide better conditions for reactive oxygen species to occur, which results in changed GSH concentration. Our experimental results, conclude that biofield energy treated cotton seeds might resist in severe drought conditions, or different unfavorable environmental conditions, and grown with high GSH level as compared with control untreated seeds of cotton, which results in final yield of cotton fibers.

Table 2. Effect of biofield energy on endogenous level of glutathione in leaves of cotton.

Group	a	b	c	d	Mean ± S.E.M.	% change
Control	0.161	0.162	0.185	0.198	0.177 ± 0.009	27.68
Treated	0.200	0.202	0.202	0.298	0.225 ± 0.024	

a, b, c, and d are four studied replicas; S.E.M.: Standard error of mean; Values are presented as millimolar (mM)

3.3. Estimation of Indole Acetic Acid (IAA) Content in Cotton Seedlings

IAA is the most common phytohormone and, physiologically active auxins. It is produced in cells of the apex and very young leaves of the plant via. several independent biosynthetic pathways, and also produced by plant-associated commensal bacteria. The concentration of auxin can also be depends upon the level of plant infections [30]. However, IAA was reported with production of longer roots and enhanced root hairs by increasing the nutrient uptake from soil [31]. Further, IAA stimulates cell elongation by altering the necessary conditions such as increase in permeability of water and osmotic contents into cell, increase the cell wall synthesis via protein synthesis, and decrease the

wall pressure. It also delays or inhibits the leave abscission, and induces flowering and fruiting. The difference in IAA concentrations in cotton seedlings before and after biofield treatment was presented in Figure 2. Control group showed IAA levels as 6.36 μg/mL, while after biofield treatment, it showed increased level as 6.83 μg/mL. The IAA concentration was calculated using rapid, determination using colorimetric principle by Salkowski test [32,33]. However, the IAA concentration was increased by 7.39%, which could be related with the improved overall growth of the plant, which leads to increase the final yield of crop. It can be assumed, that biofield treatment on cotton seeds, might enhance the biosynthetic pathways of IAA synthesis, or may inhibit the growth of phytopathogens, which results in increased level of IAA in biofield treated group.

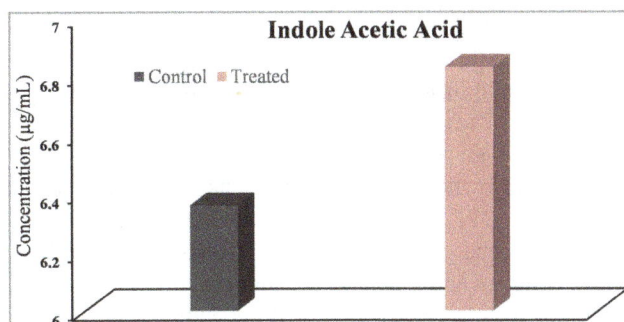

Figure 2. Concentration of indole acetic acid (IAA) in cotton (Gossypium hirsutum L.).

3.4. DNA Fingerprinting

Ninety-nine markers out of the 100 simple sequence repeat (SSR) markers were screened on the 2 samples of cotton amplified bands of the expected molecular weight. Ninety-four markers were categorized as "monomorphic", while five showed polymorphism between the treated and untreated lines. The polymorphism detected with these five markers could be classified into two categories. One was true allelic variation, while the other was additional band or a shift in molecular weight. One of the tested marker (JESPR-206) had not amplify in any fragment. The level of true polymorphism between the treated and untreated samples was evaluated at 4.0% (4/99). Overall, results suggest that biofield treatment resulted in polymorphism in biofield treated cotton seeds as compared with the control, which suggest that biofield energy might act and alter the genetic relatedness of cotton species.

Several studies suggest that the polymorphic DNA is responsible to give information about the ideal genetic markers, due to its selectively neutral nucleotide sequence and distinct genomes pattern [34]. Microsatellite or SSR markers have emerged as the most versatile and popular genetic marker in the plant systems. Because of their co-dominant, multiallelic nature, and hyper-variability, they are the leading markers for fingerprinting, conservation genetics, plant breeding, and phylogenetic studies. Microsatellites are more variable and informative than

Restriction Fragment Length Polymorphism (RFLP), Random Amplified Polymorphic DNA (RAPD) and Amplified Fragment Length Polymorphisms (AFLPs) [35].

Plants are reported with high level of plasticity, as compared with animals, which suggest better adaptive capability in their DNA due to the environmental responses. The changes are visible and can be reflected in various morphological characteristics, such as better canopy, length of root and shoot, etc. along with genetic alterations. Biofield treatment on agricultural crops such as *Withania somnifera*, *Amaranthus dubius*, tomato, etc. were recently reported as an alternate method to improve the growth of crop, with improved overall agronomical characteristics (*i.e.* leaf, stem, flower, seed setting, immunity parameters, chlorophyll content, etc.) [14,15]. Our experimental results are supporting with the studies on *in vitro* growth germination of seeds, plant growth, and development using magnetic or electromagnetic fields [36,37]. According to Rakosy-Tican et al. magnetic field influence the growth of potato and wild *Solanum* species. They have reported significant stimulation of leaf growth, even at biochemical level, the quantity of chlorophyll a and b and carotenoids were increased by more than two-fold [38]. Biofield treatment suggests the mechanism based on energy fields, through bioelectromagnetics and biophysical fields that plays a major role in cellular structure and function of the human body [39]. Thus, the human body emits the electromagnetic waves in the form of bio-photons and moving electrically charged particles. Similarly, biofield treatment on cotton seeds might enhance catalase activity as reported with improved GSH and IAA level, which may promotes the germination. Another hypothesis suggests that biofield energy can cause mitotic incoherence against standard cytogenetic benchmarks, or may have indirect effect on plant mitosis by altering the cytosol condition. Therefore, biofield energy treatment, which might be considered as low electromagnetic fields, can improve the germination rate, increased level of important phytohormones, and overall growth of cotton plant.

4. Conclusions

In conclusion, biofield energy treated cotton seeds resulted in enhanced germination rate by 20% as compared with the control. Further, the length of the shoot and root of cotton plant after biofield energy treatment was reported to be increased as compared with the control. Moreover, GSH level (*i.e.* a biochemical marker for immunity level) in plant cell of biofield treated cotton seeds was increased by 27.68%, which suggest increased immunity of cotton plants after biofield treatment. The IAA concentration after biofield treatment was increased by 7.39%, which may inhibit the growth of phytopathogens, and hence improved the overall growth of the plant. Polymorphism was detected between treated and untreated samples of cotton seeds. The percentage of polymorphism observed between treated and untreated samples was 4%; which could be a notable data in support of

biofield treatment on cotton. Based on study outcome, Mr. Trivedi's biofield energy could be used as better alternate approach to improve the overall yield of agricultural crops in near future.

Abbreviations

SSR: Simple sequence repeat;
IAA: Indole acetic acid;
ROS: Reactive oxygen species;
ASC–GSH: Ascorbate–glutathione;
PCR: Polymerase chain reaction;
NHIS: National Health Interview Survey;
NCHS: National Center for Health Statistics;
CDC: Centers for Disease Control and Prevention;
CAM: Complementary and Alternative Medicine.

Acknowledgements

Authors are thankful for DNA LandMarks Inc., Montreal, Canada for conduction SSR DNA fingerprinting assay of cotton, and thanks to Monad Nanotech Pvt. Ltd., Mumbai, India for their valued inputs for growth analysis and related parameters. Authors are grateful to Trivedi science, Trivedi testimonials and Trivedi master wellness for their support throughout the work.

References

[1] Abdellatif KF, Khidr YA, Mansy YM, Lawendey MM, Soliman YA (2012) Molecular diversity of Egyptian cotton (*Gossypium barbadense* L.) and its relation to varietal development. J Crop Sci Biotechnol 15: 93-99.

[2] Zhang T, Qian N, Zhu X, Chen H, Wang S, et al. (2013) Variations and transmission of QTL alleles for yield and fiber qualities in upland cotton cultivars developed in China. PLoS One 8: e57220.

[3] Kim HJ, Triplett BA (2001) Cotton fiber growth in planta and *in vitro*. Models for plant cell elongation and cell wall biogenesis. Plant Physiol 127: 1361-1366.

[4] Ruan YL, Llewellyn DJ, Furbank RT (2003) Suppression of sucrose synthase gene expression represses cotton fiber cell initiation, elongation, and seed development. Plant Cell 15: 952-964.

[5] Nickell LG (1982) Plant growth regulators: Agricultural uses, Springer, New York.

[6] Wareing PF, PhillipsI DJ (1970) The control of growth and differentiation in plants. Pergamon Press Ltd., New York, USA.

[7] Lopez-Bucio J, Hernandez-Abreu E, Sanchez-Calderon L, Nieto-Jacobo MF, Simpson J, et al. (2002). Phosphate availability alters architecture and causes changes in hormones sensitivity in the *Arabidopsis* root system. Plant Physiol 129: 244-256.

[8] Grill D, Tausz M, De Kok LJ (2001) Significance of glutathione in plant adaptation to the environment. Handbook of plant ecophysiology, Dordrecht: Kluwer.

[9]   http://www.indiantextilemagazine.in/uncategorized/factors-inf
luencing-cotton-production/

[10] Sances F, Flora E, Patil S, Spence A, Shinde V (2013) Impact of
biofield treatment on ginseng and organic blueberry yield.
AGRIVITA J Agri Sci 35: 22-29.

[11] Lenssen AW (2013) Biofield and fungicide seed treatment
influences on soybean productivity, seed quality and weed
community. Agricultural Journal 8: 138-143.

[12] Barnes PM, Bloom B, Nahin RL (2008) Complementary and
alternative medicine use among adults and children: United
States, 2007. Natl Health Stat Report 10: 1-23.

[13] Shinde V, Sances F, Patil S, Spence A (2012) Impact of biofield
treatment on growth and yield of lettuce and tomato. Aust J
Basic Appl Sci 6: 100-105.

[14] Nayak G, Altekar N (2015) Effect of biofield treatment on plant
growth and adaptation. J Environ Health Sci 1: 1-9.

[15] Patil SA, Nayak GB, Barve SS, Tembe RP, Khan RR (2012)
Impact of biofield treatment on growth and anatomical
characteristics of *Pogostemon cablin* (Benth.). Biotechnology
11: 154-162.

[16] Moron MS, Depierre JW, Mannervik B (1979) Levels of
glutathione, glutathione reductase and glutathione
S-transferase activities in rat lung and liver. Biochim Biophys
Acta 582: 67-78.

[17] Tang YW, Bonner J (1947) The enzymatic inactivation of
indoleacetic acid. I. Some charasteristics of the enzyme
contained in pea seedlings. Arch Biochem 13: 11-25.

[18] Zhu LF, Zhang XL, Nie YC (2003) Analysis of genetic
diversity in upland cotton (*Gossypium hirsutum* L.) cultivars
from China and foreign countries by RAPDs and SSRs. J Agric
Biotechnol 11: 450-455.

[19] Xiao J, Wu K, Fang DD, Stelly DM, Yu J, et al. (2009) New
SSR Markers for Use in Cotton (*Gossypium* spp.) Improvement.
J Cotton Sci 13: 75-157.

[20] Rohlf FJ (2000) NTSYSpc: Numerical Taxonomy System, Ver.
2.10q, Exeter, Setauket, NY, USA.

[21] Brar ZS, Singh N, Deal JS (2002) Influence of plant spacing
and growth modification practices on yield and its attributing
characters of two cotton cultivars (*Gossipium hirsutum* L.,).
Journal of Research 39: 181-183.

[22] Graig WC, David B, Steve MB (2000) Analysis of cotton yield
stability across population densities. Agron J 92: 128-135.

[23] Leonard OA, Pinckard JA (1946) Effect of various oxygen and
carbon dioxide concentrations on cotton root development.
Plant Physiol 21: 18-36.

[24] Iturbe-Ormaetxe I, Escuredo PR, Arrese-Igor C, Becana M
(1998) Oxidative damage in pea plants exposed to water deficit
or paraquat. Plant Physiol 116: 173-181.

[25] Grill D, Esterbauer H, Hellig K (1982) Further studies on the
plants by catalase inhibitors. Plant Physiol 79: 1044-1047.

[26] Loggini B, Scartazza A, Brugnoli E, Navari-Izzo F (1999)
Antioxidative defense system, pigment composition, and
photosynthetic efficiency in two wheat cultivars subjected to
drought. Plant Physiol 119: 1091-1100.

[27] Tausz M, Sircelj H, Grill D (2004) The glutathione system as a
stress marker in plant ecophysiology: Is a stress-response
concept valid? J Exp Bot 55: 1955-1962.

[28] Noctor G, Queval G, Mhamdi A, Chaouch S, Foyer CH (2011)
Glutathione. Arabidopsis Book 9: e0142.

[29] Wagner U, Edwards R, Dixon DP, Mauch F (2002) Probing the
diversity of the Arabidopsis glutathione S-transferase family.
Plant Mol Biol 49: 515-532.

[30] Yamada T (1993) The role of auxin in plant-disease
development. Annu Rev Phytopathol 31: 253-273.

[31] Datta C, Basu P (2000) Indole acetic acid production by a
Rhizobium species from root nodules of a leguminous shrub
Cajanus cojan. Microbiol Res 155: 123-127.

[32] Gordon SA, Weber P (1951) Colorimetric estimation of
indoleacetic acid. Plant Physiol 26: 192-195.

[33] Salkowski E (1885) Ueber das verhalten der
skatolcarbonsa¨ure im organismus. Z. Physiol Chem 9: 23-33.

[34] Bretting PK, Widrlechner MP (1995) Genetic markers and plant
genetic resource management. John Wiley & Son Inc. Canada.

[35] He GH, Meng RH, Newman M, Gao GQ, Pittman RN, et al.
(2003) Microsatellites as DNA markers in cultivated peanut.
BMC Plant Biol 3: 3-11.

[36] Hirota N, Nakagawa J, Koichi K (1999) Effects of a magnetic
field on the germination of plants. J Appl Phys 85: 5717-5719.

[37] Yano A, Hidaka E, Fujiwara K, Iimoto M (2001) Induction of
primary root curvature in radish seedlings in a static magnetic
field. Bioelectromagnetics 22: 194-199.

[38] Rakosy-Tican L, Aurori CM, Morariu VV (2005) Influence of
near null magnetic field on *in vitro* growth of potato and wild
Solanum species. Bioelectromagnetics 26: 548-557.

[39] Schwartz GE, Simon WL, Carmona R (2007) The energy
healing experiments: Science reveals our natural power to heal.
(1stedn), Atria Books.

Herbaceous Species Diversity in Kanawa Forest Reserve (KFR) in Gombe State, Nigeria

Abba Halima Mohammed[1], Sawa Fatima Binta Jahun[2], Gani Alhassan Mohammed[2], Abdul Suleiman Dangana[2]

[1]Biology Unit, School of Basic and Remedial Studies, Gombe State University, Gombe, Nigeria
[2]Department of Biological Sceinces, Abubakar Tafawa Balewa University, Bauchi, Nigeria

Email address:

halimamohammedabba77@gmail.com (A. H. Mohammed), halimamohammedabba@yahoo.com (A. H. Mohammed)

Abstract: The study was conducted between 2009 and 2011 in Kanawa Forest Reserve (KFR) in order to determine the impact of anthropogenic pressures and environmental changes of the herbaceous species. The project area was divided into six sites following the variety of land forms in the forest and three transects measuring 100m were laid within each site. Point Centered Quarter (PCQ) sampling method was used. Data obtained were analyzed for relative density, relative frequency, and importance value index. A total of (35) species were identified in KFR out of which 16 species belonged to grasses within three families and 19 genera. The family Poaceae had the highest number, 10 species, the Cyperaceae had 5 species, while the Typhaceae had only 1. 19 species belonged to forbs within fourteen families and 16 genera. The families Asteraceae and Leguminosae: Fabaceae had 3 species each: Rubiaceae had 2 species each. Acanthaceae, Capparidaceae, Euphorbiaceae, Nyctaginaceae, Zygophyllaceae, Portulacaceae, Polygonaceae, Solanaceae, Onagraceae, Labiatae and Commelinaceae all had 1 species each. Simpson's index of diversity was 0.998 and Shannon-Wiener Index was 4.57. This condition indicates complex vegetation. Chi square and its related statistics showed significant positive associations between site I and IV, II and IV, V and VI. Only site I and IV, II and IV, V and VI were significantly negatively associated at ($P < 0.05$). The species with the lowest importance value indices were:- Pennisetum pedicellatum, Polygonum senegalense, Vetiveria nigrinata, Zornia glochidiata. These plants therefore require more efforts on conservation.

Keywords: Herbaceous, Biodiversity, Kanawa Forest Reserve (KFR) and Inventory

1. Introduction

According to Hornby (2001), herbs are usually small tender plants, lacking of woody stems above ground. They occur in a wide variety of forms and leaf structures, including annuals, biennials and perennials, and broad-leaved plants (forbs) as well as the grasses. Herbaceous species also constitutes an important vegetative component of the environment due to its diverse ecological importance, including the maintaining of the structure and function of forests (Iwara et al, 2014). In particular they have the role of being an habitat for wild array of animals, the base for complex food web, stabilizing the soil, preventing soil erosion. Furthermore they have an economic relevance such as sources of fodder, food, fuel and medicines, aesthetic and cultural values for vast number of people throughout the world (Abdullahi, 2011). Some herbaceous species in KFR such as *Euphorbia hirta, Zornia glochidiata, Mucuna prureins, Kyllinga tenuifolia, Cyperus sphacelatus, Acanthospernum hispidium, Physallis angulata, Portulaca oleraceae, Ludwigia hyssopifolia* are used for medicinal purposes. Both human and animal ailments are treated through the use of these local vegetables. In most instances, these plant species are considered specific for a particular illness but they have occasionally mixed usages. Other species such as *Zornia glochidiata, Chloris pilosa, Digitaria horizontallis, Mucuna prureins, Pennisetum pennicelatum, Commelina erecta, Eragrostis ciliaris* are used for forage/fodder. *Vetivera nigrinata* and *Andropogan gayanus* are also raw material for the production fuel. *Typha domingensis* is an invasive herb that has a strong and possibly expanding presence in the wetter habitats of the reserve. Herbaceous species however plays a major role in improving water penetration into soils and adding organic matter that improves moisture holding capacity and plant growth. It indeed plays important functions relating to productivity and environmental health by reducing the velocity of runoff (Department of Environment and resource management, 2010). Notwithstanding its small stature, herbaceous species has huge ecological significance of mediating carbon dynamics and energy flow and influencing the cycling rates of essential nutrients, including N, P, K and Mg (Gilliam, 2007). The herb community of tropical forests is very little known, with few studies addressing its structure quantitatively. Even with this scarce body of information, it is clear that the herbs are a rich group, comprising 14 to 40% of the species found in total species counts in tropical forests. This stratum remains an underappreciated aspect of forest ecosystems (Gilliam, 2007).The quantitative plant diversity inventories are the fundamental tool for conservation and management of tropical forests (Campbell 1994), but as far as KFR are concerned they are limited and quantitative inventories of understory species are still lacking. The

documentation and classification of this unique and often neglected vegetation community may enable efforts to be made for biological conservation (Jennings et al., 2009: Ahmad and Ehsan, 2012).The problem is that due to over collection and climate change several species are fast disappearing.The herbaceous flora of KFR is one of the diverse among the savanna vegetation types of Northern Nigeria and this could mainly be due to climatic, edaphic and human impact. Although studies have been undertaken in other forest reserves in Nigeria, Kanawa Forest Reserve (KFR) (since establishment between 1940 and 1945 has not received such scientific attention. It therefore became imperative to take stock of the status of the herbaceous vegetation of Kanawa forest with a view to deriving appropriate conservation measures.The aim of the study was to obtain an inventory of the herb species of (KFR), in order to determine their relative importance and diversity.

2. Materials and Methods

2.1. The Study Area

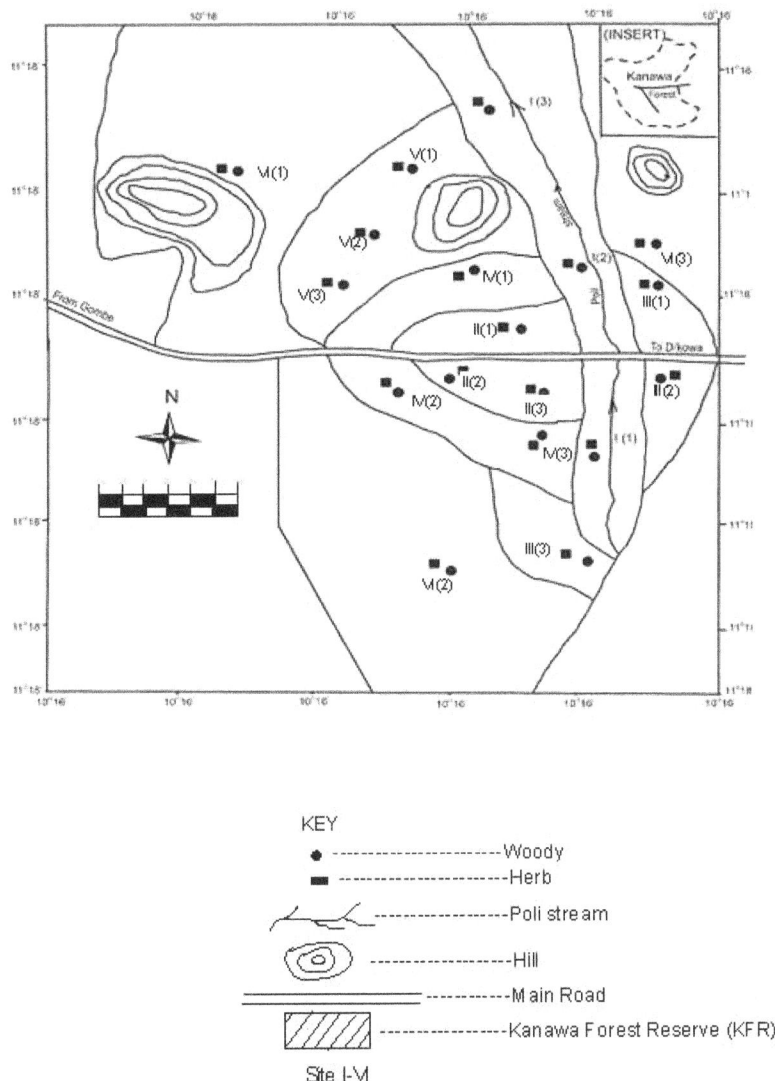

Figure 1. Map showing the study sites and location of sampling points in Kanawa Forest Reserve.

KFR is located in the North Eastern part of Nigeria in Yamaltu/Deba Local Government Area of Gombe State and it lies in the Southern part of the Sudan Savanna between latitude 10o16'N and 10o 18' 30''N longitude 11o 18'10'' E. and 11o 22' 09'' E . and at an altitude of 336-390 mamsl. The size of the forest was 41 hectares (Gombe Native Authority, 1945) and it was gazetted as a forest reserve in 1953. At present the Kanawa Forest Reserve is presently 53 hectares due to the acquisition of surrounding farmlands by Gombe State Government.

The terrain is generally undulating, with dendritic shallow V-shaped drainage (Samaila, 2011). The soil types were loamy sand, sandy and sandy loam types (Abba *et al.*, 2013)

The climate is characterized by two distinct seasons: The rainy season one that lasts from April through October and the dry ones that goes from November to March. The rainfall pattern is bimodal with an annual total which ranges from 658- 923mm. The mean annual temperature ranges from 32.2 to 32.80C. Six sites were used for vegetation sampling following the selecting criteria of topography, soil, and vegetation types. The vegetation of (KFR) is a mosaic made up of dense Sudan savanna vegetation especially around the hilly part of the reserve (Site VI). Marshy(Site IV), riparian (Site I), lowland rainforest vegetation near the Poli stream (Site II) grassland with tall grasses (Site III) and thorn vegetation in the drier part of the forest (Site V). In each site, three sampling locations were identified (Fig 1).

2.2. Sampling Procedure and Methodology

The point centred quarter (PCQ) method was employed (Cottam and Curtis 1956; Dix, 1971; Abdullahi, 2010). Three transects (1, 2, 3) of 100m long each were laid per site. There were 10 sampling points per transect giving a total of 30 sampling points. On each sampling point four quarters were demarcated and the nearest herb plant in terms of distance from the point to the center of the herb plant. Readings were recorded in meters within each quarter. This was done for all the sampling points on each transect.

2.3. Data Collection and Analysis

In each sampling point all nearest living herb species encountered were listed. In each site, all nearest living herb plant were measured with measuring tape in each quarter and recorded. Species encountered were identified on site with the help of field guides and floras and texts (Blair, 1976; Lowe and Stanfeild,1974; Akobundu and Agyakwa, 1998). Also morphological characteristics involving fruits, flowers, leaves and stem, bark and sap were used for identification. The specimens were collected and compared with herbarium specimens of the Biological Sciences Programme, Abubakar Tafawa Balewa University, Bauchi, Nigeria. Nomenclature of the species follows Hutchinson and Dalzeil (1972).

Data obtained were quantitatively analyzed for Relative density (RD), Relative frequency (RF) following Curtis (1959), and the relative values were summed up to obtain Importance Value index (IVI) (Misra,1968, Dallmier,1992; Sabogal,1992; Abdullahi, 2010). The formulae used to calculate RD, RF and (IVI) are as follows:-

RD = (number of individuals of a species)/ (total number of individuals of all species) x 100.

RF= (frequency of one species)/ (sum of all frequencies) x 100

IVI = sum of (RF+ RD) /2

The value of IV1 may range from 0 to 200%. Dividing it by 2 will result in a figure that ranges from 0 to 100%. This value is referred to as the importance percentage.

Species diversity was calculated using Simpson's index (Simpson, 1949) with this formula:

D =[N (N-1)] / [summation of n (n-1)], where D is the diversity index, N is the total number of individuals of all plant species found, n is the number of individuals of a particular species.

Species diversity index was also calculated using the Shannon-Weiner index (Shannon, 1948)

$$H' = -\sum_{i=1}^{R} \text{Pi in Pi}$$

Where H' is the = Shannon –Weiner diversity index

Pi= is the proportion of each species in the sample.

R = represents the total number of species

Chi-Square analysis was calculated using the following formulae:

$\chi2$= n (ad-bc) 2 / (a+b) (c+d) (a+c) (b+d) , while the correlation coefficient was obtained as follows: i (Causton, 1988)

r = ad-bc/ $\sqrt{}$ {(a+b) (c+d) (a+c) (b+d)} -ii

From the above equations;

n = number of species in the data set

a = number of species common to sites A and B

b = number of species present in site A but not B

c = number of species present in site B but not A

d = number of species absent in site A and B

χ= chi-square

r = correlation coefficient

3. Results and Discussion

A total of (35) species of herbs were identified out of which 16 species belonged to grasses within 3 families and 19 genera and 19 species belonged to forbs within 14 families and 16 genera within an area of 53 hectares, of which 10 species were encountered in the riparian (site I), 8 in lowland rainforest vegetation near Poli stream (Site II), 9 in grassland with tall grasses (Site III), 8 in Marshy (Site IV), 7 in thorn vegetation in the drier part of the forest (Site V), and 13 in the dense Sudan savanna vegetation especially around the hilly part of the reserve (Site VI) respectively. A total of 14 families were recorded for forbs

(Table l). In the study area, the families Asteraceae and Leguminosae: Fabaceae had 3 species each: Rubiaceae had 2 species each. Acanthaceae, Capparidaceae, Euphorbiaceae, Nyctaginaceae, Zygophyllaceae, Portulacaceae, Polygonaceae, Solanaceae, Onagraceae, Labiatae and Commelinaceae all had l species each. A total of three families were recorded for grasses (Table 2). The family Poaceae had the highest number, 10 species. The family Cyperaceae had 5 species; Typhaceae had only l species.

From the above results, the families that had the most common herb species were Poaceae and Cyperaceae. The Poaceae family is one of the most widely distributed and abundant groups and important family of the earth's flora. Their success results from their tolerance of grazing herbivores and fire, their varied means of reproduction, and their versatility in photosynthesis. These families were also known to be native species in most savannah-woodland mosaics in Africa and more typical of the Sudano-Sahelian zone. A similar study was carried out by Atiku and Bello (2011) at Wassaniya Forest Reserve of Sokoto State (Nigeria), using the point centered quarter method. Another investigation was carried out in Yankari Game Reserve (YGR), in the Savanna-Woodlands of Northern Nigeria by Abdullahi (2010). The number obtained in the present study was higher than that of Wassaniya Forest Reserve probably because of the size, location, soil composition, soil types and climatic factors which gave rise to a higher number of species. It was however lower than that of Yankari Game Reserve, because of the statute of Yankari as a game reserve and the moist–warm climate, geological and topographic nature of the area.

Table 3, reveals the relative density of herbs. The highest value in the reserve was *Typha domingensis* (37.50%) revealing that it was the most abundant species in the forest because *Typha* species were often among the first wetland plants to colonize areas of newly exposed wet mud, with their abundant wind dispersed seeds which provides a greater reproductive capacity. Buried seeds could survive in the soil for long periods of time and they germinate best with sunlight and fluctuating temperatures. The plants also spread by rhizomes, forming large, interconnected stands with extensive root systems which provides a rapid growth and makes them invasive and proliferative.

The lowest value was recorded on *Polygonum senegalense* (2.50%), probably due because of the continous use for removal of ecto parasites from livestock and also orally administered to cattle for managing unstated diseases. The herb species *Hyptis suaveolens* had the highest relative densities (15.00%) at site I and II probably because *Hyptis* shows strategy for better survival and establishment. It exhibit vigorous growth in the forest. Fenner (1985) reported that a number of vegetative characters are helpful in the heavy proliferation of *Hyptis* such as small seed size, prolific seed production forming persistent propagule bank within short period, seed dimorphism, autogamic and allogamic mode of reproduction, good proliferation from the perrenating

rootstock, probable allele chemicals and presence of essential oil conferring resistance to it against variety of pathogens. Factors favoring the growth of *Hyptis* is the small seed size, which easily penetrates into cracks or small openings in the soil. Reduction in size along with large number of seeds has also been associated with predator avoidance as reported by Fenner (1985). Seed dimorphism in *Hyptis* is helpful in its germination across a range of temperature conferring year round seed germination.

Typha domingensis had the highest relative densities (15.00%) at site I and *Kyllinga tenuifolia* at site IV. Those species were high because of their extensive root system and they often exclude other plants with their dense canopy. The species *Vetiveria nigritana* had the lowest relative density (4.17%) at site VI, because the roots have been used continuously for their fragrance and are woven into fragrant-smelling mats and fans by the locals living around the reserve. Lastly, *Vetiveria nigritana* is beneficial to poor farmers because it provides traditional cost effective ways to fight infections and diseases. The plant also contains active ingredients used in traditional medicine and as botanical pesticide.

Euphorbia hirta (8.16 m2/ ha) had the highest absolute density in the reserve, hence the most densely populated species. This was because it grows on all type of soils and it reproduces by seeds up to 3,000 seeds per plant. The plant is dispersed in an active way by projection or in a passive way by the activity of ants. Herbivours also play a key role in long distance dispersal of propagules (Milton and Dean, 2001) enhancing local colonization processes and plant diversity (Olff and Ritchie, 1998). The lowest value was recorded on *Zornia glochidiata* (0.600 m2/ ha), because the species is considered a good fodder plant for animals and leaves are sources of food and medicine. *Chloris pilosa* and *Chamaecrista mimosoides* had the highest absolute density (5.64 m2/ ha) in site V. The two species belongs to Poaceae and Leguminosae: Fabaceae family which is known to be propagated by seeds. The lowest absolute density (0.56 m2/ ha) was recorded on *Digitaria horizontalis* at site I, because it was highly grazed upon as sources of fodder by animals. This study was consistent with the findings of (Abdullahi *et al.,* 2009; Atiku and Bello, 2011) at YGR, Bauchi and Wassaniya Forest Reserve, Sokoto State, Nigeria.

The species with the highest absolute density in the reserve was *Mucuna pruriens* (46.36 m2/ ha), hence the most densely populated. This shows that this herb had the highest number of species per unit area in the reserve. This high number could be because it was rapidly dispersed by seeds and the associated processes of germination, seedling establishment and survival were rapid. This finding was consistent with that of (Harper's, 1977).

Achyranthes aspera (8.54 m2/ ha) was the species with the least number in the forest. It is highly sought for by humans as sources of food and medicine. *Abrus precatorius, Ipomoea asarifolia, Luffa cylindrica* and *Mucuna pruriens* had the highest absolute density in site l while the lowest values (3.13%) was recorded on *Combretum racemosum* and *Abrus precatorius*. *Mucuna pruriens* (22.50) had the highest relative densities at site III while the lowest value was recorded on *Achyranthes aspera* (7.50).

Relative density was high because site III contains loamy sand, minimum amounts of moisture content, high organic matter content, total nitrogen and available phosphorus, with high amounts of exchangeable bases.

The relative frequency of herbs for KFR is presented in Table 4. *Kyllinga tenuifolia* and *Cyperus sphacelatus* had the highest grass frequency (80%) in the reserve hence the most frequent grass species. The high frequency could probably be because of the presence of seeds that could easily be dispersed together with their rapid regeneration abilities. The lowest frequency in the reserve was recorded on *Acanthospermum hispidium*, *Fimbristylis ferruginea*, *Fimbristylis meliacea*, *Pennisetum pedicillatum*, *Physalis angulata*, *Polygonum senegalense*, *Vetivera nigrinata*, *Zornia glochidiata*. This could be because they were highly grazed upon by animals. The herb species *Commelina erecta* and *Hyptis suaveolens* had the highest frequencies at site II and *Kyllinga tenuifolia* had the highest frequency at site IV. The species *Digitaria horizontalis*, *Ludwigia hyssopifolia*, *Physalis angulata*, *Tridax procumbens*, *Chloris pilosa* had the lowest frequencies at site I. *Chloris pilosa*, *Vernonia cineria*, *Portulaca oleraceae*, *Euphorbia hirta* had the lowest frequencies at site II. *Digitaria horizontalis*, *Kyllinga tenuifolia*, *Fimbristylis ferruginea*, *Fimbristylis meliaceae*, *Vetiveria nigritana*, *Eragrostis ciliaris*, *Eragrostis tremula*, and *Spermacose viticilata* had lowest frequencies at site III. *Tribulus terrestris* had lowest frequencies at site IV. *Euphorbia hirta*, *Tridax procumbens*, *Commelina erecta*, *Polygonum senegalense*, *Boerhavia diffusa*, *Cenchrus biflorus*, *Pennisetum pedicillata*, *Alysicarpus vaginalis* had lowest frequencies at site V. *Commelina erecta*, *Eragrostis tremula*, *Acanthospermum hispidium*, *Dactyloctenium aegyptiaca*, *Boerhavia diffusa*, *Cenchrus biflorus*, *Alysicarpus vaginalis*, *Zornia glochidata* had the lowest frequencies at site VI. These low values could be due to human impact and their importance as sources of medicine, forage /fodder and food.

Cyperus sphacelatus and *Kyllinga tenuifolia* had the highest relative frequencies (41.91%) in the reserve. This shows that the species had random distributions because the environmental conditions and resources were consistent and their modes of dispersal are sporadic. The lowest relative frequencies (5.26%) were recorded on *Polygonum senegalense*. *Euphorbia hirta*, *Hyptis suaveolens* and *Typha domingensis* had highest relative frequencies at site I. *Cyperus sphacelatus* (17.65%) had the highest relative frequency at site III. *Vernonia cineria*, *Portulaca oleracea* and *Euphorbia hirta* had lowest relative frequencies of (4.17%) at site II. All this showed clumped distribution of species mainly because water only reaches the surface of the terrain in specific places and there is just enough to support their growth. This study was also in line with the works of (Abdullahi and Sanusi, 2006) and of Atiku and Bello (2011) who studied herbaceous species in Yankari

game reserve, Bauchi State and Wasaniya Forest Reserve, Sokoto State, Nigeria.

From Table 5, the herb species with the highest Importance Value Index (IVI) in the whole forest reserve was *Typha domingensis* (43.81%). The herb species *Scleria verrucosa* (3.63%) had the lowest IVI in the reserve. The herb species *Typha domingensis* had the highest IVI of (15.83%) at site I while the lowest was recorded for *Scleria verrucosa* (3.63%) at site V. The importance value, or the importance percentage, gives an overall estimate of the influence of importance of a plant species in the community.

The IVI is imperative at comparing the ecological significance of species and it indicates the extent of dominance of a species in the structure of a vegetation stand (Curtish and McIntosh, 1951; Rosenzweig, 1995: Abdullahi, 2010). IVI is also a reasonable measure to assess the overall significance of a species since it takes into account several properties of the species in the vegetation. The IVI was calculated as per Curtis and McIntosh (1951). The importance value indices IVI of the herbaceous species were generally low in the study area. The low IVI values could be due to different species with few individuals represented in each herbaceous species. This is consistent with the works of (Abdullahi, 2010). All these species were characteristic of Sudan Savanna vegetation types. The grasses were shorter than in the Guinea Savanna zones, and they had different characters. Most of the grasses were annuals because of drought-stress in the long dry period. A number of perennials grow vigorously too (Broekhuyse and Allen, 1984; Vierich and Stoop, 1990).

The results of Simpson's index of diversity (1-D) was (0.998), while Shannon- Weiner index of Diversity for the study area was (4.57) for the herb species. The values were high indicating a more complex and healthy community with greater variety of species allowing for more species interactions. This will ensure greater system stability, and indicating good environmental conditions. The high diversity could also be due to the high fertility related parameters and high moisture contents in some sites. It could also be due to the position of Kanawa Forest as a forest reserve.

A similar study was carried out by Abdullahi (2010) where he reported Simpson's index of diversity for herbs in Yankari National Park as (0.998). Richard *et al.*, (2011) also reported Simpson's index of diversity as (0.957) for the Miombo woodland of Bereku Forest Reserve, in Tanzania. These values were higher than those of the present study due to size and higher enforcement of protection laws.

The Shannon-Wiener Index of diversity (H') was reported by Richard et al., (2011) to be 4.27 for the Miombo Woodlands of Bereku Forest Reserve and 2.76 in Khadimnagar National Park of Bangladesh as reported by Sobuj and Rahman (2011). The value of 4.27 is lower than the 4.57 obtained in the present study. The higher value observed in this study may be due to riparian vegetation found along the Poli stream that transverse part of the KFR along with its concomitant high fertility and variation in the ecotypes within the KFR.

The results of the association, similarity and correlation between site (Table 6) revealed that site III and VI had the highest association values (10.7905), while the lowest value (0.064815) was recorded in site I and II. Three positive associations at site I and IV, II and IV, V and VI. Only site I and IV, II and IV, V and VI were significantly negatively associated at ($p < 0.05$). Sites I and III (23.58), had higher sum of chi-square (Sx). The Sx (sum of chi-square) and Sr (sum of correlation coefficient) values were similarly less than the total number of species in the data set (35) and > greater than 1(one) except for three positive relationships depicting weak relationships and heterogeneity of the flora of KFR. According to Kumar *et al.,* (2006) species distribution in areas with high diversity is always heterogeneous.

Table 1. *Families and species of forbs identified at Kanawa Forest Reserve (KFR), Gombe State, Nigeria during 2010-2011.*

S/No	Family	Species
1	Acanthaceae	*Nelsonia canescens* (Lam) Sprengel
		Acanthospermum hispidum DC
2	Asteraceae	*Tridax procumbens* L.
		Vernonia cineria Schreb
3	Capparidaceae	*Cleome viscosa* L.
4	Commelinaceae	*Commelina erecta* L
5	Euphorbiaceae	*Euphorbia hirta* L.
		Alysicarpus vaginalis (L.) DC
6	Leguminosae: Fabaceae	*Zornia glochidata* Rehb ex DC
		Chamaecrista rotundifolia (Pers.)
7	Labiatae	*Hyptis suaveolens* L.
8	Nyctaginaceae	*Boerhavia diffusa* L.
9	Zygophyllaceae	*Tribulus terrestris* L.
10	Onagraceae	*Ludwigia hyssopifolia* (G.Don) Exell
11	Portulacaceae	*Portulaca oleraceae* L.
12	Polygonaceae	*Polygonum senegalense* Meisn Senegalensis (Meisn) sojak
13	Rubiaceae	*Spermacose verticillata* (L.)
		Oldenlandia corymbosa L.
14	Solanaceae	*Physalis angulata* L.

Table 2. *Families and species of grasses identified at Kanawa Forest Reserve (KFR), Gombe State, Nigeria during 2010-11.*

S/No	Family	Species
1	Cyperaceae	*Cyperus sphacelatus* Rottb
		Fimbristylis ferruginea (L.)
		Fimbristylis meliaceae Vahl
		Kyllinga tenuifolia Steaud
		Scleria verrucosa Willd
2	Poaceae	*Andropogan gayanus* Kunth
		Cenchrus biflorus Roxb
		Chloris pilosa Shumach
		Dactyloctenium aegyptium Linn P.Beauv
		Digitaria gayana (Kunth) Stapf ex A.Chev
		Digitaria horizontalis Willd
		Eragrostis ciliaris (L) R.Br
		Eragrostis tremula Hochst ex Steaud
		Pennisetum pedicillatum Trin
3	Typhaceae	*Vetiveria nigrinata* (Benth) Sap
		Typha domingensis (Pers) Steaud

Table 3. *Relative and absolute density of herbs identified at Kanawa Forest Reserve (KFR), Gombe State, Nigeria.*

| | | Absolute density (m2/ha) | | | | | | | Relative density (%) | | | | | | |
| | | Site | | | | | | | Site | | | | | | |
S/No	Species	I	II	III	IV	V	VI	TOTAL	I	II	III	IV	V	VI	TOTAL
1	Acanthospermum hispidium						1.00	1.00						8.33	8.33
2	Alysicarpus vaginalis					2.60	1.00	3.60					5.00	8.33	13.33
3	Andropogan gayanus					3.04	0.90	3.94					5.83	7.50	13.33
4	Boerhavia diffusa					3.47	0.90	4.37					6.67	7.50	14.17
5	Cenchrus biflorus					1.30	1.20	2.50					10.00	10.00	20.00
6	Chamaecrista rotundifolia					5.64	1.00	6.64					10.83	8.33	19.16
7	Chloris pilosa	0.72	1.51			5.64		7.87	8.33	7.50			10.83		26.66
8	Cleome viscosa				3.08			3.08				12.50			12.50
9	Commelina erecta		2.18			3.90	1.00	7.08		10.83			7.50	8.33	26.66
10	Cyperus sphacelatus		1.68	2.22	3.08			6.98		8.33	11.67	12.50			32.50
11	Dactyloctenium aegyptium					4.34	0.90	5.24					8.33	7.50	15.83
12	Digitaria gayana				2.46			2.46				10.00			10.00
13	Digitaria horizontalis	0.55		1.55				2.10	5.00		8.33				13.33
14	Eragrostis ciliaris		1.58				1.30	2.88		8.33				10.83	19.16
15	Eragrostis tremula		1.58				1.00	2.58		8.33				8.33	16.66
16	Euphorbia hirta	1.38	2.01			4.77		8.16	12.50	10.00			9.68		32.18
17	Fimbristylis ferruginea		1.58					1.58		8.33					8.33
18	Fimbristylis meliacea		1.58					1.58		8.33					8.33
19	Hyptis suaveolens	1.7	3.02					4.73	15.00	15.00					30.00
20	Kyllinga tenuifolia		1.68	1.58	3.69			6.95		8.34	8.34	15.00			31.66
21	Nelsonia canescens					5.64		5.64					10.33		10.33
22	Ludwigia hyssopifolia	0.92		1.90				2.82	8.33		10.00				18.33
23	Oldenlandia corymbosa	1.01						1.01	9.17						9.17
24	Pennisetum pedicellatum					5.20		5.20					3.33		3.33
25	Physalis angulata	0.72						0.72	8.33						8.33
26	Portulaca oleraceae		1.51		3.08			4.59		7.50		12.50			20.00
27	Polygonum senegalense					1.30		1.30					2.50		2.50
28	Scleria verrucosa			1.90				1.90			10.00				10.00
29	Spermacose verticillata		0.95				1.23	2.18			5.00			10.00	15.00
30	Tribulus terrestris	1.10			2.46			3.56	10.00			15.00			15.00
31	Tridax procumbens	0.72	2.51			4.77		8.00	8.34	12.00			9.17		29.51
32	Typha domingensis	1.65	1.51		3.69			6.85	15.00	12.50		10.00			37.50
33	Vernonia cineria		2.51	1.74	3.08			7.33		12.50	9.17	12.50			34.17
34	Vetiveria nigrinata		0.79					0.79			4.17				4.17
35	Zornia glochidiata						0.60	0.60						5.02	5.02

Table 4. Frequency, relative frequency of herbs identified at Kanawa Forest Reserve (KFR), Gombe State, Nigeria.

| | | Frequency (%) | | | | | | | Relative frequency (%) | | | | | | |
| | | Site | | | | | | | Site | | | | | | |
S/No	Species	I	II	III	IV	V	VI	TOTAL	I	II	III	IV	V	VI	TOTAL
1	Acanthospermum hispidium						10	10						6.67	6.67
2	Alysicarpus vaginalis					10	10	20					5.27	6.67	11.94
3	Andropogan gayanus					20	20	40					10.53	13.3	23.86
4	Boerhavia diffusa					10	10	20					5.26	667	11.93
5	Cenchrus biflorus					10	10	20					5.26	6.67	11.93
6	Chamaecrista rotundifolia					20	10	30					10.53	6.67	17.20
7	Chloris pilosa	10	10			20		40	5.56	4.17			10.53		20.26
8	Cleome viscosa				20	10		30			11.76				11.76
9	Commelina erecta		40			10	10	60		16.67			5.26	6.67	28.60
10	Cyperus sphacelatus		30	30	20			80		12.50	17.65	11.76			41.91
11	Dactyloctenium aegyptium				30		10	40					15.79	6.67	22.46
12	Digitaria gayana				20			20			11.76				11.76
13	Digitaria horizontalis	10		10				20	5.56		5.88				11.44
14	Eragrostis ciliaris			10		20		30			5.88			13.3	18.86
15	Eragrostis tremula			10			10	20			5.88			6.67	12.55
16	Euphorbia hirta	30	10		10			50	16.67	4.17			5.26		26.10
17	Fimbristylis ferruginea			10				10			5.88				5.88
18	Fimbristylis meliacea			10				10			5.88				5.88
19	Hyptis suaveolens	30	40					70	16.64	16.65					33.29
20	Kyllinga tenuifolia		30	10	40			80		12.50	5.88	23.53			41.91
21	Nelsonia canescens					20		20					10.53		10.53
22	Ludwigia hyssopifolia	10		20				30	5.56		11.76				17.32
23	Oldenlandia corymbosa	20						20	11.11						11.11
24	Pennisetum pedicellatum				10			10					5.26		5.26
25	Physalis angulata	10						10	5.56						5.56
26	Polygonum senegalense				10			10					5.26		5.26
27	Portulaca oleraceae		10		20			30		4.17		11.76			15.93
28	Scleria verrucosa			20				20			11.76				11.76
29	Spermacose verticillata			10		20		30			5.88			13.3	19.21
30	Typha domingensis	30	20		20			70	16.67	8.33		11.76			36.76
31	Tribulus terrestris	20			10			30	11.11			5.91			17.02
32	Tridax procumbens	10	40		10			60	5.56	16.67			5.26		27.49
33	Vernonia cineria		10	20	20			50		4.20	11.76	11.76			27.72
34	Vetiveria nigrinata			10				10			5.88				5.88
35	Zornia glochidiata					10		10						6.67	6.67

Table 5. *Importance value index for herbs identified at Kanawa Forest Reserve (KFR) Gombe State, Nigeria.*

		Importance value index (%)						
		Site						
S/No	Species	I	II	III	IV	V	VI	TOTAL
1	*Acanthospermum hispidium*						7.50	7.50
2	*Alysicarpus vaginalis*					5.13	7.50	12.63
3	*Andropogan gayanus*					8.18	10.42	18.60
4	*Boerhavia diffusa*					5.96	7.08	13.04
5	*Cenchrus biflorus*					7.63	8.33	15.96
6	*Chamaecrista rotundifolia*					10.68	7.50	18.18
7	*Chloris pilosa*	6.94	5.83			10.68		23.45
8	*Cleome viscosa*				12.13			12.13
9	*Commelina erecta*		13.75			6.38	7.50	27.63
10	*Cyperus sphacelatus*		10.72	7.68	12.13			30.53
11	*Dactyloctenium aegyptium*					12.08	7.08	19.16
12	*Digitaria gayana*				10.88			10.88
13	*Digitaria horizontalis*	5.28		7.11				12.39
14	*Eragrostis ciliaris*			7.11			12.08	19.19
15	*Eragrostis tremula*			7.11			7.50	14.61
16	*Euphorbia hirta*	14.58	7.08			7.21		28.87
17	*Fimbristylis ferruginea*			14.22				14.22
18	*Fimbristylis meliacea*			14.22				14.22
19	*Hyptis suaveolens*	10.83	14.71					25.54
20	*Kyllinga tenuifolia*		10.42	7.11	19.26			36.79
21	*Nelsonia canescens*					10.68		10.68
22	*Ludwigia hyssopifolia*	6.94			10.88			17.82
23	*Oldenlandia corymbosa*	10.14						10.14
24	*Pennisetum pedicellatum*					4.30		4.30
25	*Physalis angulata*	6.94						6.94
26	*Portulaca oleraceae*				12.13			12.13
27	*Polygonum senegalense*					3.88		3.88
28	*Scleria verrucosa*			3.63				3.63
29	*Spermacose verticillata*			5.44			11.67	17.11
30	*Tribulus terrestris*	15.58			7.94			25.52
31	*Tridax procumbens*	6.94	14.58			7.22		28.74
32	*Typha domingensis*	15.83	14.58		13.40			43.81
33	*Vernonia cineria*		8.33	10.47	12.13			30.93
34	*Vetiveria nigrinata*			5.02				5.02
35	*Zornia glochidiata*						5.83	5.83

Table 6. *Association, similarity and correlation between sites for herbaceous species at Kanawa Forest Reserve (KFR), Gombe State, Nigeria.*

Site	Chi-square	Correlation	ad/bc	Sx	Sr
Site1/11	0.064815	-0.043033	-	34.93519	1.043033
Site1/111	1.268116	0.190347	-	33.73188	1.190347
Site1/1V	1.495726	0.206725	+	33.50427	0.793275
Site1/V	7.304348	-0.456832	-	27.69565	1.456832
Site1/V1	0.305944	-0.093495	-	34.69406	1.093495
Site11/111	0.396873	-0.106486	-	34.60313	1.106486
Site11/1V	3.201963	0.302464	+	31.79804	0.697536
Site11/V	5.410628	-0.393179	-	29.58937	1.393179
Site11/V1	6.127946	-0.418431	-	28.87205	1.418431
Site111/1V	0.782547	-0.149528	-	34.21745	1.149528
Site 111/V	1.752653	-0.223776	-	33.24735	1.223776
Site 111/V1	10.79051	-0.555248	-	24.20949	1.555248
Site 1V/V	2.887944	-0.287250	-	32.11206	1.28725
Site 1V/V1	1.155266	-0.181680	-	33.84473	1.18168
Site V/V1	3.512149	0.316776	+	31.48785	0.683224

4. Conclusion

Kanawa Forest Reserve is dominated by the following herb species: *Cyperus sphacelatus-Hyptis suaveolens-Typha domingensis* complex. The presence of *Typha* being an invasive herb species could also be a threat to the native species within the reserve. In order to protect the forest reserve from further encroachment from human activies such as harvesting the herbs for medicinal purposes, and animals foraging on the grasses, the reserve need to be fenced and additional guards be obtained. It was also concluded that the species with the lowest importance value indices were:- *Pennisetum pedicellatum, Polygonum senegalense, Vetiveria nigrinata, Zornia glochidiata.* These plants therefore require more efforts on conservation.

References

[1]   H.M. Abba; F.B.J. Sawa; A.M. Gani . and S. D. Abdul (2013). Study of Kanawa Forest Reserve (KFR). Soil analysis. Unpublished field work for Ph.D thesis.

[2]   Abdullahi, M.B (2010). Phytosociological Studies and Community Rural ApppraisalTowards Biodiversity Conservation in Yankari Game Reserve, Bauchi State, Nigeria . An unpublished Ph.D Thesis. Abubakar Tafawa Balewa University, Bauchi, Nigeria, pp 99 .

[3]   Abdullahi M.B (2011). An investigation into the Herbaceous plant formations Of Yankari Game Reserve Bauchi, Nigeria. Botany Research Journal 4(3) pp 29-34.

[4]   Abdullahi, M.B and Sanusi S.S (2006). A phytosociological survey of the herbaceous Vegetation of the Gaji flood Plains.Yankari National Park, Bauchi.Nigerian Journal of Botany,19:pp 61-67.

[5]   Abdullahi, M.B, S.S Sanusi, S.D, Abdul and F.B.J. Sawa (2009). An assessment of the herbaceous species vegetation of yankari game reserve, Bauchi, Nigeria. American Eurasian Journal of Agricultural and Environmental Science 6:pp20-25.

[6]   Ahmad, S. S. and Ehsan, H. (2012) Analyzing the herbaceous flora of LohiBher Wildlife Park under variable environmental stress. Pak. J. Bot., 44(1):pp11-14.

[7]   Akobundu, I.O. and C.W Agyakwa. (1998). A Hand Book of West African Weeds.International Institute of Tropical Agriculture, IITA, Ibadan.pp 1-.557.

[8]   Atiku, M, and Bello, A.G (2011). Diversity of herbaceous plants in Wassaniya Forest Reserve of Sokoto State, Nigeria.Forestry Association of Nigeria, Conference paper.pp 438-443

[9]   Blair, R.A (1976). A Field Key to the Common Genera of Nigerian Grasses.Part 1 and 2. University Press Ibadan.pp76 .

[10] Broekhuyse, J.T and A.M. Allen (1984). Farming Systems Research on the Northern Mossi Plateau.Journal of Human biology Organs.47:pp 330-342.

[11] Campbell, DG (1994). Scale and pattern of community structure in Amazonian rainforest in Edwaeds P J. May RM.Web NR (eds) Larger scale ecology and conservation biology. Blackwell, Oxford.

[12] Causton, DR (1988). Introduction to Vegetation Analysis: Principles, Practice and Interpretation. Unwin Hyman Ltd, London.pp 342.

[13] Cottam, G and Curtis J.T (1956). The use of distance measurements in phytosociological sampling.Ecology 37(3)pp460.

[14] Curtis, J.T and McIntosh, R.P (1951). An Upland continuum in the Praire- Forest Border region of Wisconsin, Ecology, 32,pp 476 – 496 .

[15] Curtis, JT (1957). The vegetation of Winsconsin: An ordination of plant communities.Uniersity of Wisconsin press Madison,Wisconsin.pp657

[16] Dallmeier, F. (Ed.). (1992). Long-term monitoring of biological diversity in Tropical Digest 11 .UNESCO, Paris.

[17] Department of Environment and resource management (2010). Fitzroy Basin Draft Water Resource Plan environmental assessment: Stage 2 assessment report, DERM, Brisbane.

[18] Dix, R. L (1971). An application of the point-centered quarter method to the samplingof grassland vegetation . Journal of Range Management. 14:63-69pp

[19] Fenner, M. (1985). Seed Ecology. Chapman and Hall, London, 151 pp.

[20] Gilliam, F. S. (2007). Ecological Significance of the Herbaceous Layer in Temperate Forest Ecosystems; Biosceince.57(10) 845 - 858pp. 2007 American Institute of Biological Sceinces.http:// dx.doi.org/ 10.164//B571007 http:// www.bioone.org/doi/full/10.1641B5710007

[21] Gombe Native Authority (1945). Preliminary report on proposed Gombe Native Authority Forest reserve . No 4 of the former Gombe Native Authority, Northern Nigeria.Now Gombe State, Nigeria.pp 1-10.

[22] Harper, J. L (1977). Population Biology of Plants, Academic Press: New York–London,

[23] Hornby, A. S. (2001). Oxford Advanced Learners Dictionary .6th Edition . Oxford University Press, Oxford, New York.pp580.

[24] Hutchinson J and Dalzeil J.M. (1954-1972). Flora of West Tropical Africa.Vol.1 - lll . 2nd Rev.Edition. Millbank, London.

[25] Iwara A.L.Offiong,R.A, Nar, G.N.Ogundele,F.O (2014). An assessment of Herbaceous Species Diversity, Density, Cover in Agoi Ekpo, Cross River State, Nigeria.International Journal of Biological Sciences. 01(01) : 21-29.

[26] Jennings, M.D., Faber-Langendoen, D., Loucks, O. L., Peet, R. K. and Roberts, D. (2009) Standards for associations and alliances of theU.S. National Vegetation Classification. Ecological Monographs, 79 (2):pp 173-199.

[27] Kumar, A., B. G Marcot and A Saxena (2006).Tree speciesDiversity and distribution patterns in tropical forests of Garo Hills.Current Science. 91 (10); pp11-25

[28] Lowe, J and Stanfeild, D.P (1974).The Flora of Nigerian Sedges (Family Cyperaceae).University press. Ibadan. Nigeria.pp 140-185.

[29] Milton, S.J. and Dean, W.R.J. (2001). Disturbance, drought and dynamics of desert dune grassland, South Africa.Plant Ecology.150:pp 37-51.

[30] Misra, R (1968). Ecology work Book.Oxford and IBH Publishing Co.Calcutta.244 pp.of grassland vegetation.Journal of Range Management.14:pp63-69

[31] Olff, H. and Ritchie, M.E (1998). Effects of herbivore on grassland plant diversity, Trends in Ecology and Evolution 13,pp 261–265

[32] Richard, A. G., Emmanuel, K. B., Canisius J. K., Emmanuel B.M., Almas, M. K and Philipina F. S (2011). Species Composition, Richness and Diversity in Mimbo Woodland of Bereku Forest Reserve, Tanzania. Journal ofBiodiversity, 2(1): (2011) pp1-7.

[33] Rosenzweig, M. (1995). Species diversity in space and time. Cambridge University Press, Cambridge.171pp

[34] Sabogal, C. (1992). Regeneration of tropical dry forests in Central America, with examples from Nicaragua. J Veg Sci 3:407–416 pp

[35] Samaila, M (2011). The geology of Kanawa and its environs, part of Gombe sheet 152 NE, Gombe State, Nigeria.An unpublished B.Sc thesis. Department of geology, Gombe State University, Gombe, Nigeria.pp 47.

[36] Shannon, C.E. (1948). A mathematical theory of communication.Bell System Technical Journal 27: 379–423 and pp623–656.

[37] Simpson E. H (1949). Measurement of diversity. Nature, 163, 688pp

[38] Sobuj, N.A and Rahman.M (2011). Assessment of plant diversity in Khadimnagar National Park of Bangladesh.Department of Forestry and Environmental Science, School of Agriculture and Mineral Sciences, Shah Jalal University of Science and Technology, Sylhet 3114, Bangladesh. International Journal of Environmental Sciences. 2 .No.1. pp 91.

[39] Vierich, H.I.D and Stoop, W.A (1990). Changes in West African Savanna Agriculture in Response to Growing Population and Continuing Low Rainfall.Agric. Ecosys. and Environ. 31:pp115- 132.

Effects of Lime-Aluminium-Phosphate Interactions on Maize Growth and Yields in Acid Soils of the Kenya Highlands

Esther Mwende Muindi[1, *], Jerome Mrema[2], Ernest Semu[2], Peter Mtakwa[2], Charles Gachene[3]

[1]Department of Crop Science, Pwani University, Kilifi, Kenya
[2]Department of Soil Science, Sokoine University of Agriculture, Morogoro, Tanzania
[3]Department of Land Resource Management, University of Nairobi, Nairobi, Kenya

Email address:
e.muindi@pu.ac.ke (E. M. Muindi)

Abstract: Soil acidity and phosphorus (P) deficiency are some of the major causes of low maize yields in Kenya. Although considerable work has been done to establish liming rates for acid soils in many parts of the world, information on the effects of the lime-Al-P interactions on maize growth and yield is limiting. A green house pot experiment was conducted at Waruhiu Farmers Training Centre, Githunguri to evaluate the effects of lime-Al-P interactions on maize growth and yield in acid soils of the Kenya highlands. Extremely acidic (pH 4.48) and strongly acidic (pH 4.59) soils were used for the study. Four lime (CaO) rates and phosphorus (Ca $(H_2PO_4)_2$ rates were used. The liming rates were: 0, 2.2, 5.2 and 7.4 tonnes ha^{-1} for extremely acidic soil and 0, 1.4, 3.2, and 4.5 tonnes ha^{-1} for the strongly acidic soil. Phosphorus applications rates were: 0, 0.15, 0.30 and 0.59 mg P kg^{-1} soil for the extremely acidic soil and 0, 0.13, 0.26, and 0.51 mg P kg^{-1} for strongly acidic soil. The experiments were a 4^2 factorial laid down in a Randomized Complete Block Design (RCBD) and replicated three times. Data collected included: plant height, number of leaves, P-uptake and maize dry matter yield. Lime-Al-P interaction significantly (P$\leq$ 0.05) increased P concentrations in maize tissues, maize height, dry matter yields. Use of 7.4 tonnes ha^{-1} in extremely acidic soils and 4.5 tonnes ha^{-1} in strongly acidic soils significantly (P$\leq$0.05) increased maize height compared to lower lime rates. Phosphorus uptake and dry matter yields did not however, vary when 7.4 tonnes ha^{-1} lime was combined with either 0.59 mg P kg^{-1} or 0.3 mg P kg^{-1} in extremely acidic soils, and 4.5 tonnes ha^{-1} was combined with either 0.51 mg P kg^{-1} or 0.26 mg P kg^{-1} in strongly acidic soils. It was, therefore, concluded that lime and P positively interact to reduce Aluminium toxicity in the soils and improve maize growth, P uptake and yields in acid soils in the Kenya highlands. However, further research is required to evaluate long term effects of the interactions on crop yields, uptake of plant nutrients under field conditions.

Keywords: Lime, Phosphorus, Aluminium, Acid Soils, Maize Yields, P-uptake

1. Introduction

In highly weathered, acid soils around the world, P, Ca, and Mg deficiencies and aluminum (Al) and manganese (Mn) toxicities are the most important nutritional and/or element disorders that limit crop yields [1,2]. Phosphorus deficiency in such soils is attributed to adsorption of P by Al/Fe oxides and hydroxides, such as gibbsite, hematite, and goethite [3]. The phosphorus is first adsorbed on the surface of clay minerals and Fe or Al oxides by forming various complexes. Non - protonated and protonated bidentate surface complexes may coexist at pH 4 to 9, while the protonated bidentate inner sphere complex is predominant under acidic soil conditions [4, 5]. Phosphorus may also be occluded in nanopores that frequently occur in Fe / Al oxides, and thereby become unavailable to plants [5].

Aluminium toxicity is a major growth limiting factor for crop production in acid soils [6]. It impedes both cell elongation and cell division leading to reduced root growth [7, 8] hence reduced ability of the plant to explore the soil volume for nutrients and water leading to nutrient and water stress. Aluminium toxicity can also interfere with active ion uptake processes across the root-cell plasma membrane [7].

Liming modifies the physical, chemical and biological characteristics of soil through its direct effect on amelioration of soil acidity [9, 10, 11]. It also plays an indirect role of mobilization of plant nutrients, immobilization of toxic heavy metals and improvement of soil structure [12]. Physical amelioration of lime occurs through flocculation of colloid particles which leads to changes in surface potential and charge densities [13] while chemical amelioration of lime occurs through increased Ca^{2+} and /or Mg^{2+} ions in the soil solution and increased soil pH, thereby reducing the activities / concentrations of Al^{3+} and Fe^{3+}, H^+, Mn^{4+} and Fe^{3+} ions in the soil solution. Liming also improves microbiological activities of acid soils, which in turn can increase dinitrogen fixation by legumes and liberate nitrogen (N) from incorporated organic materials.

Although liming of acid soils has been shown to have a variable effect on P sorption capacity, increased sorption on limed soils is attributed to formation of active x-ray amorphous Al hydroxyl polymers, which actively sorb more P than Al^{3+} [14].The active Al hydroxyl polymers formed can also coat the surfaces of minerals, thereby affecting their surface charge characteristics [15] and P sorption of the soils.

Liming an acid soil to above about pH 5.5 has also been reported to increase plant growth [16]. This positive growth response to lime has been attributed to amelioration of Al-toxicity and / increased P availability [17, 18]. However, high rates of lime, which increase the pH values above 6.5, have been reported to depress plant growth [16, 19]. The decrease in yields have been attributed by [16] to three possible reasons, namely, reduction of infiltration due to formation of smaller soil aggregates, micronutrient deficiencies at higher pH and induced P deficiency due to formation of insoluble calcium phosphate (Ca-P) compounds. However, [20] attributed the yield decrease to Al reactions in the soil where the level of exchangeable Al at a given pH reflected the reactivity of the aluminium surfaces which, in turn, governed the solubility of P.

Kenyan soils, similar to other agricultural soils of the tropics, have low available P, high Al concentrations and high P-fixation capacities [21, 22, 23] attributable to extensive weathering and dominance of Al, Fe, and Mn oxyhydroxides and 1:1 layer silicates and extensive leaching of the basic cations and high concentration of Al^{3+} and Fe^{3+} in the soil solution and on the cation retention sites and the soil colloids. Under these situations, an appropriate combination of lime and P is an important strategy for improving field crops yield in highly weathered soils. Several workers have tested and documented the effect of liming on nutrient availability in acid soils of the western highlands of Kenya [22, 24, 25]. However, information on the interaction of lime, Al and P to ensure optimal availability of P as assessed by plant uptake in Kenya highlands soils is limiting. The objective of this study was, therefore, to evaluate the effects of lime-Al-P interaction on P uptake, maize growth and yield.

2. Materials and Methods

2.1. Experimental Layout, Design and Crop Husbandry

A greenhouse pot experiment was carried out at Waruhiu Farmers' Training Centre, Githunguri, Kiambu County. Two composite soil samples representing extremely acid (pH 4.0-4.5) and strongly acidic (pH 5.0-5.5) soils, as described by [9] were used in the study. The experiments were a 4^2 factorial laid down in a Randomized Complete Block Design (RCBD) and replicated three times. The treatments were lime application rates and phosphorus application rates. Liming rates were chosen to obtain 0, 30, 70 or 100% reduction in amounts of M KCl-extractable Al originally present in the soil while the phosphorus levels added were: 0, 0.5, 1 or 2 times the standard phosphorus requirement (SPR) of the tested soils (Table 1). Burnt lime (CaO) containing about 21% calcium oxide was used in this study while triple superphosphate fertilizer [Ca $(H_2PO_4)_2$] as used as P- source.

Table 1. Actual amounts of phosphorus and lime added in the acid soils.

	Lime added (tones ha^{-1})				P added (g P kg^{-1} soil)			
Soil	0%	30%	70%	100%	0 SPR	0.5 SPR	1 SPR	2 SPR
Extremely Acidic	0	2.2	5.2	7.4	0	0.15	0.30	0.59
Strongly Acidic	0	1.4	3.2	4.5	0	0.13	0.26	0.51

SPR-Standard phosphate requirement

Five kilogramme composite soil samples for both extremely and strongly acid soils were measured and put in nine litre plastic pots. The different lime levels were measured and incorporated into the soil samples by thoroughly mixing and incubating at field capacity for a period of 21 days. Water was added every 2 days to compensate for evaporative losses and the soils were remixed thoroughly. After incubation, soils from each liming level were air-dried, sieved and returned to the plastic pots. Various P levels were then added and the soils re- incubated at the same conditions for 14 days after which they were air dried and used for plant growth studies.

A plant growth study was conducted in a greenhouse. Lime-P treated soils (4 kg) were mixed with Calcium

Ammonium Nitrate (CAN) at the rate of 50 kg N ha^{-1} then placed in plastic containers with the container lids placed underneath to obtain any leachate. Each treatment was replicated three times. The potted soils were moistened to field capacity with water and three maize seeds planted in each pot. Nduma maize variety was used. After 14 days, the plants were thinned to one plant per pot.

2.2. Laboratory Analysis

Soil physiochemical analysis and P adsorption were determined before application of treatments. Soil pH, exchangeable aluminum, CEC, and particle size distribution were analyzed as described by [26]. Extractable P was determined by dry ashing techniques as described by [27] while phosphorus sorption capacities of the soils were evaluated as described by [28] and the P adsorbed data for the two soils fitted into the linearized form of the Langmuir equation. The lime requirements of the soils were calculated using the equation of [29]. The equation aims at reducing the % Al saturation to a level that is commensurate with crop Al tolerance, and is given as: Lime required ($CaCO_3$ equiv.) tones ha^{-1} = 1.8[Al - RAS (Al + Ca + Mg) /100] where Al = cmol kg^{-1} soil in the original exchange complex, RAS = Required percentage Al saturation, Ca = cmol kg^{-1} soil in the original exchange complex, Mg = cmol kg^{-1} soil in the original exchange complex. A RAS value of 20% was used. Soil characterization data in Table 2 below was used for lime requirement determinations.

2.3. Crop Growth Data Collection

Crop growth data collected included: plant height, number of leaves, and dry matter yield. Maize height and number of leaves were measured weekly from crop emergence until 35 days after emergence. Thirty five days after germination, the maize plants were cut and the above ground parts weighed, then oven dried at 70^0C up to constant weights. Soil from each pot was also thoroughly mixed and subsamples taken to the soil science laboratory at the University of Nairobi for physical and chemical analysis.

2.4. Statistical Analysis

Data obtained were subjected to Analysis of Variance (ANOVA) using the GenStat statistical package [30] and treatment effects were tested for significance using the F-test at 5% level of significance. Means were ranked using Duncan's New Multiple Range Test. Dependency tests were also conducted to find out if there was a relationship between the various variables used.

3. Results

3.1. Initial Soil Physical and Chemical Characteristics

The tested soils were acidic with pH <5.5 (Table 2). Exchangeable Aluminium levels for both soils were > 2 cmol kg^{-1} and % Aluminium saturation > 20%. Extractable was low while CEC was <15 cmol kg^{-1} soil. The tested soils had clay texture.

Table 2. Physiochemical properties of the two soils before pot experiment.

	Extremely acidic	Strongly acidic
pH	4.48	4.59
Exch. Al (cmol kg^{-1})	3.85	3.90
OC (%)	1.75	1.83
P (mg kg^{-1})	10.50	13.50
CEC	10.82	11.68
Al Saturation (%)	55.82	49.66
%Clay	56.32	50.00
% Silt	21.00	17.00
% Sand	22.68	33.00
Textural class	Clay	Clay

3.2. The Effect of Lime-Al-P Interaction on Growth of Maize

The effect of lime-Al-P interaction on maize height varied with levels of P and lime used (Table 3). Lime rates resulting into 100% reduction in AL^{3+} in the soils were observed to significantly (P$\leq$ 0.05) promote the highest maize plant heights in both extremely and strongly acidic soils.

Table 3. Effect of lime-Al-P interactions on maize height (cm) in acid soils of the Kenya Highlands (averaged over 5 sampling period).

	Phosphorus levels applied (mg P kg^{-1})							
	Extremely acidic soils				Strongly acidic soils			
Lime to give	0	0.15	0.3	0.59	0	0.13	0.26	0.51
0% reduction in Al^{3+}	91.7a	108.0a	113.0a	116.0a	89.7a	102.0a	108.7a	103.8a
30% reduction in Al^{3+}	97.8a	118.3a	123.5a	132.2b	97.0a	110.0b	114.6a	121.4b
70% reduction in Al^{3+}	103.7a	138.3b	136.0b	132.2b	102.3a	118.0b	121.0b	122.0b
100% reduction in Al^{3+}	115.6b	152.9c	207.2c	158.2c	112.9b	129.7c	131.3c	132.0c
% CV	3.2	3.2	3.2	3.2	2.0	2.0	2.0	2.0

Values followed by the same letter(s) on the same column are not significantly different at P ≤ 0.05.

Phosphorus use was observed to significantly (P$\leq$ 0.05) increase the number of leaves per maize plant in the acid soils (Table 4). Use of 0.3 mg P kg^{-1} gave the highest number of leaves in extremely acidic soils while 0.26 mg P kg^{-1} gave the highest plant height in strongly acidic soils.

Table 4. Effect of lime-Al-P interactions on number of leaves per maize plant in acid soils of the Kenya Highlands (averaged over 5 sampling period).

	Phosphorus levels applied (mg P kg^{-1})							
	Extremely acidic soils				Strongly acidic soils			
Lime to give	0	0.15	0.3	0.59	0	0.13	0.26	0.51
0% reduction in Al^{3+}	21.0a	26.7b	29.3c	26.3b	20.7a	21.3a	26.3b	23.7c
30% reduction in Al^{3+}	22.0a	27.7b	30.3c	27.7b	21.3a	24.0b	27.0c	24.3b
70% reduction in Al^{3+}	22.7a	27.7b	31.0c	27.7b	21.7a	24.0b	28.7c	24.3b
100% reduction in Al^{3+}	23.0a	27.7b	31.0c	28.7b	22.7a	27.3a	31.7b	27.0b
% CV	3.7	3.7	3.7	3.7	4.7	4.7	4.7	4.7

Values followed by the same letter(s) within a row are not significantly different at P ≤0.05.

Average height of maize plants growing under different lime rates in the extremely acidic soils was not significantly different until the 4th week after germination (Figure 1). On the other side, the heights of the maize plants in strongly acidic soils were significantly different from the 3rd week after emergence (Figure 2).

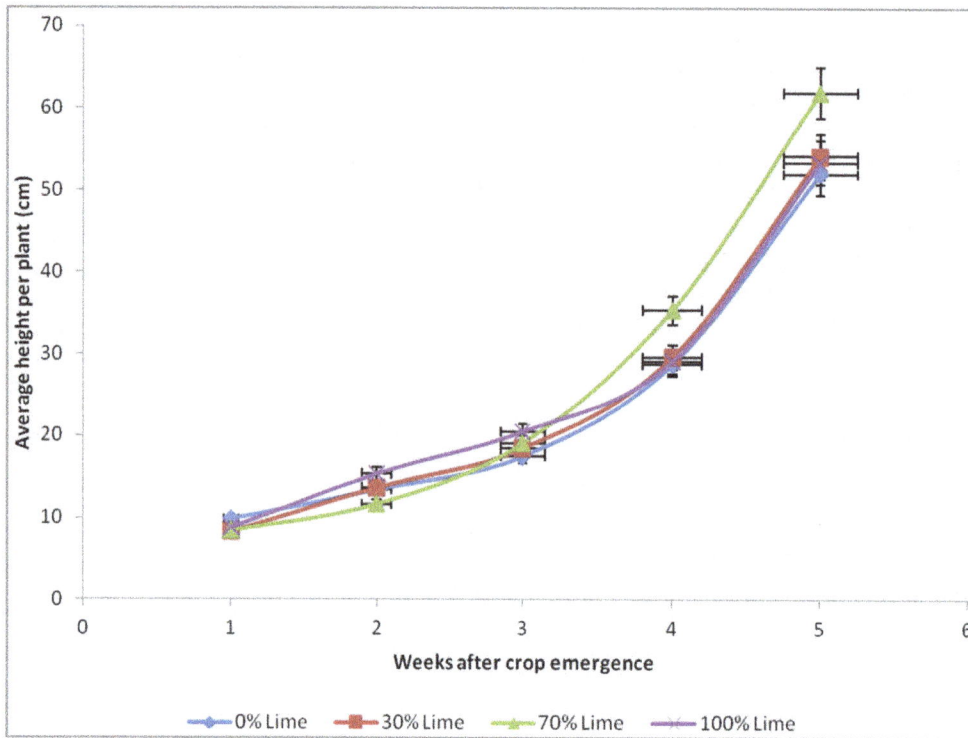

Figure 1. *Effects of liming on heights of maize plants planted in extremely acidic soils (LSD bars inserted).*

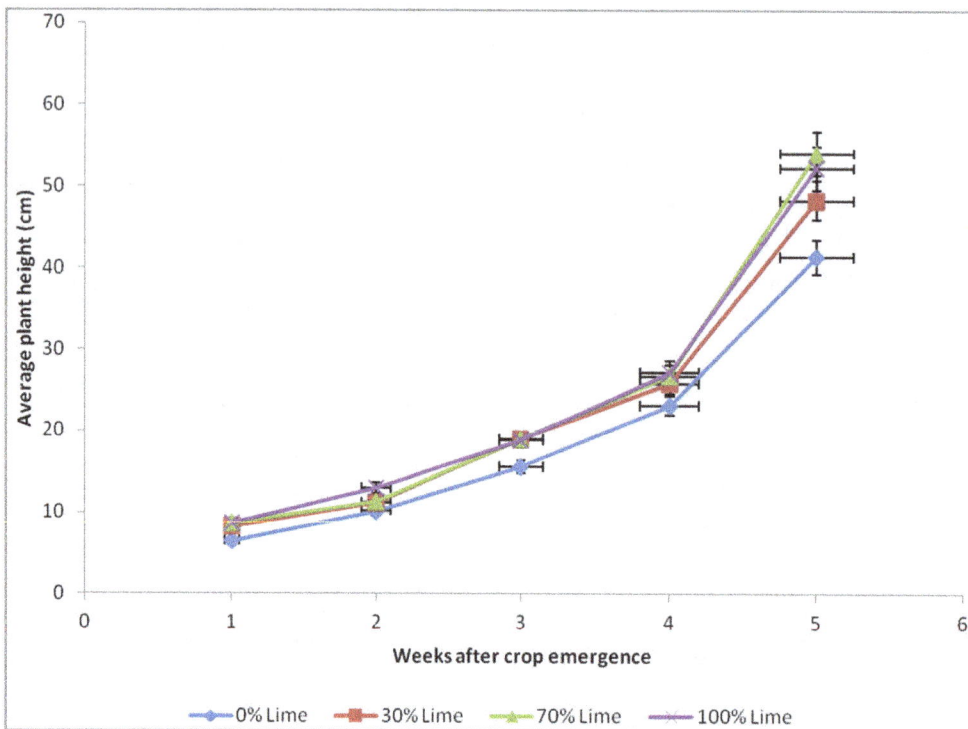

Figure 2. *Effects of liming on heights of maize plants planted in strongly acidic soils (LSD bars inserted).*

Average numbers of leaves from maize plants grown under different lime rates in extremely acidic soils were significantly different from the 4th week after emergence (Figure 3). On the contrally, number of leaves from plants grown in strongly acidic soils were significantly different from 2nd week after emergence (Figure 4).

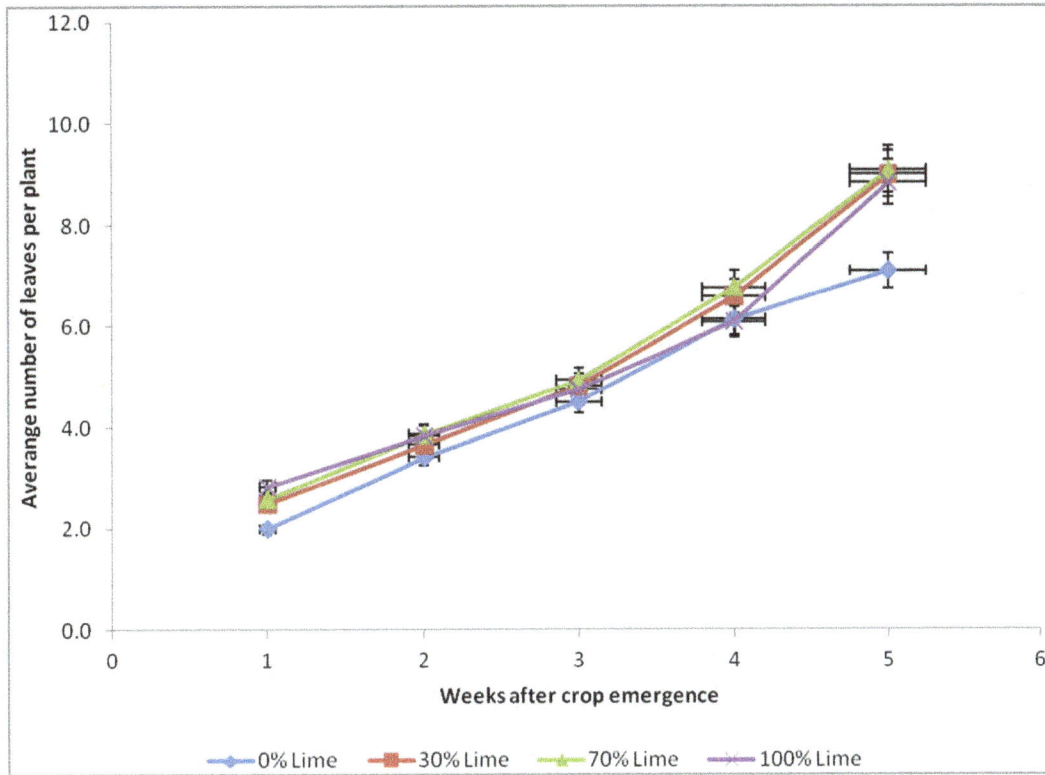

Figure 3. Effects of liming on number of leaves from maize planted in extremely acidic soils (LSD bars inserted).

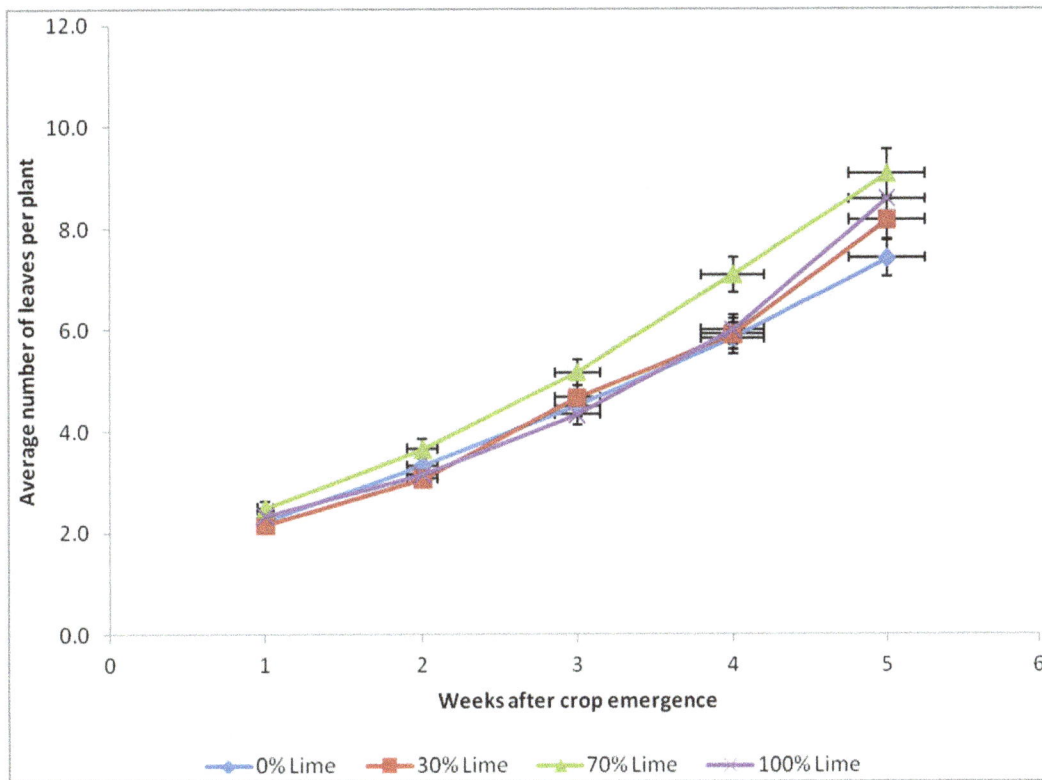

Figure 4. *Effects of liming on maize plant leaves in strongly acidic soils (LSD bars inserted).*

Maize leaf area index was significantly different from 3rd week after emergence under different lime rates in extremely acid soils (Figure 5). Similar trends were observed in strogly acid soils (Figure 6).

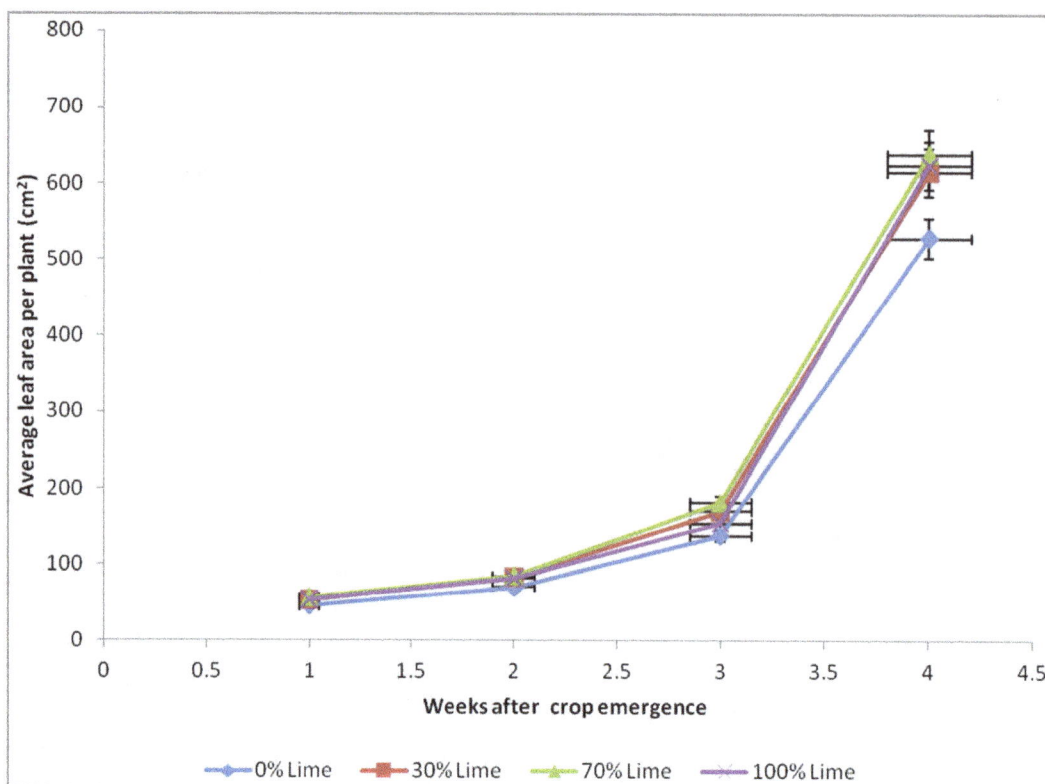

Figure 5. *Effects of liming on leaf area index of maize planted in extremely acidic soils (LSD bars inserted).*

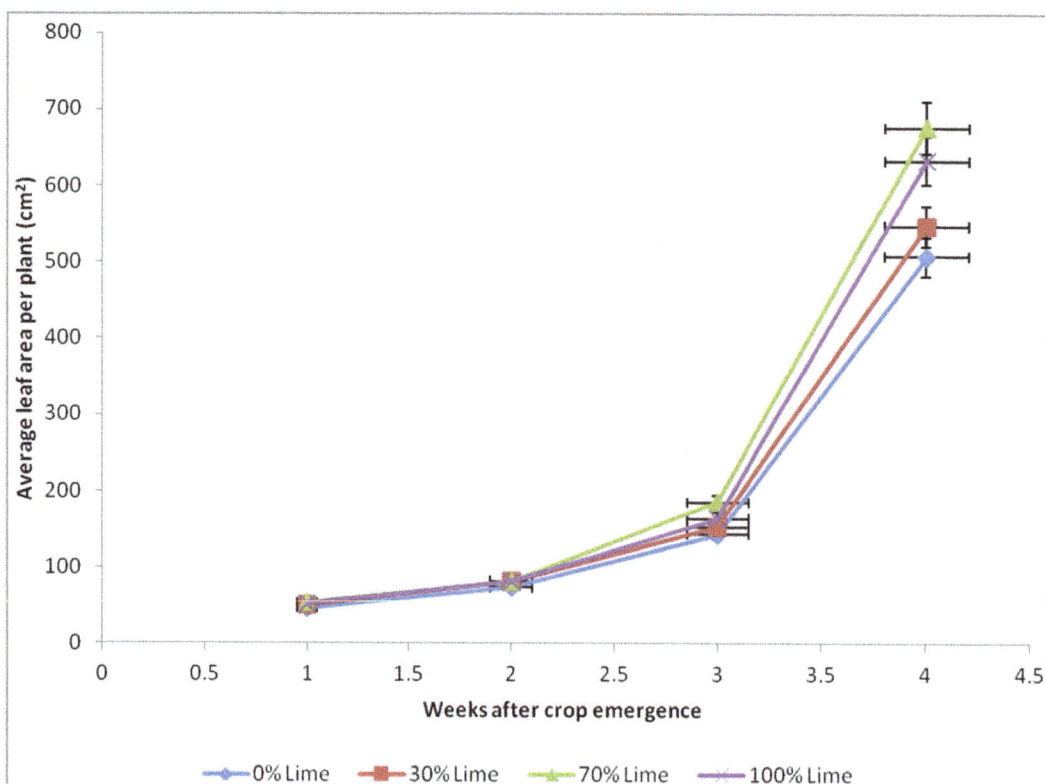

Figure 6. *Effects of liming on leaf area index of maize planted in strongly acidic soils (LSD bars inserted).*

Phosphorus concentrations in plant tissues 35 days after emergence were observed to increase significantly with lime-Al-P interactions (Table 5). The P concentration increase was significantly dictated by reduction of Al concentrations in the soil in the order: 0 % reduction of Al^{3+} = 30 % reduction of Al^{3+} < 70 % reduction of Al^{3+} ≤ 100 % reduction of Al^{3+} in extremely acidic soils and, 0 % reduction of Al^{3+} = 30 % reduction of Al^{3+} < 70 % reduction of Al^{3+} = 100 % reduction of Al^{3+} in strongly acidic soils.

Table 5. *Effects of lime-Al-P interactions on P concentration (%) in maize tissues 35 days after emergence.*

	Phosphorus levels applied (mg P kg⁻¹)							
	Extremely acidic soils				Strongly acidic soils			
Lime to give	0.00	0.15	0.3	0.59	0.00	0.13	0.26	0.51
0% reduction in Al^{3+}	0.17a	0.26a	0.29a	0.31a	0.16a	0.23a	0.28a	0.30a
30% reduction in Al^{3+}	0.20b	0.27a	0.30a	0.31a	0.19a	0.23a	0.30a	0.32a
70% reduction in Al^{3+}	0.30c	0.35b	0.45b	0.47b	0.37b	0.40b	0.45b	0.49b
100% reduction in Al^{3+}	0.37d	0.43c	0.51c	0.53c	0.34b	0.41b	0.47b	0.49b
% CV	3.50	3.50	3.50	3.50	2.90	2.90	2.90	2.90

Values followed by the same letter(s) on the same column are not significantly different at $P \leq 0.05$.

Lime-Al-P interactions significantly increased maize dry matter yield (Table 6).

It was observed that liming to achieve 100% reduction in Al^{3+} significantly ($P \leq 0.05$) produced the highest dry matter yields in extremely acidic soils.

Table 6. *Effect of lime-Al-P interactions on maize dry matter (tonnes ha⁻¹) 35 days after emergence.*

	Phosphorus levels applied (mg P kg⁻¹)							
	Extremely acidic soils				Strongly acidic soils			
Lime to give	0	0.15	0.3	0.59	0	0.13	0.26	0.51
0% reduction in Al^{3+}	6.3a	29.7b	49.7c	39.4d	11.0a	23.0b	43.3c	32.2b
30% reduction in Al^{3+}	7.4a	31.2b	54.1c	49.8c	13.6a	29.1b	48.2c	40.5e
70% reduction in Al^{3+}	15.0b	32.6b	54.7c	54.4c	14.5a	33.4c	58.8d	42.3e
100% reduction in Al^{3+}	21.8c	39.1c	73.3d	59.0e	17.8a	41.1c	59.3d	44.1e
% CV	2.7	2.7	2.7	2.7	4.8	4.8	4.8	4.8

Values followed by the same letter(s) on the same column are not significantly different at $P \leq 0.05$.

4. Discussion

4.1. Chemical and Physical Characteristic of the Soils

As per the rating suggested by [31], the soils had low levels of CEC (<15 cmol kg⁻¹), Ca (<4.0 cmol kg⁻¹) and P (<10 bicarbonate extractable P) and high levels of exchangeable aluminium (>2.0 cmolkg⁻¹) and Al saturation (>20%) implying that they were of low fertility status. The high levels of Al and Al saturation are considered to be toxic to maize plants [26, 32]. According to [9], the soils were strongly to extremely acidic with pH of 4.48-4.58. Such acid soils with high Al^{3+} ions, low bases and CEC are characteristic of highly weathered soils, which have lost most of the basic cations through the process of leaching [32]. As a result, their high levels of Fe and Al sesquioxides may lead to high P fixation, resulting in low available P [32, 33, 34, 35]. Additionally, the acidity could be attributed to the mineralogy of the parent materials [35] because most of these soils are developed from non calcareous parent materials such as syenites, phololites, trachytes and nepholites which are acidic in nature [36].

4.2. The Effect of Lime-Al-P Interactions on Growth of Maize

The significant (P≤ 0.05) increase in maize growth after lime application can be attributed to reduced aluminium toxicity which inhibit root growth by reducing cell elongation and cell division hence reduced main axis and lateral root formation [6]. The reduction of Al toxicity by liming occurs through precipitation of soluble and exchangeable Al as hydroxyl- Al species; the positively charged monomeric $AlOH_2^+$ and $Al(OH)_2^+$ species may polymerize to form both large and small positively charged polynuclear complexes which become sorbed to clay mineral and organic matter surfaces around the root zone [37] hence improved ability of the plants to explore the soil volume for nutrients and water

The significant (P≤ 0.05) increase of P concentrations in plant tissues after lime application can be attributed to reduced adsorption-precipitation reaction between Al and P at the root surface and in the root free space or ' P spring effect' of lime [11, 13]. According to [13], liming may increase plant uptake in soils high in exchangeable and soluble Al by decreasing Al, rather than by increasing P availability per se. This may be due to reduced interference of root cell-plasma membrane by Al, hence improved root growth, allowing a greater volume of soil to be explored [7, 8]. Once P is taken up by plants, it plays a great role in energy transfer processes including, photosynthesis which leads to biomass accumulation [2, 38]. The positive relation between shoot P and biomass have been reported among *sesbania* seedlings in Kenyan acid soils [39].

5. Conclusions

Lime-Al-P interactions significantly increased P- uptake, maize growth and dry matter yield. Application of lime that leads to 100% reduction in concentration of Al ions in the soil (7.4 tones ha⁻¹ for extremely acid soils and 4.5 tones ha⁻¹ for strongly acid soils) significantly reduced the Al ions in the soils compared to other liming amounts. Combined application of 7.4 tonnes ha⁻¹ lime with either 0.59 mg P kg⁻¹ or 0.3 mg P kg⁻¹ in extremely acidic soils, and 4.5 tonnes ha⁻¹ with either 0.51 mg P kg⁻¹ or 0.26 mg P kg⁻¹ in strongly acidic soils did not however, significantly vary in their effect on improving plant height, P uptake and dry matter yield. Further studies are therefore, required to ascertain long term, optimal lime-Al-P interactions under field conditions.

Acknowledgment

The authors express their appreciation to Waruhiu- farmers training centre for provision of research facilities and the Alliance for Green Revolution Africa (AGRA), for funding the research work.

References

[1] Fageria, N. K. (1994). Soil acidity affects availability of nitrogen, phosphorns, and potassium. Better Crops International. 10:8-9.

[2] Brady, C. N. and Weil, R. R. (2008). *The Nature and Properties of Soils,* 14th Ed; Pearson Prentice Hall, New Jersey. 975pp.

[3] Parfitt RL (1978) Anion adsorption by soils and soil minerals. Advances in Agronomy 20: 323–359

[4] Luengo, C., Brigante, M., Antelo, J., Avena, M. (2006). Kinetics of phosphate adsorption on goethite: comparing batch adsorption and ATR-IR measurements. *Journal of Colloidal Interface Sciences* 300: 511–518

[5] Arai, Y. and Sparks, D.L. (2007) Phosphate reaction dynamics in soils and soil minerals: a multiscale approach. *Advances in Agronomy* 94: 135–179

[6] Foy, C.D. (1988). Plant adaptation to acid, aluminium-toxic soils. Communication in Soil Science and Plant Analysis 19: 959–987.

[7] Kochian, L.V. (1995). Cellular mechanisms of aluminium toxicity and resistance in plants. *Annual Review of Plant Physiology and Plant Moleculer Biology* 46:237–260.

[8] Haynes, R. J. (1982). Effects of liming on phosphate availability in acid soils. *Plant and Soil* 68(3):289-308.

[9] Kanyanjua, S. M., Ireri, L., Wambua, S., Nandwa, S. M. (2002). Acid soils in Kenya: Constraints and remedial options. KARI Technical Note, 11:24.

[10] The, C., Calba, H., Zonkeng, C., Ngonkeu, E.L.M., Adetimirin, V.O. (2006). Response of maize grain yield to changes in acid soil characteristics after soil amendment. *Plant and Soil* 284:45-57.

[11] Fageria KF, Baligar VC, Jones CA. (2010). Growth and mineral nutrition of field crops. 3rd Edition. CRC Press. New York. London. 586Pp.

[12] Haynes, R. J. and Naidu, R. (1998). Influence of lime, fertilizer and manure applications on soil organic matter content and soil physical conditions: a review. *Nutrient Cycling in Agro ecosystems 51:123-137.*

[13] Bolan, N.S., Adriano, D.C., Curtin, D. (2003). Soil acidification and liming interactions with nutrient and heavy metal transformation and bioavailability. *Advances in Agronomy* 78: 215-272

[14] Roborage, W.P., and Corey, R.B. (1979). Adsorption of phosphate by hydroxyl aluminium species on a cation exchange resin. *Soil Science Society of America Journal* 43:481-487.

[15] Sims, J.T. and Ellis, B.G. (1983). Changes in phosphorus adsorption associated with aging of aluminum hydroxide suspensions. *Soil Science Society of America Journal 47*: 912-916.

[16] Kamprath, E.J. (1970). Exchangeable aluminum as a criterion for liming leached mineral soils. *Soil Science Society of America Proceedings* 34: 252-254.

[17] Sanchez, P.A. and Uechara, G. (1980). Management considerations for acid soils with high phosphorus fixation capacity. In: *The role of phosphorus in Agriculture.* (Edited by Khasawneh, F. E., Sample, E. C., Kamprath, E. J.) American Society for Agronomy, Wisconsin. pp.471-514.

[18] Kisinyo, P.O., Opala, P.A., Gudu, S.O., Othieno, C.O., Okalebo, J.R., Palapala, V., Otinga, A.N. (2014a). Recent advances towards understanding and managing Kenyan acid soils for improved crop production. *African Journal of Agricultural Research* 9(31):2397-2408.

[19] Maier, N.A., McLaughlin, M.J., Heap, M., Butt, M., Smart, M.K. (2002). Effects of current- season application of calcite lime and phosphorus fertilization on soil pH, potato growth, yield, drymatter content, and cadmium concentration. *Communications in Soil Science and Plant Analysis* 33(13-14): 2145-2165.

[20] Sumner, M.E. (1979). Interpretation of foliar analyses for diagnostic purposes. *Agronomy Journal* 71:343-348.

[21] Obura, P. A. (2008). Effects of soil properties on bioavailability of aluminium and phosphorus in selected Kenyan and Brazilian soils. Ph. D Thesis, Purdue University, USA. pp. 1-57.

[22] Kisinyo, P.O., Gudu, S.O., Othieno, C.O., Okalebo, J. R., Opala, P.A., Maghanga, J.K., Agalo, J.J., Ngetich, W.K., Kisinyo, J.A., Osiyo, R.J., Nekesa, A.O., Makatiani, E.T., Odee, D.W., Ogola, B.O. (2012). Effects of lime, phosphorus and rhizobia on *Sesbania sesban* performance in a Western Kenyan acid soil. *African Journal of Agricultural Research* 7(18):2800-2809.

[23] Muindi, E.M., Mrema J.P., Semu E., Mtakwa P.W., Gachene C.K., Njogu M.K. (2015). Phosphorus adsorption and its relation with soil properties in acid soils of Western Kenya. *International Journal of Plant and Soil Science, 4(3):203-211.*

[24] Nekesa, A.O. (2007). Effects of Mijingu phosphate rock and agricultural lime in relation to maize, groundnut and soybean yield on acid soils of western Kenya. M.Phil. Thesis. Moi University, Eldoret, Kenya. pp. 1-64.

[25] Kisinyo P.O., Othieno C.O., Gudu S.O., Okalebo, J.R., Opala, P.A., Ng'etich, W.K., Nyambati, R.O., Ouma, E.O., Agalo, J. J., Kebeney, S.J., Too, E.J., Kisinyo. J.A., Opile, W.R. (2014b). Immediate and residual effects of lime and phosphorus fertilizer on soil acidity and maize production in Western Kenya. *Experimental Agriculture* 50(1):128-143.

[26] Okalebo, J.R., Gathua, K.W., Woomer, P.L. (2002). Laboratory methods of soil analysis: A working manual (2nd ed.). TSBR-CIAT and SACRED Africa, Nairobi, Kenya. 88pp.

[27] Mechlich, A.A., Pinkerton, R. W., Kempton, R. (1962). Mass analysis methods for soil fertility evaluation. Ministry of Agriculture, Nairobi. pp.1-29.

[28] Fox, R.L. and Kamprath, E.G. (1970). Phosphate sorption isotherms for evaluating the phosphate requirements of soils. *Soil Science Society of America Proceedings* 34: 902-907.

[29] Cochrane, T. T., Salinas, J. G., Sanchez, P. A. (1980). An equation for liming acid mineral soils to compensate crop aluminium tolerance. *Tropical Agriculture (Trinidad)* 57: 33-40.

[30] GenStat. (2010). The GenStat Teaching Edition. GenStat Release 7.22 TE, Copyright 2008, VSN International Ltd.

[31] Landon, J.R. (1984). *Booker tropical soil manual: A handbook for soil survey and agricultural land evaluation in the tropics and sub tropics.* Longman, New York. 450pp.

[32] Landon, J. R. (1991). *Booker Tropical Soils Manual: A Handbook for Soil Survey and Agricultural Land Evaluation in the Tropics and Subtropics.* John Wiley and Sons, New York. 465pp.

[33] Buresh, R. J., Smithson, P. C., Hellums, D. T. (1997). Building soil phosphorus capital in Africa. In: *Replenishing soil fertility in Africa.* (Edited by Buresh, P. J. *et al.*) Soil Science Society of America Special Publication, Madison. pp. 111-149

[34] Sanchez, P.A., Shephard, K.D., Soule, M.J., Place, F.M., Buresh, R.J., Izac, A.N., Mokwunye, A.U., Kwesiga, F.R., Ndiritu, C.G., and Woomer, P.L. (1997). Soil fertility replenishment in Africa: An investment in natural resource capital. In: *Replenishing soil fertility in Africa, SSSA Special Publication* (Edited by Buresh, R.J., Sanchez, P.A., Calhoun, F.) SSSA, Madison. pp. 1-46.

[35] Van Straaten, P. (2002). *Rocks for crops: Agro minerals for sub-Saharan Africa.* ICRAF, Nairobi. pp. 25-28.

[36] Sombroek, W. G., Braun, H. M. H., van de Pouw. (1982). Exploratory soil map and agro-climatic zone map of Kenya. Scale 1:1000, 000. Exploratory soil survey report No. E1. Kenya Soil Survey, Nairobi. pp. 1-78.

[37] Stol R.J., Van Helden A.K., de Bruyn P.L. (1976) Hydrolysis precipitation studies of aluminium (III) solutions. 2. A kinetic study and model. Journal of Colloid Interface Sci ence57: 115–131

[38] Xie, P., Niu, J., Gan, Y., Gao, Y., Li, A. (2015). Optimizing phosphorus fertilization promotes dry matter accumulation and P remobilization in oilseed flax. *Crop Science* 54(4):1729-1736.

[39] Gudu, S. O., Kisinyo, P. O., Makatiani, E. T., Odee, D.W., Esegu, J. F. O., Chamshama, S. A. O., Othieno, C. O., Okalebo, J. R., Osiyo, R. J. and Owuoche, J.O. (2009). Screening of Sesbania for tolerance to aluminum toxicity and symbiotic effectiveness with acid tolerant rhizobia strains in Western Kenya acid soils. *Experimental Agriculture* 45:417-427.

18

Chemical Composition, Bio-Diesel Potential and Uses of *Jatropha curcas* L. (*Euphorbiaceae*)

Temesgen Bedassa Gudeta

Department of Biology, School of Natural Sciences, Madda Walabu University, Bale-Robe, Ethiopia

Email address:

tasgabifenet@gmail.com

Abstract: This review paper focuses some basic aspect of the taxonomic, biology, cultivation, chemical composition, bio-diesel potential, medicinal values and uses of Jatropha curcas Linn. *The genus Jatropha is distributed throughout the tropics and sub-tropics growning in marginal lands and is a potential biodiesel crop worldwide. Due to its adaptability to marginal soils and environments the cultivation of Jatropha curcas is frequently mentioned as the best option for producing biodiesel. The seed oil can be used as a feed stock for biodiesel. Alternatively Jatropha oil is used in soap, glue or dye industry. The seed cake is rich in nitrogen and phosphorus and can be used as manure.* Ash from the roots and branches of *Jatropha curcas* L. is used as cooking salt, and as lye in dyeing. The dark blue dye extracted from the bark of Jatropha is a useful dye. *The plant parts and its oil along with its latex used for different reasons such as pesticides, anti-inflammatory activities, wound healing, lighting (lamp), bio-gas production, fertilizer and other purposes.* The objective of this review paper focuses some basic aspect of the taxonomic, biology, cultivation, chemical composition, bio-diesel potential, medicinal values and uses of Jatropha curcas Linn.

Keywords: Chemical Composition, *Jatropha curcas*, Biodiesel, Uses

1. Introduction

Jatropha curcas Linn is commonly known as 'physic nut' is a non-food bioenergy plant and *currently* considered as alternative substitute to fossil [1]. It is perennial shrub belongs to *Euphorbiaceae* family same as rubber and cassava trees [2, 3]. Originally, *Jatropha curcas* was native tree in South America and was induced to Thailand about 200 years ago by Portuguese who produced soap from Jatropha oil. Generally, Jatropha tree is 3-6 meter tall, smooth grey bark, having latex and heart green leaf. *Jatropha curcas* L. or physic nut is a drought resistant large shrub or small tree, producing inedible oil containing seeds [4]. It is the commonest specie found in Nigeria, but many species exist in different parts of the world. It is a multipurpose, drought resistant tree and can be cultivated in areas of low rainfall [5]. *Jatropha curcas* L. is a suitable plant for quick and efficient domestication compared with other woody species [18]. Names used to describe the plant vary per region or country. It is most commonly known as "Physic nut". In Zimbabwe it is known as "Mufeta/

mujirimono" to mean it 'oil tree' [5]. In Nigeria it is known as "binidazugu/cinidazugu" and "lapa lapa" in Hausa and Yoruba languages respectively [6, 7]. At present, the varieties being used to established plantations in Africa and Asia are inedible [8]. Due to its toxicity, *J. curcas* oil is not edible and is traditionally used for manufacturing soap and medicinal [4]. Although there have been few and increasing research investigations in the previous works about *J. curcas* L., they have been based on little evidence-based information concerning chemical composition, bio-diesel potential and uses of the plant all the required information at one place in detail and consolidated form. There are many knowledge gaps concerning the chemical composition and potential applications of this plant. So, in view of such concerns of this plant, this review aims to provide an up-to-date overview of the chemical composition, bio-diesel potential and uses of different parts from *J. curcas* L.,which could be significant in providing insights for present and future research aimed at both ethnopharmacological validation of its popular use, as well as its exploration as a new source of herbal drugs and/or bioactive natural products along with its bio-diesel potentiality.

1.1. Taxonomic and Botanical Description

The Euphorbiaceae family, which is considered one of the largest families of Angiosperms, covers about 7,800 species distributed in approximately 300 genera and 5 subfamilies worldwide [9]. These species occur preferentially in tropical and subtropical environments [10, 11]. Among the main genera belonging to this family, there is Jatropha L., which belongs to the subfamily Crotonoideae, Jatropheae tribe and is represented by about 200 species. This genus is widely distributed in tropical and subtropical regions of Africa and the Americas [10]. The name "Jatropha" is derived from the Greek words "jatros," which means "doctor" and "trophe," meaning "food," which is associated with its medicinal uses. The leaves have significant variability in their morphology from green to pale green, alternate to sub opposite, and three- to five-lobed with a spiral phyllotaxis [1]. Flowers of *Jatropha curcas* produce nectar and are scented. The nectaries are hidden in the corolla and only accessible to insects with a long proboscis or tongue. The sweet, heavy perfume at night and greenish yellow colour of the flowers suggest that they are pollinated by moths.

Table 1. Taxonomic Classification J. curcas Linn.

Kingdom:	Plantae
Subkingdom:	Angiosperms
Infrakingdom:	Streptophyta
Superdivision:	Embryophyta
Division:	Tracheophyta
Subdivision:	Spermatophytina
Class:	Magnoliopsida
Superorder:	Rosanae
Order:	Malpighiales
Family:	Euphorbiaceae
Subfamily:	Crotonoideae
Tribe:	Jatropheae
Genus:	Jatropha
Species:	Jatropha curcas Linn

Source: [72]

In inflorescences, the female flowers open one or two days before the male ones or at the same time as the earliest male flowers. Male flowers last only one day. Seed never sets in indoor cultivation unless the flowers are pollinated by hand. Plants raised from seed are more resistant to drought than those raised from cuttings, because they develop a taproot. Fruit development from flowering to seed maturity takes 80–100 days. Plants from cuttings produce seeds earlier than plants grown from seed. Full production is achieved in the 4th or 5th year [12].

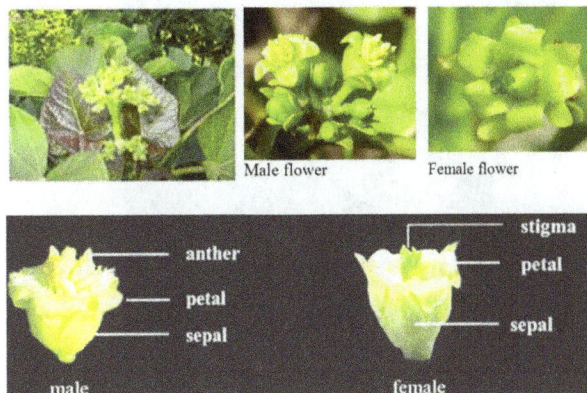

Figure 1. Flower of J. curcas L.; Sources: [13, 14].

Male and female flowers are produced on the same inflorescence, averaging 20 male flowers to each female [15]. The petiole length ranges from 6.1–23.1 mm. The inflorescence can be formed in the leaf axil. Plants are monoecious and also possess hermaphroditic flowers occasionally. Flowers are formed terminally, individually, with female flowers usually slightly larger occurring in hot seasons. Where continuous growth, an unbalance of pistillate or staminate flower production results in a higher number of female flowers. More female flowers mean more fruits. Fruits are produced in winter when the shrub is leafless, or it may produce several crops during the year if soil moisture is good and temperatures are sufficiently high. Each inflorescence yields a bunch of approximately 10 or more ovoid fruits. A three bi-valved cocci is formed after the seeds mature and the fleshy exocarp dries [1]. The seeds are mature when the capsule changes from green to yellow. The whole genome of *J. curcas* L was sequenced by *Kazusa DNA Research Institute*, Chiba Japan in October 2010 [16]. It was reported, somatic chromosome numbers were counted from root-tip cells of four individuals per population and all had 2n=22 chromosomes, corresponding to the diploid level (x=11) in which all the plant populations were found diploid [17]. This lack of variation in chromosome numbers contrasts with the high variability in other characteristics such as seed size, weight, and oil contents due to environment and genetic interaction [18].

Figure 2. J. curcas L with ripen fruits; Source: [19].

Figure 3. J. curcas L with ripen fruits; Source: [19].

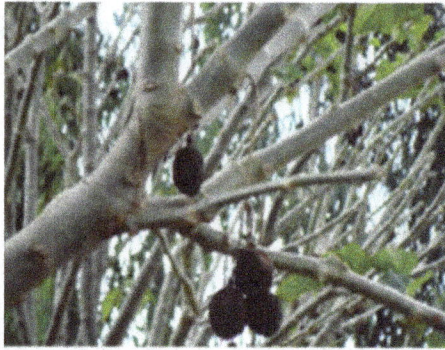

Dried fruits of J. Curcas L. Source: [5]

Figure 4. (a) Fresh and (b) dried seeds of J. Curcas L.; Source: [19].

1.2. Ecological Requirements

Jatropha curcas grows almost anywhere, even on gravelly, sandy and saline soils. It can thrive on the poorest stony soil. It can grow even in the crevices of rocks. The leaves shed during the winter months (dry season) form mulch around the base of the plant [20]. The organic matter from shed leaves enhance earth-worm activity in the soil around the root-zone of the plants, which improves the fertility of the soil. Regarding climate, *Jatropha curcas* is found in the tropics and subtropics and likes heat, although it does well even in lower temperatures and can withstand a light frost. Its water requirement is extremely low and it can stand long periods of drought by shedding most of its leaves to reduce transpiration loss. Jatropha is also suitable for preventing soil erosion and shifting of sand dunes.

1.3. Pests and Diseases

The seed oil, extracts of *J. curcas* seeds and phorbol esters from the oil have been used to control various pests, often with successful results. In Gabon, the seeds, ground and mixed with palm oil, are used to kill rats. The oil has purgative properties, but seeds are poisonous; even the remains from pressed seeds can be fatal [21]. It is popularly reported that pests and diseases do not pose a significant threat to jatropha, due to the insecticidal and toxic characteristics of all parts of the plant. However, incidence of pests and diseases such as collar rot, leaf spots, root rot and damping-off, may be controlled with a combination of cultural techniques (for example, avoiding waterlogged conditions) and fungicides are widely reported [2]. In Nicaragua (*Pachycoris klugii*) and India (*Scutellera nobilis*), causes flower fall, fruit abortion and seed malformation. Other serious pests include the larvae of the moth *Pempelia morosalis* which damages the flowers and young fruits, the bark-eating borer *Indarbela quadrinotata*, the blister miner *Stomphastis thraustica*, the semi-looper *Achaea janata*, and the flower beetle *Oxycetonia versicolor*

[20]. Termites may damage young plants. Carefully and judiciously adding an insecticide to the planting pit may be advisable if problems are endemic the use of pesticides is not necessary, due to the pesticidal and fungicidal properties of the plant. An insecticide may be included as a precaution against termites [22].

1.4. Cultivation

J. curcas L. can be planted by two common methods; seed or seedling propagation and the cutting method [23]. It was reported that vegetative propagation can be achieved by stem cuttings, grafting, budding and by air layering techniques [24]. The investigation leads to the recommendation that cuttings should be taken preferably from juvenile plants and treated with 200 microgram per liter of indol butric acid IBA (rooting hormone) to ensure the highest level of rooting in stem cuttings. These vegetative methods have potential for commercial propagation of these plants and yields faster results than multiplication by seeds [20]. The plant can grow in wastelands and grows on almost any terrain, even on gravelly, sandy and saline soils [25, 57]. Mycorrhizae have been observed on the roots; they promote growth, especiallly where phosphate is limiting [12]. Complete seed germination is achieved within nine days. Adding manure during the germination has negative effects during that phase, but is favorable if applied after germination is achieved [26].

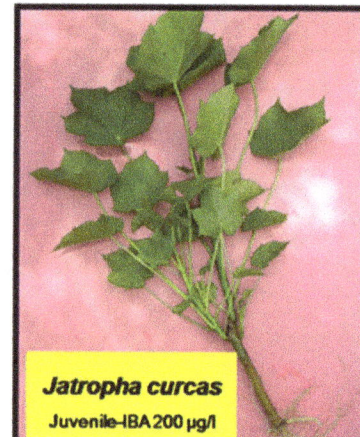

Figure 5a. Propagation of J. curcas by stem cutting.

Figure 5b. Propagation of J. curcas by grafting.

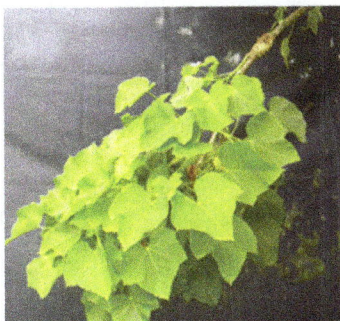

Figure 5c. Propagation of J. curcas by air layering techniques; Source: [20].

Jatropha curcas thrives on a mere 250 mm of rain a year, and only during its first two years does it need to be watered in the closing days of the dry season. Ploughing and planting are not needed regularly, as this shrub has a life expectancy of approximately forty years. Jatropha is planted at densities ranging from 1 1010 to 2500 plants per hectare. Yield per tree is likely to increase with wider spacing but with a decline in yields per hectare [18]. Spacing decisions should be based on the environment, i.e. how it affects competition among trees for water, light and nutrients. Semi-arid, low-input systems should use wider spacing such as 3.0 x 2.0, 3.0 x 2.5 or 3.0 x 3.0 metres. Alternate planting in succeeding rows will minimize mutual shading. In addition, consideration should be given to access. At least 2.5 m between trees allows easier passage for fruit pickers, while a 5-metre alley at every fourth row facilitates access by carts. Planting holes of 30–45 cm wide and deep should be prepared and organic matter incorporated before planting [20]. In the case of vegetative propagation of *J. curcas,* the method uses 40-50 cm long cuttings. Unlike seedlings, cuttings are planted during the dry season mostly two to three months prior to the commencement of rainy season. This is mainly because the plant has so much water that it can decompose if planted during the rainy season [5]. This contains resonance with thin which the jatropha cuttings have a thin layer of wax that prevents the easy evaporation of water hence they have to be planted early to lose some water [12].

1.5. Growth and Development

Growth in *J. curcas* is intermittent and sympodial which is a zigzag pattern of growth. Dormancy is induced by fluctuations in rainfall, temperature and light. Not all plants respond simultaneously; in a hedge plants without leaves may be found besides ones full of green leaves. The potential lifespan of *J. curcas* is 30–50 years [12].

Figure 6. Sympodial i.e. a zigzag pattern of growth in which the main plant stem develops from a series of lateral branches. Source: [5].

2. Chemical Composition of *Jatropha curacas* L. Components

2.1. Fruit Shell

The fruit shell describes the fruit pericarp, while the seed consists of the inner kernel and the outer husk or seed coat [27]. As stated in Farmers Handbook prepared in Kenya when the fruit is crushed by a simple hand tool, the shells and seeds are separated [14]. The shell is mechanically removed from the fruit in the first step during oil extraction. The chemical analysis of *Jatropha curcas* L. shell has shown that it is made up of 34%, 10% and 12% cellulose, hemicellulose and lignin, respectively. Volatile matter, ash and fixed carbon content of the shell have been shown to be 69%, 15% and 16%, respectively [22, 27]. These results show that *J. curcas* L. shells have very high ash content. This has an influence on the type of conversion technology that can be used to obtain energy from the shells. Jatropha shell ash fuses at temperatures above 750oc. Depending on the magnitude of the ash content, the available energy of the fuel is reduced proportionally [28]. At these high temperatures the ash reacts to form a slag, which can reduce plant throughput in combustion equipment. It was found that the caloric value of *J. curcas* L. shells is 11.1MJ kg-1 [29]. It was also found that the chemical composition of this plant shells seems to suggest that it is a good feedstock for biological conversion and for briquetting. Several conversion technologies have been studied using *J. curcas* L. shells as an energy feedstock [22]. These include briquetting and combustion [22], pyrolysis [30] and bio-methanation [29].

2.2. Seeds

The matured seeds contain 21% saturated fatty acids and 79% unsaturated fatty acids [31], and they yield 25%–40% oil by weight [32]. Additionally, the seeds contain other chemical compounds, such as saccharose, raffinose, stachyose, glucose, fructose, galactose, and protein. As stated by [18]. On average the seeds contain about 30-45% viscous oil which varies depending on where the jatropha is planted and the care it receives, water and nutrients [23].

2.3. Seed Husks

According to [27] the seed kernel contains predominantly crude fat oil and protein while the seed coat (husk) contains mainly fibre. Figure 8, shows that seed contains about 42% seed husks (seed coat) and 58% kernels, part of the seed in the seed coat. Although there is limited information in literature about the use of *J. curcas* L. seed husks for energy purposes. Analysis of the husks by [22] and [33] showed that the husks contained 4% ash, 71% volatile matter and 25% fixed carbon. The calorific value of the husks is 16 MJ kg-1, comparable to that of wood [33]. The physical properties of husks (for example, bulk density of 223 kg m-3) make them amenable to briquetting. The briquettes can be used as an energy source via combustion. It was also stated that *gasification* is a mature commercial energy conversion technology that can be used with *J. curcas* L. seed husks. In this study, it was found that the *syngas* calorific value and concentration of carbon monoxide, along with gasification efficiency increased with the increase in gas flow rate. This study showed that seed husks can be successfully used as feedstock for open core down draft gasifier [20, 25, 33].

2.4. Seed Cake

One liter of J. curcas L. oil comes from about 4 kg of seed which will give about 3 kg of the seed cake, a pressed seed cake, left after oil extraction [2].

Jatropha curcas L. seed cake makes an excellent organic fertilizer with a high nitrogen content similar to, or better than, chicken manure with macronutrient contents such as Nitrogen% (4.4-6.5), P% (2.1-3.0), K% (0.9-1.7), Ca% (0.6-07) and Mg% (1.3-1.4) [34].

Figure 7. Pressed seed cake.

Jatropha curcas L. seed cake contains mainly proteins and carbohydrates. The cake is made up of the seed husks (42%) and kernel [27]. Based on the extraction efficiencies and the average oil content of the whole seed, press-cake can contain 9-12% oil by weight [18]. This oil influences the gross energy value of press cake, which is about 18.2 MJ kg-1 [18]. Production of biogas by anaerobic digestion of *Jatropha* press-cake has been demonstrated [22, 35].

2.5. Seed Oil

The oil content of *J. curcas* seed generally lies in the range of 25–40 percent [32]. This can be extracted by heat, solvents or by pressure. The seed oil is potentially the most valuable end-product of *J. curcas*.

Figure 8. Harvested fruits products of J. curcas L. and their fractions by weight. Sources: [27, 32, 39].

The plant seed oil chemically consists of triacylglycerol with linear fatty acid chain (unbranched) with/without double bonds [19]. With related to this, two interesting forms of diterpenoids, namely Diterpenoid Curcuson A and Diterpenoid Curcuson C, having aromatic qualities are found in this plant. This is a great prospect for the plant in perfumes and other cosmetics industries. The oil is almost all stored in the seed kernel, which is accounted for around 58% of the seed [4]. Similar result was obtained as the seed of this plant consists of 40-42% of husk and 58-60% of kernel by mass [36]. This compares well to groundnut kernel (42 percent), rape seed (37 percent), soybean seed (14 percent) and sunflower seed (32%). The seed yield can attain 4 tonnes of seed/ha/year while the oil yield can reach 1590kg/ha [37, 38].

The removal of shell from Jatropha seed can improve oil yield and quality during the oil extraction process in biodiesel production. A Jatropha shelling machine was evaluated to analyse its efficiency to separate between the seed and the shell [39]. 63.16% oil content from Malaysian *J. curcas* seed kernels was determined using Hexane [40]. High oil content of *J. curcas* indicated the plant is suitable as non-edible vegetable oil feedstock in oleochemical industries such as biodiesel, fatty acids, soap, fatty nitrogenous derivatives, surfactants and detergents. In addition to a high oil content (64.4%), there are also lipid classes such as dominant triacylglycerol lipid species (88.2%) and appreciable percentage composition of fatty acids such as linoleic acid (47.3%) after extraction from samples of *Jatropha curcas* seeds obtained from markets in five different towns in Nigeria; as indicated below in Table 2 and 3 [41].

Table 2. Percentage composition of lipid classes of J. curcas seed oil.

R. No	Composition	Percentage
1	Unsaponifiable lipids	3.8
2	Stereo esters	4.8
3	Triglycerols	88.2
4	Free fatty acids (FFA)	3.4
5	Diacylglycerols	2.5
6	Sterols	2.2
7	Monoacylglycerols	1.7
8	Polar lipids	2.0

Table 3. Fatty acid composition of J. curcas seed oil.

R. No	Composition	Percentage
1	Palmitic acid (C16:0)	11.3
2	Stearic acid (C18:0)	17.0
3	Oleic acid (C18:1)	12.8
4	Linoleic acid (C18:2)	47.3
5	Arachidic acid (C20:1)	4.7
6	Arachidoleic acid (C20:0)	1.8
7	Behenic acid (C22:0)	0.6
8	(C24:0)	44

Sources: [20-22, 41]

2.6. Roots of J. curcas L. Contents

Ash from the roots and branches of *Jatropha curcas* L. is used as cooking salt, and as lye in dyeing. The dark blue dye extracted from the bark of Jatropha is a useful dye. HCN and *rotenone* are present in the roots of *J. curcas* L. [42]. In China isolated thirteen compounds from the roots of *Jatropha curcas* L. combining the determination of physico-chemical constants and spectral analyses (IR, 1H-NMR, 13C-NMR, EIMS, FABMS), the structures of the compounds were identified as 5α-stigmastane-3, 6-dione (1), nobiletin (2), β-sitosterol (3), taraxerol (4), 2S-tetracosanoic acid glyceride-1(5),5-hydroxy-6,7-dimethoxycoumarin (6), jatropholone A (7), jatropholone B (8), 6-methoxy-7-hydroxycoumarin (9), caniojane (10), 3-hydroxy-4-methoxybenzaldehyde (11), 3-methoxy-4-hydroxybenzoic acid (12) and daucosterol (13) [43]. Among them, compound 5 is a new compound which has never been reported in China and abroad, compound 1, 2, 9, 10, 11, 12 were first time isolated from the plant, 7 and 8 are a pair of stereoisomers which can be inverted in dilute basic solution. 10 is a diterpenoid containing peroxide bridge. The bark of *Jatropha curcas* L. contains a wax composed of a mixture of 'melissyl alcohol' and its melissimic acid ester [21]. Leaf juice stains red and marks linen an indelible black. The 37% tannin found in bark is said to yield a dark blue dye; latex also contains 10% tannin and can be used as marking ink. Ashes from the roots and branches are used in the dyeing industry, and pounded seeds in tanning in [43, 21].

3. Bio-diesel Potential of *Jatropha curcas* L. Oil

The oil is mainly used as biodiesel for energy. The process through which the glycerin is separated from the biodiesel is known as transesterification. Glycerin is another by-product from Jatropha oil processing that can add value to the crop. Transesterification is a simple chemical reaction that neutralizes the free fatty acids present in any fatty substances in Jatropha. A chemical exchange takes place between the alkoxy groups of an ester compound by an alcohol. Usually, methanol and ethanol alcohol are used for the purpose. The reaction occurs by the presence of a catalyst, usually sodium hydroxide (NaOH) or caustic soda and potassium hydroxide (KOH), which forms fatty esters (e.g., methyl or ethyl esters), commonly known as biodiesel. It takes approximately 10% of methyl alcohol by weight of the fatty substance to start the transesterification process [28, 44]. The production of jatropha biodiesel is a chemical process whereby the oil molecules (triglycerides) are cut to pieces and connected to methanol molecules to form the jatropha methyl ester. An alkali normally sodium hydroxide (caustic soda) is needed to catalyze the reaction. Glycerine (glycerol) is formed as a side product. Methanol is normally used as the alcohol for reasons of cost and technical efficiencies. Sodium hydroxide is dissolved in methanol to form sodium methoxide, which is then mixed with jatropha oil. The glycerine separates out and is drained off. The raw biodiesel is then washed with water to remove any remaining methanol and impurities [1; 20]. Typical proportions used in the reaction are:

Inputs:
100 units of jatropha oil
10-15 units of methanol
0.5-2 units of NaOH catalyst
Outputs:
100 units of biodiesel
10-15 units of glycerine
The figure 9 below shows the chemical process for methyl ester, biodiesel. The reaction between the fat or oil and the alcohol is a reversible reaction, so the alcohol must be added in excess to drive the reaction towards the right and ensure complete conversion.

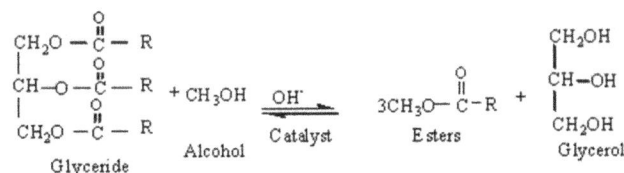

Figure 9. *Chemical process for methyl ester, biodiesel. Source: [20].*

Here the biodiesel is produced in the form of methyl ester where as glycerine is the byproduct of it [45]. Concerning the production methyl ester (biodiesel) there are two techniques to produce biodiesel from crude *Jatropha curcas* seed oil (CJSO) [46]. The first one is alkali base catalyzed transesterification process in which the presence of high concentration of free fatty acids, FFA, (15%) reduced the yield of methyl esters. The second is two-step pretreatment process in which the high level of JCSO was reduced to less than 1%. A two-stage transesterification process was selected to improve the methyl ester yield. Biodiesel may be used as partial blends (e.g. 5 percent biodiesel or B5) with mineral diesel or as complete replacements (B100) for mineral diesel. In general, B100 fuels require engine modification due to the different characteristics of biodiesel and mineral diesel. Solvent action may block the fuel system with dislodged residues, damage the hoses and seals in the fuel system, or cause cold filter plugging, poorer performance due to the lower heating value of biodiesel, some dilution of the engine lubricating oil, and deposit build-up on injectors and in combustion chambers [47]. It is generally accepted by engine manufacturers that blends of up to 5 percent biodiesel should cause no engine compatibility problems. Higher blends than this may void manufacturers' warranties. Jatropha biodiesel has proven to conform to the required European and USA quality standards. Jatropha biodiesel generally exceeds the European standard. For every 1 litre of biodiesel, 79 millilitres of glycerine are produced, which is equivalent to around ten percent by weight. The raw glycerine contains methanol, the sodium hydroxide catalyst and other contaminants, and must be purified to create a saleable product. Traditional low volume/high-value uses for glycerine are in the cosmetic, pharmaceutical and confectionary industries, but new applications are being sought as productionshifts to high volume/low value. Glycerine is used in the production of fuel, plastics and antifreeze. Pure jatropha oil may be used directly in some diesel engines, without converting it into biodiesel. The main problem is that jatropha oil has higher viscosity than mineral diesel, although this is less of a problem when used in the higher temperature environment of tropical countries. The following are the available options for using jatropha oil in diesel engines [9, 20, 48].

Indirect-injection engines: Some indirect-injection diesel engines of older design, such as the Lister single cylinder engines, can use jatropha oil without any problems. These engines, made in India, require no modification other than an appropriate fuel filter. In fact the higher oxygen content of the jatropha oil can deliver greater power under maximum load than diesel. These engines can be run on jatropha oil, biodiesel,

mineral diesel or a blend [5, 20, 45].

Two-tank system: The power unit may be modified to a twotank system. This is effectively a flex-fuel power unit which may run on mineral diesel, any blend of biodiesel or on vegetable oil [45]. The problem of cold starting with the more viscous vegetable oil is avoided by starting and stopping the engine using diesel or biodiesel and then switching tanks to run on the oil when it reaches the critical temperature. Detergents in the mineral diesel prevent the build-up of carbon deposits and gums in the pump and on the fuel injectors. Switching between fuels may be manual or automatic [5, 20].

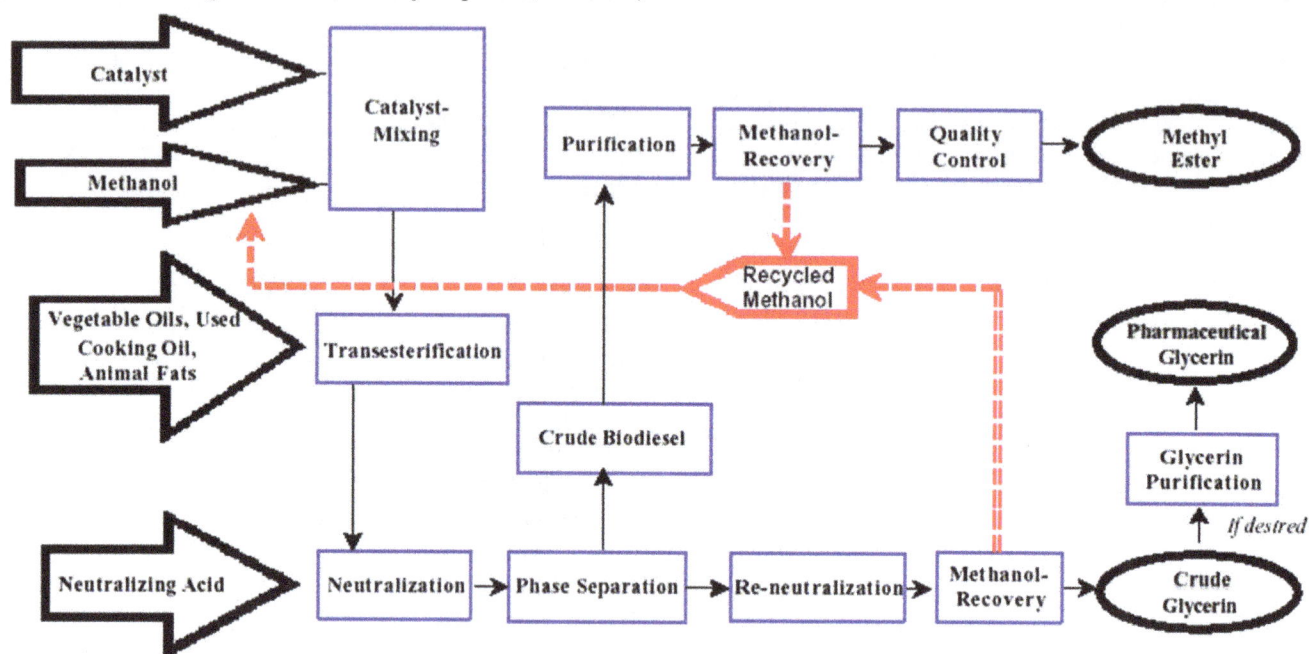

Figure 10. *General Schematic diagram for Production of Bio-diesel from J. curcas and other inputs; Source: [45].*

Single-tank vegetable oil system: A single-tank vegetable oil system uses fuel injectors capable of delivering higher pressures to overcome the high oil viscosity, stronger glow plugs, a fuel pre-heater and a modified fuel filter. A number of manufacturers produce engines that use these single and two-tank technologies. The addition of proprietary organic solvents to the vegetable oil is sometimes recommended to improve engine performance. The long-term viability of these systems in terms of engine performance and reliability remains to be fully assessed. The oil must be of a quality satisfactory for long-term performance of engines run on jatropha oil. Although fresh jatropha oil is low in free fatty acids, it must be stored in closed, dry, cool conditions. The presence of particles and phosphorous in the oil can block filters and cause engine wear [5, 9, 20, 25]. Phosphorous content is lower when the oil is pressed at temperatures less than 60°C. The oil should be well filtered (five microns) to remove contaminants and its water content kept as low as possible to reduce corrosion and wear in the engine, and avoid build up of microbial growth in the fuel delivery system [49]. Jatropha oil has been found adequate for use as a crankcase engine lubricant in Lister-type diesel engines. Crude jatropha

oil is relatively viscous, more so than rapeseed. It is characteristically low in free fatty acids, which improves its storability, though its high unsaturated oleic and linoleic acids make it prone to oxidation in storage. The presence of unsaturated fatty acids (high iodine value) allows it to remain fluid at lower temperatures. Jatropha oil also has a high cetane (ignition quality) rating. The low sulphur content indicates less harmful sulphur dioxide (SO_2) exhaust emissions when the oil is used as a fuel. These characteristics make the oil highly suitable for producing biodiesel [1, 25].

4. Uses of *Jatropha curcas* L

4.1. As Substitute of Fossil Diesel

It was extensively studied that seed oil from the plant is used as an alternative stationary engine or transportation fuel [18]. This is due to its potential to substitute fossil diesel. The calorific value of seed oil is 39MJ kg-1 which is higher than anthracite coal and comparable to crude oil [29]. Raw oil has been used as a substitute for petro-diesel both in modified and unmodified diesel engines. The use of raw oil in diesel engines was also studies in detail which has not shown satisfactory results due to its high viscosity [50]. High viscosity of raw oil causes problems in its use in diesel engines. These include reducing the fuel atomization and increasing fuel spray, which would be responsible for engine deposits, injector coking, piston ring sticking and thickening of lubricating oil [50]. Despite the problems caused by its high viscosity for use in diesel engines, raw oil can have some other energy uses. It has been used in slow-speed stationary diesel engines such as pumps and generators with success [51]. Tests with low heat rejection diesel engine showed that use of JCL oil results in higher brake specific energy consumption lower brake thermal efficiency (BTE), higher exhaust gas temperature and lower NO_x emissions than fossil diesel [52]. The reduction in NO_x emission has environmental benefits. Pre-heating and blending raw oil with fossil diesel are techniques that have also been used to improve the use of raw *J. curcas* L as a fuel [18]. Both techniques have the effect of reducing the viscosity of the seed oil.

4.2. As Source of Biogas

Biogas has been produced from fruit shells. In addition, trials showed that seed husks can be used as a feedstock for a gasification plant [35, 18]. *J. curcas* L fruit shells and seed husks can be used for direct combustion. Since the shells make up around 35–40 percent of the whole fruit by weight and have a calorific value approaching that of fuelwood, they could be a useful by-product of jatropha oil production. As shown in Table 4 the calorific values of *Prosopis juliflora* (a fuelwood species of semi-arid areas) and jatropha fruit shells are similar. However, four times the volume of fruit shells is required to equal the heating value of fuel wood, due to their lower bulk density. The seed cake has a high energy content of 25MJ kg-1. Experiments have shown that some 60 percent more biogas was produced from jatropha seed cake in anaerobic digesters than from cattle dung, and that it had a higher calorific value [20, 27, 35].

Table 4. The calorific values of Prosopis juliflora (a fuelwood species of semi-arid areas) and jatropha fruit shells.

	Wood (Prosopis juliflora)	Biomass Briquettes	Jatropha fruit shell	Jatropha seed husk
Bulk density Kg/m³	407	545	106.18	223.09
Ash content% dm	1.07	8.77	14.88	3.97
Caloric value Kcal/kg	4018	4130	3762	4044

Adopted from: [27]

Seed husks have a higher heating value and greater bulk density which makes them more valuable than the fruit shells as a combustible fuel. However, the technology required to separate the seed husk from the kernel is more suited to large processing plants than small rural industry. The fruit shells can be dried and ground to a powder and formed into fuel briquettes. A trial found that 1 kg of briquettes took around 35 minutes for complete combustion, giving temperatures in the range of 525°C–780°C [22]. The ash left after combustion of jatropha shell briquettes is high in potassium, which may be applied to crops or kitchen gardens. The fruit shells and seed husks also can be left around jatropha trees as mulch and for crop nutrition. For jatropha grown on degraded land, this has clear advantages because nutrient re-cycling – through returning the seed cake to the plantation – is unlikely to happen, due to the effort required and the higher utility to be gained from applying the seed cake to high-value crops [47-49].

4.3. For Varnish Preparation

Oil is used to prepare varnish after calcination with iron oxides. Hardened physic nut oil could be a satisfactory substitute for tallow or hardened rice bran oil. In Europe it is used in wool spinning and textile manufacture. Along with burnt plantain ashes, oil is used in making hard homemade soap [19, 21].

4.4. Human Consumption

Jatropha can be toxic when consumed due to the seeds contain toxic Phorbol esters. However, a non-toxic variety of Jatropha is reported to exist in Mexico and Central America, [53, 54]. This variety is used for human consumption after roasting the seeds/nuts, and "the young leaves may be safely eaten, steamed or stewed." They are favored for cooking with goat meat, said to counteract the peculiar smell. This non-toxic variety of Jatropha could be a potential source of oil for human consumption, and the seed cake can be a good protein source for humans as well as for livestock [55, 56].

4.5. Jatropha curcas L Oil in Household Activities

Traditionally, it is used for the manufacture of candles and soap, as lamp oil and as fuel for cooking. It is a poor lubricant as it dries quickly. Throughout the tropics and warm subtropics *J. curcas* is increasingly planted for bio-fuel purposes. The oil is either used directly in adapted engines powering local grain mills, oil presses, water pumps and small generators, or first refined by trans-esterification with methanol or ethanol to produce regular fuel suitable for high-performance diesel engines [20, 25-26].

4.5.1. Light / Lamp

The oil is put in a small container and a wick is dipped in through the top lid to make a lamp. Alternatively, the jatropha seed can be joined with a thin wire and when the top seed is lit it can slowly burn, producing light. This method is commonly used especially when a household does not have money to buy another alternative light source. The problem of jatropha oil's high viscosity also applies to lamp design. A lamp with a floating wick offers one solution to the oil's poor capillary action. This allows the wick to be kept as short as possible, with the flame just above the oil. The oil lamp requires a very short wick so that the flame is very close to the oil surface. It requires periodic cleaning of the wick to remove carbon deposits. Ordinary kerosene lamps may be modified to lower the wick, but the oil level has to be maintained at a constant level and the wick again needs frequent cleaning. There is anecdotal evidence that using a jatropha oil lamp deters mosquitoes [22-25]. It can also be used as an illuminant in lamps as it burns without emitting too much smoke [54].

Figure 11a. Oil lamp extracted from jatropha oil.

Figure 11b. Special, Kakute stove works by jatropha oil.

4.5.2. Cooking Fuel

Raw oil can also be used as a substitute for kerosene in lamps and cooking stoves. However, these will need to be modified to account for high oil viscosity and low absorbance [51]. There are clear advantages to use plant oil instead of traditional biomass for cooking. These include health benefits from reduced smoke inhalation, and environmental benefits from avoiding the loss of forest cover and lower harmful Green House Gas, GHG, emissions, particularly carbon monoxide and nitrogen oxides. The high viscosity of jatropha oil compared to kerosene presents a problem that necessitates a specially designed stove. In Zimbabwe, Mutoko district, Kakute, an NGO provided out-grower farmers with cooking stoves called 'Kakute stove' that uses jatropha oil and it proved to be beneficial to local people [47, 70].

4.6. Soap Making

Jatropha oil has a very high saponification value and is ideal for soap manufacture. Soap manufactured from non-petroleum sources is gaining increasing popularity especially in the European countries. Jatropha oil has a very high saponification value and is ideal for soap manufacture. Soap manufactured from non-petroleum sources is gaining increasing popularity. Jatropha soap is made by adding a solution sodium hydroxide (caustic soda soda) to jatropha oil. This simple technology has turned soap making into a viable small-scale rural enterprise appropriate to many rural areas of developing countries. The glycerin that is a by-product of biodiesel can be used to make soap, and soap can be produced from Jatropha oil itself. It will produce a soft, durable soap, and the rather simple soap making process is well adapted to household or small-scale industrial activity. In many countries, Jatropha oil is known for its usage for soap making [13, 19-20]. It has been used commercially for soap manufacture for decades, both by large and small soap producers. Soap produced from jatropha is sold as a medical soap, effective in treating skin ailments. In former times Portugal imported Jatropha seeds from Cape Verde Islands to produce soap [54]. In India, Nepal and Zimbabwe the prize of tallow or the prize of Jatropha and other plant oils is at least 2.5 times the selling prize of diesel. Obviously, selling Jatropha oil for soap making is far more profitable in these countries than using it as a diesel or kerosene substitute [68]. It was stated that in GUANTANAMO, Cuba - Farming families in El Oro community, located in the semi-arid region of this eastern Cuban province, have been successful in making soap from the extracted oil from the *Jatropha curcas* tree [12, 19].

4.7. As a Source of Fertilizer

The residue from the biogas digester can be used further as a fertilizer. Where cow dung is used for household fuel, as in India, the seed cake can be combined with cow dung and cellulosic crop residues, such as seed husks, to make fuel briquettes. The seed cake of the plant can be used for biomass feedstock to power electricity plants, or as biogas or high quality organic fertilizer [18, 34, 65].

4.8. Jatropha as Fodder

Jatropha also contains curcasin, arachidic, linoleic, myristic, oleic, palmitic, and stearic acids and curcin. Curcin and phorbol ester aretoxic compounds contained in the Jatropha meal. However, the meal can be suitable for animal feed after a detoxification process [66]. As livestock feed Jatropha seed cake is high in protein – 58.1 percent by weight compared to soy meal's 48% and would be a valuable livestock protein feed supplement if it were not for its toxicity. Currently, removal of toxins is not commercially viable. Using non-toxic varieties from Mexico could make greater use of this potentially valuable by-product, but even these varieties may need treatment to avoid sub-clinical problems that could arise with long-term feeding of jatropha seed cake to livestock [55]. Curcin and phorbol ester aretoxic compounds contained in the

Jatropha meal. However, the meal can be suitable for animal feed after a detoxification process [66]. Due to the fact that the seeds of *J. curcas* contains some toxins compounds such as those reported by [8] a protein (curcin) and phorbol-esters (diterpenoids), the detoxification or complete removal of phorbol esters is found to be important before its use in industrial or medicinal applications [71]. The major toxin phorbolester is not vulnerable to heat, but can be hydrolyzed to less toxic substances extractable by either water or ethanol [69]. Heat treatment, alkaline hydrolysis and solvent extraction for the detoxification of the *Jatropha curcas* seed cake [56]. Despite the toxicity of *J. curcas*, edible varieties are also known to exist in Mexico which are not currently being exploited [8].

4.9. Medicinal Properties of Jatropha Plant

Jatropha species are used in traditional medicine to cure various ailments in Africa, Asia, and Latin America or as ornamental plants and energy crops [3]. Several known species from genus Jatropha have been reported for their medicinal uses, chemical constituents, and biological activities such as *Jatropha curcas*, *Jatropha elliptica*, *Jatropha gossypiifolia*, and *Jatropha mollissima*, among others [58]. Although the leaves are toxic when consumed, the green pigment that comes out of the leaves and the latex that comes from the stem can be used to stop bleeding wounds on both humans and livestock. Apart from being used as a live fence, the plant is used as a repellant agent. Some people believe that jatropha protects the home from evil spirits and snakes. In Mutoko, Zimbabwe, witchcraft is a common social phenomenon hence jatropha is believed to have the power to repel witches and bad omens [5].

The latex of Jatropha contains jatrophine, an alkaloid which is believed to have anti-cancerous properties. It is also used as an external applicant for skin diseases, rheumatism, livestock sores, piles and as an antidote for certain snake-bites The dark blue dye extracted from the bark of Jatropha is a useful dye. Jatropha oil cake is rich in nitrogen, phosphorous and potassium and can be used as organic manure. Jatropha leaves are used as food for the tusser silkworm. The seeds are considered anti helmintic in Brazil, and the leaves are used for fumigating houses against bed bugs. In addition, the ether extract shows antibacterial properties against Staphylococcus aureus and Escherichia coli [13]. Medically it is used for diseases like cancer, piles, snakebite, paralysis, dropsy etc [1]. Protein, constituting 18.2% of the mass of Jatropha curcas seed, is mainly made up of curcin; curcin toxicity is similar with that of ricin in Ricinuscommunis seeds and crotin in *Croton tiglium* seeds [59]. Curcin is also named *Jatropha curcas* ribosome inactivating protein (RIP), is a RNA n-glycosidase that can cause inactivation of eukaryotic ribosomes and inhibition of protein syntheses [60]. It can effectively inhibit the in vitro proliferation of human gastric cancer cells (SGC-7901), murine myeloma cells (Sp20), and human hepatoma cells [61]. The full-length cDNA sequence (GenBank Accession Number AY069946) and gene sequence (GenBank Accession Number AF469003) of curcin have been cloned in *Escherichia coli* cells

and expressed into bioactive mature proteins with conserved protein domains [62]. Other researcher screened for two kinds of protein from Jatropha curas seeds. One of them is a 28kD anticancer protein, which suppresses the activity of ribosome and is thereby curcin [20, 63].

The roots, bark, leaves and seeds of *J. curcas* are used for medicinal purposes. *J. curcas* seeds are characterized by its anti inflammatory, and anti-swelling effects. *J. curcas* seed oil can be used as a laxative, and is also widely used for treating various skin diseases, and for pain relief, including rheumatic diseases. Overdose of *J. curcas* seed oil could lead to diarrhea and gastroenteritis. *J. curcas* seeds have been found to exhibit significant anti-cancer activity. The seeds also possess value for use as an industrial oil, prevention of pests and plant diseases; it is also shows promising pharmaceutical value. J. curcas is also an excellent species for domestication in hot and barren mountainous lands at river valley areas. With advances in basic research and further industrialization of J. curcas domestication, it is expected that J. curcas will hold an exceptional economical value to China [60-63].

The latex has a widespread reputation for healing wounds, as a haemostatic and for curing skin problems; it is applied externally to treat infected wounds, ulcers, ringworm, eczema, dermatomycosis, scabies and sarcoptic mange in sheep and goats. The latex is discharged from the bark when the plant is cut and it is white and watery. The latex of Jatropha contains jatrophine, an alkaloid which is believed to have anti-cancerous properties. It is also used as an external applicant for skin diseases, rheumatism, livestock sores, piles and as an antidote for certain snake-bites Upon drying, the initially viscous latex forms an airtight film, resembling that produced by collodion. The latex has a styptic effect and is used against pains and stings of bees and wasps. Dried and pulverized root bark is made into poultices and is taken internally to expel worms and to treat jaundice [1, 5, 20]. Leaves are also applied on wounds and in decoction they are used against malaria in Mali and Madagascar, while in Benin and Réunion a decoction is taken against hypertension. The leaf sap is used externally to treat haemorrhoids in Benin and Madagascar. In Guinea Bissau a hot water extract of the leaves is taken orally to accelerate secretion of milk in women after childbirth. Fresh stems are used as chew sticks to strengthen the gums, and to cure bleeding, spongy gums or gum boils. A decoction of the roots is a cure for diarrhoea and gonorrhoea. In Madagascar a decoction of the leaves and roots is taken to treat malaria. *Jatropha curcas* is also used in the preparation of arrow poison and in the Philippines the bark is used to prepare a fish poison. The seeds are often a source of accidental poisoning, both in animals and humans [1, 21].

Leaf sap yields a black dye or ink that is said to be indelible; the bark yields a dark blue dye, which, however, is not fast. *Jatropha curcas* is widely cultivated in the tropics as a living fence, for erosion control, demarcation of boundaries and for protection of homesteads, gardens and fields against browsing animals. In Madagascar and elsewhere in Africa it serves as a support for vanilla, black pepper and yams. The seed cake left after oil extraction is too toxic to be used as animal feed, but constitutes a valuable organic fertilizer rich in nitrogen. Some accessions of *Jatropha curcas* found e.g. in Mexico are almost free of toxins and the seed cake from such selections would provide a nutritious feedstock on account of the high protein content. Their seeds are sometimes boiled or roasted and eaten as a snack and young leaves as a vegetable. Jatropha oil has molluscicidal properties against the vector snails of the *Schistosoma* parasite that causes bilharzia. The emulsified oil has been found to be an effective insecticide against weevil pests and house flies, and an oil extract has been found to control cotton bollworm and sorghum stem borers [13, 31].

Table 5. Parts of J. curcas L. to be used.

Parts of *J. curcas* L. to be used		Uses	References
Fruits hulls		-Combustibles	[25]
		- Green Manure	[21]
		-Biogas production	[18,34,65]
Seeds	Seed oil	Soap production	[19]
		Fuel	[26]
		Insecticide	[2, 18]
		Medicinal uses	[64]
	Seed cake	Fertilizer	[18, 34, 65]
		Biogas production	[18, 34-35]
		Fodder (from non toxic varieties)	[66]
	Seed shell	Combustibles like fruit hulls	[25]
		To develop Eri Silkworm	[13]
Leaves		Medicinal uses	[21]
		Anti-inflammatory substance	[19]
Latex		Contains wound healing protease (Curcain)	[21]
		Medicinal Uses	[1]

The use of oil extracts as an insecticide, molluscicide, fungicide and nematicide. These potential uses have yet to be commercialized. As previously mentioned, the oil is widely used as a purgative in traditional medicine. Extracts from grounded *J. curcas* seeds exhibit insecticidal activity in snails carrying Schistosoma worms [18]. Jatropha leaves are used as food for the tusser silkworm. The seeds are considered anti helmintic in Brazil, and the leaves are used for fumigating houses against bed bugs. The ether extract shows antibacterial properties against Staphylococcus aureus and Escherichia coli. Curcin, seed oil and ethanol extracts from J. curcas seeds and investigated the insecticidal activity against Lipaphis erysimi [62-64]. The results showed that curcin had no significant insecticidal effects on *L. erysimi*; however, *J. curcas* seed oil showed high toxicity, and ethanol extracts showed a higher pesticidal activity than seed oil. The authors found that 2.02 g/L of ethanol extracts exhibited significant biological control of Lipaphis erysimi; biocontrol efficacy was as high as 72.11% after 7 days of treatment. The effects of curcin was tested as fungicidal with Magnaporthe grisea, Pestalotia funerea, Rhizoctonia solani and Sclerotinia sclerotiorum [67]. The results showed that 5 ug/mL can significantly inhibit the growth of fungal mycelium and spore germination. When the concentration was increased to 50 ug/mL, there was no significant M. grisea spore formation and the inhibition rate of P. funerea reached as high as 83.8%. Following treatment, mycelial cells showed evident

shrinkage and deformation under light microscope; SDS-PAGE analysis further confirmed decreased protein band intensity in mycelial cells. The results demonstrated that inhibition of both mycelial growth and sporulation may be due to an inhibitory effect of curcin on protein synthesis. It was reported that by providing physical barriers, jatropha can control grazing and demarcate property boundaries while at the same time improving water retention and soil conditions [2, 18, 25].

5. Conclusion

The objective of this review paper focuses some basic aspect of the taxonomic, biology, cultivation, chemical composition, bio-diesel potential, medicinal values and uses of Jatropha curcas Linn. The entire *J. curcas* plant can be used for medicine. It can also be used to refine bio-diesel. These benefits cannot be overlooked in the growing need for clean and efficient biofuels. *J. curcas* are also semi-succulent plants, shedding leafs during prolonged arid conditions, and can easily adapt to arid- and semi-arid ecological systems. They will not compete with crops for arable land, since their requirements for soil nutrition are very low. Taken together, cultivation of Jatropha curcas, particularly in areas that suffer severe soil and water loss, has ecological significance for both soil and water conservation. The ecological effect of large-scale cultivation of *J. curcas* on natural requires evaluation. *J. curcas* seeds exhibit toxicity and can be used to develop new lead compounds. The probability of developing new medical compounds with high efficiency and low toxicity directly from plants is low. Actions on *J. curcas* seeds to remove its toxicity should allow drug discovery of its active constituents. J. curcas seeds possess significant economic value, but research on its chemical composition, bio-diesel potentiality, and study on its diverse uses remains rare. Further reseach studies that can improve the production, oil content yield, chemical composition, combustibility, bio-diesel potentiality and decrease the toxicity of the plant will provide a clearer understanding of its practical benefits.

Acknowledgement

I would like to acknowledge Madda Walabu University for providing related reading materials in the library and regular availability of internet connection in the campus.

References

[1]   Nahar, K. and Ozores-Hampton, M. (2011). Jatropha: An Alternative Substitute to Fossil Fuel. (IFAS Publication Number HS1193). Gainesville: University of Florida, Institute of Food and Agricultural Sciences.

[2]   Srinophakun P., Saimaneerat A., Sooksathan I., Visarathanon N., Malaipan S., Charernsom K. Chongrattanameteekul W., 2011. Integrated Research on *J. curcas* Plantation Management. World Renewable energy Congrence, Linkoping, Sweden.

[3] Grover A., Kumari M., Singh S., Rathode S. S., Gupta S. M., Pandey P., Gilotra S., Kumar D., Arif M., Ahmed Z., 2013. Analysis of Jatropha curcas transcriptome for oil enhancement and genic markers. Physiol Mol Biol Plants (January–March 2014) 20(1):139–142 DOI 10.1007/s12298-013-0204-4 (Short Communication).

[4] Jongschaap, R. E. E., Corr´ e, W., Bindraban, P. S. and Brandenburg, W. A.; 2007. Claims and Facts on Jatropha curcas L., Global Jatropha curcas evaluation, breeding and propagation programme; Report 158, 42 pages, Plant Research International Wageningen UR.

[5] Mubonderi J., 2012. Jatropha: the broom of poverty; myth or reality? A critical analysis of the Zimbabwean jatropha programme in Mutoko district.

[6] Blench, R. 2003. Hausa names for plants and trees. Available at http://www.org/odi/staff/r. Printout February 27, 2016

[7] Blench, R. 2007. Hausa names for plants and trees. http://www.rogerblench.info/RBOP.htm Printout March 5, 2016 Accessed at http://www.rogerblench.info/Ethnoscience%20data/Hausa%20plant%20names.pdf

[8] King AJ, He W, Cuevas JA, Freudenberger, Ramiaramanana D and Graham IA. 2009. Potential of Jatropha curcas as a source of renewable oil and animal feed: Review paper. Journal of Experimental Botany.10.1093/jxb/erp025.

[9] Grass M., 2009. Jatropha curcas L.: Visions and Realities. Journal of Agriculture and Rural Developmen in the tropics and Subtropics Volume 110, No. 1, 2009, pages 29–38.

[10] Webster G. L., 1994. "Classification of the euphorbiaceae," Annals of the Missouri Botanical Garden, vol. 81, pp. 3–143.

[11] Alves M. V., 1998. "Checklistdasesp´ecies de Euphorbiaceae Juss. ocorrentes no semi´arido pernambucano, Brasil," Acta Botancia Brasileira, vol. 12, no. 3, pp. 485–495.

[12] Henning, R. K. (2009) The Jatropha System An integrated approach of rural development. Rothkreuz 11, D-88138 Weissensberg, Germany.

[13] Lozano J. A. D., 2007. Botanical Characteristics of Jatropha curcas L. Jatropha Project in Mexico.

[14] DEGJSP, 2012. 'Farmers Handbook: advice on growing Jatropha curcas in East Africa'. DEG Jatropha support Program (DEGJSP). June 2012. Pipal Ltd. Nairobi.

[15] Oliveira H. de, A. C. P. Juhász, S. Pimenta, B. O. Soares, Batista Morais de Lourdes, D., Rabello, 2009. Floral biology and artificial polinization in physic nut in the north of Minas Gerais state, Brazil [Biologia floral e polinização artificial de pinhão-manso no norte de Minas Gerais] Pesquisa Agropecuaria Brasileira, 44(9): 1073–1077 (in Portuguese).

[16] DNA Res., 1994. "Sequence Analysis of the Genome of an Oil-Bearing Tree, Jatropha curcas L".Oxford Journals, DNA Research, Kazusa DNA Research Institute. 2010-12-08. Retrieved 2016-02-23.

[17] Dahmer, N.; N. and Wittmann, M. T. S. and Dias, L. A. dos S. (2009) Chromosome numbers of Jatropha curcas L.: an important agrofuel plant. Crop Breeding and Applied Biotechnology, 9 (4). pp. 386-389.

[18] Achten, W. M. J., Verchot, L., Franken, Y. J., Mathijs, E., Singh, V. P., Aerts, R., and Muys, B. (2008). Jatropha Bio-diesel Production and Use. Biomass and Bioenergy, 32, 1063-1084.

[19] Warra A. A., 2012. Cosmetic potentials of physic nut (Jatropha curcas Linn.) seed oil: A review. Department of Biochemistry, Kebbi State University of Science and Technology, American Journal of Scientific and Industrial Researc Science, http://www.scihub.org/AJSIR   ISSN: 2153-649X, doi:10.5251/ajsir.2012.3.6.358.366.

[20] Arif M. and Ahimed Z., 2009. Biodiesel, J. curcasL. (a promising Source). Govt. of India, Ministry of Defense.

[21] Orwa C, A Mutua, Kindt R, Jamnadass R, S Anthony. 2009 Agroforestree Database:a tree reference and selection guide version 4.0 (http://www.worldagroforestry.org/sites/treedbs/treedatabases.asp)

[22] Singh, Neelu & Kumar, Sushil, 2008. Anti termite activity of Jatropha curcas Linn. biochemicals. Journal of Applied Sciences and Environmental Management, Vol. 12, No. 3, 2008, pp. 67-69.

[23] Jingura, R. M., Matengaifa, R., Musademba, D., and Musiyiwa, K. (2011) Characterization of land types and agro-ecological conditions for production of Jatropha as a feedstock for biofuels in Zimbabwe. Biomass and Bioenergy, Vol. 35, pp. 2080-2086.

[24] Prakash G. K. 2006. Department of Forestry, Indira Gandhi Agricultural University Raipur (C. G.)M.Sc. Forestry Thesis "Vegetative propagation of Jatropha, Karanj and Mahua by Stem cuttings, Grafting, Budding and Air layering".

[25] Brittaine R., 2010. Jatropha: A Smallholder Bioenergy Crop. FAO, Food and Agriculture Organization of the United Nations. The Potential for Pro-Poor Develop- ment. Rome, Italy: Integrated Crop Management Vol. 8.

[26] Satish Lele, 2007. Developmentof the Jatropha cultivation and bio-fuel production system. [Internet] http://www.svlele.com/jatropha_plant.htm.

[27] Abreu, F., 2009, Alternative by-products from Jatropha. Download, http://www.ifad.org/events/jatropha/harvest/f

[28] McKendry, P., 2002, Energy production from biomass (part 1): overview of biomass. Bioresource Technology, Vol. 83, pp. 37-46.

[29] Sotolongo, J. A., Beatón, P., Diaz, A., de Oca, S. M., del Valle, Y., Pavón, S. G. and Zanzi, R, 2009. Jatropha curcas L. as a source for the production of biodiesel: A Cuban experience, Download http://hem.fyristorg.com/zanzi/paper/W2257.pdf.

[30] Manurung, R., Wever, D. A. Z., Wildschut, J., Venderbosch, R.H., Hidayat, H., van Dam, J. E. G., Leijenhorst, E. J., Broekhuis, A. A. and Heeres, H. J., 2009, Valorisation of J. curcas L. parts: Nut shell conversion to fast pyrolysis oil. Food andBioproducts Processing, Vol. 87, pp. 187-196.

[31] Gubitz, G. M., Mittelbach, M., Trabi, 1999. Exploration of tropical oiloil seed plant J. curcas L. Bioresource technology 67, 73-82.

[32] Deng X, Fang Z, Liu YH, 2010. Ultrasonic transesterification of Jatropha curcas L. oil to biodiesel by a two-step process. Energy Conversion Management.

[33] Vyas, D. K., Singh, R. N., Srivastava, N. S. L., and Narra M., 2009. SPERI experience on holistic approach to utilize all parts of Jatropha curcas fruit for energy. Renewable Energy, Vol. 33, pp. 1868-1873.

[34] Patolia, J. S., A. Ghosh., J. Chikara, D. R. Chaudharry, D. R. Parmar, and H. M. Bhuva. 2007. "Response of Jatropha curcas L. Grown on Wasteland to N and P Fertilization." Paper presented at the FACT Seminar on Jatropha curcas L. Agronomy and Genetics, March 26–28, Wageningen. Article No. 34.

[35] Staubmann, R., Foidl, G., Foidl, N., Gübitz, G. M., Lafferty, R. M., Arbizu, V. M. V. and Steiner, W, 1997. Biogas production from *Jatropha curcas* press-cake. *Applied Biochemistry and Biotechnology*, Vol. 63-65, pp. 457-467.

[36] Pandey, V. C., Singh, K., Singh, J. S., Kumar, A., Singh, B., and Sing, R. P. (2012). Jatropha Curcas: A Potential Biofuel Plant for Sustainable Environmental Development.

[37] Pradhan, R. C., Naik, S. N., Bhatnagar, N., and Vijay, V. K. (2010). Design, Development and Testing of Hand-operated Decorticator for Jatropha Fruit. *Applied Energy*, 87, 762-768.

[38] Silitonga, A. S., Atabani, A. E., Mahlia, T. M. I., Masjuki, H. H., Irfan, A. B., and Mekhilef, S. (2011). A Review on Prospect of Jatropha Curcas for Biodiesel in Indonesia. *Renewable and Sustainable Energy Reviews*, 15, 3733-3756.

[39] Lim, B. Y., Shamsudin, Baharudin B. T. H., R., Yunus, R., 2014. The Performance of a Jatropha Fruit Shelling Machine and the Future Improvement. Universal Journal of Applied Science 2(7): 233-237.

[40] Akbar E, Yaakob Z., Kamarudin SK, Ismail M, and Salimon J. 2009. Characteristic and Composition of *J. Curcas* Oil seed from Malaysia and its Potential as Biodiesel Feedstock. European Journal of Scientific Research. Vol. 29 No.3 pp 396-403.

[41] Adebowale KO and Adedire CO. 2006. Chemicalcomposition and insecticidal properties of the underutilized *Jatropha curcas* seed oil. *African Journal of Biotechnology* Vol 5 No.10 pp 901-906.

[42] Morton JF (1981). Atlas of medicinal plants of middle America: Bahamas to Yucatan. Charles C. Thomas, Springfield, USA; 1420.

[43] Ling-yi K., Zhi-da M., Jian-xia Sh. and Rui F., 2015. Chemical Constituents from Roots of *J. curcas*. The Institute of Botany, the Chinese Academy of Sciences Acta Botanica Sinica Volume 38 Issue 2.

[44] Ibeto, C. N., A. U. Ofoefule, and H. C. Ezugwu. 2011. "Analytical Methods for Quality Assessment of Biodiesel from Animal and Vegetable Oils." Trends in Applied Sciences Research 6: 537–553.

[45] Folaranmi J., 2013. Production of Biodiesel (B100) from *J. curcas* oil Using Sodium Hydroxide as Catalyst. Hindawi Publishing Corporation Journal of Petroleum Engineering.

[46] Berchmans H. and Hirat S., 2008. Biodiesel production from crude Jatropha curcas L. seed oil with a high content of free fatty acids. Science Direct; Bioresource Technology 99 (2008) 1716–1721. Available online at www.sciencedirect.com.

[47] Van Gerpen, J. (2005). Biodiesel processing and production: biodiesel processing and production. Fuel Processing Technology 86, 1097-1107.

[48] Francis, G., Edinger, R., Becker, K. (2005). A concept for simultaneous waste land reclamation, fuel production, and socio-economic development in degraded areas in India: need,

potential and perspectives of Jatropha plantations. Natural Resources Forum 29, 12-24.

[49] De Jongh, J. & Adriaans, T., 2007. 'Jatropha oil quality related to use in diesel engines and refining methods', Technical report, Arrakis, Ingenia.

[50] Shahid, E. M. and Jamal, Y., 2008. A review of biodiesel as a vehicular fuel. Renewable and Sustainable Energy Reviews, Vol. 2, No. 9, pp. 2484-94.

[51] Tomomatsu Y. and Brent S., 2007. Jatropha curcas biodiesel production in Kenya Economics and potential value chain development for smallholder farmers.

[52] Prasad D. M. R., Izam A. and Md, 2012. Plant of medical benefits Journal of Medicinal Plants Research Vol. 6(14), pp. 2691-2699.

[53] Makkar HPS, Becker K, Sporer F, Wink M (1998). Studies on nutritive potential and toxic constituents of different provenances of Jatropha curcas. J. Agric. Food Chem., 45: 3152-3157.

[54] Misra M. and Misra N. A., 2010 Jatropha: The Biodiesel Plant Biology, Tissue Culture and Genetic Transformation – A Review. Int. J. Pure Appl. Sci. Technol., 1(1) (2010), pp. 11-24. ISSN 2229 – 6107. Available online at www.ijopaasat.in.

[55] Makkar, H. P. S., Becker, K. (2009). *Jatropha curcas*, a promising crop for the generation of biodiesel and value-added coproducts. European Journal of Lipid Science and Technology 111, 773-787.

[56] Aregheore, E. M., Becker, K., Makkar, H. P. S. (2003). Detoxification of a toxic variety of *Jatropha curcas* using heat and chemical treatments, and preliminary nutritional evaluation with rats. South Pacific Journal of Natural Science 21, 50-56.

[57] Wani, S., Sreedevi, T. K. & Marimuthu, S. 2008. Pro-poor biodiesel initiative for rehabilitating degraded drylands. In: *International Consultation on Pro-poor Jatropha Development*. 10–11 April 2008, Rome, IFAD (available at http://www.ifad.org/events/jatropha/).

[58] Sabandar C. W., Ahmat N., Jaafar F. M., and Sahidin I., 2013. Medicinal property, phytochemistry and pharmacology of several Jatropha species (Euphorbiaceae): a review, Phytochemistry, vol. 85, pp. 7–29, 2013.

[59] Stirpe F, Pession-Brizzi A, Lorenzoni E, 1976. Studies on the proteins from the seeds of Croton tiglium and Jatropha curcas [J]. Biochem J, 156: 1—6.

[60] Lin J, Yan F, Tang L, 2002. Isolation, purification and functional investigation on the N-glycosidase activity of curcin from the seeds of Jatropha curcas [J]. High Techn Lett, 12 (11): 36—40.

[61] Lin J, Yan F, Tang L, 2003a. Anti-tumor effects of curcin from seeds of *J. curcas*. Acta Pharmacol Sin, 24: 241—246.

[62] Lin J, Chen Y, Xu Y, 2003b. Cloning and expression of curcin, a ribosome-inactivationg protein from the seeds of *J. curcas*. Acta Bot Sin, 45: 858—863.

[63] Chen Y, Wei Q, Tang L, 2003. Proteins in vegetative organs and seeds of Jatropha curcas L. and those induced by water and temperature stress [J]. Chin J Oil Crop Sci, 25: 98—104.

[64] Li J., Yan F, Wu FH, 2004. Insecticidalactivity of extracts from Jatropha curcas seed against Lipaphiserysimi [J]. Acta Phytophyla Sin, 31: 289—293.

[65] Ghosh, A., D. R. Chaudhary, M. P. Reddy, S. N. Rao, J. Chikara, and J. B. Pandya. 2007. "Prospects for Jatropha Methyl Ester (Biodiesel) in India." Int. J Environ. Stud. 64: 659–674.

[66] Gaur, S., B. Viswanathan, R. Seemamahannop, S. Kapila, and N. L. Book. 2011. "Development and Evaluation of an Efficient Process for the Detoxification of Jatropha Oil and Meal."

[67] Wei Q, Liao Y, Zhou LJ, 2004. Antifungal activity of curcin from seeds of J. curcas. Chin J Oil Crop Sci, 26: 71—75.

[68] Openshaw K. 2000. A review of *J. curcas*: an oil plant of unfulfilled promise. *Biomass and bioenergy* Vol. 19 pp1-15.

[69] Usman LA, Ameen OM, Lawal A and Awolola GV. 2009. Effect of alkaline hydrolysis on the quantity of extractable protein fractions (prolamin, albumin,globulin and glutelin) in J. curcas seed cake. African Journal of Biotechnology. Vol. 8.

[70] Marjolein and Romijn, 2010. The Jatropha Biofuels Sector in Tanzania 2005-9: Evolution Towards Sustainability? Eindhoven Centre for Innovation Studies (ECIS), School of Innovation Sciences, Eindhoven University of Technology, The Netherlands.

[71] Goel, G., Makkar, H. P. S., Francis, G. and Becker, K. 2007. Phorbol esters: structure, occurrence and biological activity. Int. J. Toxicol. 26, 279–288.

[72] http://www.itis.gov/servlet/SingleRpt/SingleRpt?search_topic =TSN&search_value=28335 *Jatropha curcas* L. Taxonomic Serial No.: 28335. Accessed on 23rd March, 2016.

19

Effect of Nitrogen and Phosphorus Fertilizer Rates on Yield and Yield Components of Barley (*Hordeum Vugarae L.*) Varieties at Damot Gale District, Wolaita Zone, Ethiopia

Mesfin Kassa[1], Zemach Sorsa[2]

[1]Soil science, Wolaita Sodo University, College of Agriculture, Wolaita Sodo, Ethiopia
[2]Plant Breeding, Wolaita Sodo University, College of Agriculture, Wolaita Sodo, Ethiopia

Email address:

mesfine2004@gmail.com (M. Kassa), zemachsorsa@yahoo.com (Z. Sorsa)

Abstract: A field experiment was conducted at Damot Gale District, Wolaita Zone, SNNPRS to evaluate the response of barely varieties to nitrogen and phosphorus fertilizer application since the response varies from location to location due to several factors. Thus, there is a need to determine specific nitrogen and phosphorus fertilizer requirement of specific variety. The barley varieties (HB1370 and Shage) were used as test crop and the experiment contained factorial combination of four levels of N/P (0/0, 23/10, 46/20, 69/30 kg ha^{-1}) and was laid out in randomized complete block design with three replications. The results from this study indicated that nitrogen and phosphorus fertilizer showed no significant effect on number of days to heading while number of fertile tillers, total biomass and yield were significantly increased by application of nitrogen and phosphorus. However, the effects of nitrogen and phosphorus were significant ($P < 0.05$) on plant height, spike length, number of seeds per spike and grain yield. In general, grain yield tended to be higher under NP 69/30 kg ha^{-1} treatment (2.02t/ha). In contrast, the lowest grain yield (0.86t/ha) was obtained from 0/0 NP treatment, although the interaction effects of nitrogen and phosphorus were significant on treatments with varieties and balanced amount of nitrogen and phosphorus. The future studies should articulate towards the studies involving more varieties, multi-location and additional rates of nitrogen and phosphorus applications, under diverse management practices such as research and farmer's field's conditions, which may facilitate fine-tuning of fertilizer recommendations.

Keywords: Fertilizer, Phenology, Growth, Yield

1. Introduction

Barley (Hordeum vulgarae L.) is cultivated in every region of Ethiopia implicating its wide ecological plasticity and physiological amplitude (Lakew *et al.*, 1996). Barley ranks fourth in worldwide production of all cereals (FAO, 2004). In Ethiopia, barley is ranked fifth of all cereals, based on area of production, but third based on yield per unit area (CSA, 2004).Within Ethiopia, the highest levels of barley consumption occur in highlands where it is widely cultivated, accounting for the bulk of the total crop harvest (Kiros, 1993). In Ethiopia, barley covers 10% of the land under crop cultivation with a yield of 1.2t ha^{-1} (CACC, 2002), where as the potential yield goes up to 6 t/ha on experimental plots (Berhane Lakew, Habtamu *et al.*, 2014; Hailu Gebre and

Fekadu Alemayehu, 1996). Several a biotic and biotic factors have contributed to this low productivity, such as poor crop management practices; the use of low yielding cultivars; the limited availability of the very few improved cultivars released; weeds, insects and diseases; and the inherently low yield potential of the prevalent local varieties (Asfaw Negassa *et al.*, 1997; Hailu Gebre and Fekadu Alemayehu, 1996; ChilotYirga, Fekadu Alemayehu and Woldeyesus Sinebo, 1998; Woldeyesus Sinebo andChilot Yirga, 2002).

This low productivity is mainly due to traditional methods of production and poor soil fertility. Poor soil fertility and low pH are among the most important constraints that threaten barley production in Ethiopia. Since the major barley producing areas of the country are mainly located in the highlands, severe soil erosion and lack of appropriate soil conservation practices in the past have resulted in soils with low fertility and pH (Grando and McPherson, 2005). Particularly deficiency of nitrogen and phosphorus is the main factor that severely reduces the yield of barely. According to Desta Beyene (1987), although soil fertility status is dynamic and variable from locality to locality, and it is difficult to end up with a blanket recommendation invariably, some soil amendment studies were undertaken at different times and places. Even though several researches have been conducted on high land areas of Ethiopia, like Bale, Wello, North Shewa and some parts of Arsi region have bimodal pattern of rainfall permitting production of barley twice a year both on summer and winter seasons (Alemayehu, 1994). The soils that are mostly under cereal cultivation in the study area are Nitisols and Vertisols (Solomon, 2006). Barley yields are increased also by phosphorus and potash fertilizers on some soils (Bulman and Smith, 1993). In addition to fertilizer rates, soil acidity also affects the productivity of the land by affecting availability of nutrients and hindering the activity of microorganisms. Also nitrogen is commonly the most limiting nutrient for crop production in the major world's agricultural areas and therefore adoption of good N management strategies often result in large economic benefits to farmers. Among the plant nutrients, nitrogen plays a very important role in crop productivity (Zapata & Cleenput, 1986; Ahmad, 1999; Miao et al., 2006; Oikeh et al., 2007; Assefa and Chekole, 2015, Worku et al., 2007). Soil fertility is one of the major production constraints in the Southern region of Ethiopia (Getahun and Tenaw, 1990). The various factors accounting for the poor soil fertility include topography, soil erosion, deforestation, population pressure, and continuous cultivation without proper soil fertility maintenance (Tenaw, 1996). This area under barley cultivation is about 123000 ha and production is estimated to be about 150000 t, with a mean yield of 1.2 t/ha in the main cropping season (MOA, 2000). In the Wolaita zone, where the experiment was conducted, barley is the fourth most important cereal after maize, sorghum and tef in area coverage (BOA, 1998). Despite its importance and cultivation of the yield and productivity of barley in the country, Region and in the Wolaita zone have low yields compared with its potential. Among the major constraints on increased production of barley are poor soil fertility, limited supply of production inputs, low prices for the produce, and undeveloped markets (Hailu Gebre, 1978). Furthermore, phosphorus has a high nutritional value, with high concentrations of nitrogen, and potassium, while the contamination by heavy metals and other toxic substances are very low (Asghar et al,. 2006).There was no research conducted concerning fertilizers rates as a result of this fact, the farmers rely on traditional practices and local cultivars. Most of the farmers in experiment site do not use NP combination above the recommended rate. Therefore, there is a need to study the effect of different NP rates on the yield and yield components of barely with the following specific objectives:

- To examine the effect of NP fertilizer on yield and yield components of barley varieties.
- To evaluate barley varieties in the study area

2. Materials and Methods

2.1. Description of the Study Site

This study was carried out at Damote Gale, Wolaita zone, which is found in the Southern Nations Nationalities and Peoples Regional State. It is located 375 km south of Addis Ababa. The study area is located on latitude of 6o 55 ' 36.1" N and longitude of 370 50'12.1" E, with an elevation of 2110 m.a.s.l.

2.2. Treatments and Experimental Design

The treatments consisted of combination of four levels of phosphorus and nitrogen (0/0, 10/23, 20/46 and 30/69 kgha-1) and barely varieties (HB-1307, and Shage).The twenty four treatment combinations were replicated three times in factorial RCBD design. A plot size of $3m^2$ with 20 cm spacing between rows and spacing of 1m between blocks and 0.5 m between plots were used. Nitrogen was applied in the form of urea, while P was applied in the form of triple supper phosphate. Nitrogen was applied in split, half during sowing and half at booting stage.

2.3. Data Collection and Analysis

The data were collected as Days to heading, plant height, spike length, number of seeds per spike, number of tiller per plots, number of non fertile tillers per plots, thousand seed weight, and total above ground biomass.

2.4. Soil Sampling and Analysis

For soil analysis, before planting soil samples were randomly taken from the experimental site at a depth of 30cm using an auger and the samples were mixed thoroughly to produce one representative composite sample of 1kg.The soil samples were air-dried and ground to pass 2 and 0.5 mm (for total N) sieves. All samples were analyzed following standard laboratory procedures as outlined by Sahlemedhin and Taye (2000). Organic carbon and total N contents of the soil were determined following the wet combustion method of Walkley and Black, and wet digestion procedure of Kjeldahl method, respectively. The available P content of the soil was determined following Olsen method. Soil texture was analyzed by Bouyoucos hydrometer method. The cations exchange capacity (CEC) of the soil was determined following the 1N ammonium acetate (pH 7) method. Ca and Mg contents were measured by using EDTA titration, the pH (1:2.5 solid: liquid ratio) of the soils was measured in water using pH meter with glass-calomel combination electrode.

2.5. Statistical Analysis

The data collected on different parameters were statistically analyzed using PROC ANOVA function of SAS program. After performing ANOVA the differences between the treatment means were compared by LSD test at 5% level of significance.

3. Result and Discussion

3.1. Physical and Chemical Properties of the Soil

Soil analysis before sowing showed that soil pH, available P, CEC and Total N found in the range of slightly acidic based on Herrera (2005) classification, however a result of some soil physical and chemical properties of the experimental site were showed that medium. The soil analysis result indicated that texture of the soil was clay loam and pH of the soil was 6.0 slightly acidic, this range is suitable for different crops have different requirements but the optimum pH range for barley is 6.0 - 7.0. Therefore, the pH of soil is suitable for barley (CLDB, 2001). Other soil chemical properties were total Nitrogen (TN %) is rated by Havlin et al. (1999) as very low (<0.1), low (0.1 to 0.15), medium (0.15 to 0.25), and high (> 0.25). According to Olsen et al. (1954) rating, P (mg Kg-1) content is: (< 3) very low, (4 to 7) low, (8 to 11) medium, (>11) high. Then available P at the site was within medium and low respectively. It might be because of this that some growth and yield components responded better to P than to N. In line with this M. Osundwa et al. (2013) reported liming of acidic soils reduces the P sorption thus increasing its availabilities. However, over all residual soil P of the treatments were categorized under very low status

3.2. Effects of Nitrogen and Phosphorus Fertilizer Rates on Plant Growth and Yield Parameter

Plant height was significantly affected by different rates of NP application (Table 1.) In line with this, Rashid et al (2007) indicated that plant height was linearly increased with increasing levels of NP fertilization. The maximum plant height (86.02cm) was recorded from application of 69/30 kg NP ha-1 and zero application of NP result the minimum plant height, which was significantly lower than the effect of other rates. Such increment of plant height along with increase of NP rate might be related to the effect of nitrogen which promotes vegetative growth as other growth factors are in conjunction with it. This result is in line with the report of Wakene et al. (2014) who stated that plant height of barely was increase with increasing rates of NP from 0/0 to 69/30 kg ha-1. The experiment by Rashid et al. (2007) was conducted in arid zone which received 182 mm rain fall during the growing period. Most of the time, arid zone soils are salt affected soils in which nutrient availability is influenced and application of P slightly increased plant height. The maximum plant height of 83.1cm was recorded at the highest NP rates of 69/30 kg ha-1and the minimum plant height of 61.3cm was recorded in NP rates of 0/0 kg ha-1. Similarly, Taye Bekele et al. (1996) and Woldeyesus Sinebo (2005) reported that the yield of barley increase with increasing N/P fertilizer application at many locations.

Interaction between varieties and Plant height was no significant (Table 1). Maximum days (100 days) were recorded from plants which received 0 kg N/P ha-1. Similar results were reported by Kernich and Halloran (1996), who

observed that nitrogen fertilizer considerably influenced duration of the pre-anthesis period and spike in barley. The probable reason might be that optimum NP supply played an essential role in plant growth and development. Maqsood et al. (2001), Kenbaev & Sade (2002), Salwa et al. (2005), Soylu et al. (2005), Arif et al. (2006) and Pervez et al. (2009) reported significant increase in plant height of wheat with application of nitrogen. In case of NP levels, maximum no of tillers (51.9) was produced with the application of 69/30 kg NP ha-1 when compared with other treatments (Table1). The probable reason might be that optimum nitrogen availability plays an essential role in plant growth whereas low or very high dose of nitrogen caused reduction in above ground vegetative growth of plant. Increase in the number of tillers of wheat due to N application was also reported by Rajput et al. (1993) and Ahmad (1999).Non significant (P<0.05) differences were observed in two verities, spike length, grains spike-1, 1000 grain weight, grain yield, straw yield, biological yield, harvest index, grain and straw N due to various levels of N and varieties (Table 2). Cantero et al., (1995), Le Gouis et al., (1999) and Oweis et al., (1999) observed similar results for grains spike-1 in barley was significantly increased with increasing N fertilization as reported by Moselhy & Zahran (2002). They further revealed that application of nitrogen fertilizer significantly increased spike length, number of grains spike-1, 1000 grain weight, grain yield and N uptake by the crop (Chaudhary & Mehmood, 1998; Bakhsh et al., 1999; Wakene et al., 2014; Tilahun et al., 2000)

Both N and P significantly influenced total biomass but their interaction effect was not significant (Table 1). The highest TBM of 3.9 t ha-1 was recorded in the treatment which received 69/30 NP kg N ha-1 though not significantly different to the TBM obtained from 0/0NP Kg N ha-1 (2.4 t ha-1) (Table 1). Nitrogen and phosphors increases vegetative growth of plants, especially at higher doses. Besides, the significant increase in spike length, number of seeds per spike, number of fertile tillers, non fertile tillers and grain yield by NP contributed for the significant increase in TBM. This is in agreement with Alam and Haider (2006) who indicated that increased nitrogen level increased total dry matter irrespective of cultivars. The highest TBM of 3.91 t ha-1 was obtained from the highest level (69/30 NP kg ha-1) though not significantly different from 3.2 t ha-1 obtained at 46/20 NP tha-1. Other authors also reported similar results from researches conducted on wheat Alcoz et al. (1993); Tilahun Geleto et al. (1996) indicating that P significantly increased TBM. On the other hand, application of NP fertilizer as well as its interaction with verities had non-significant effect on growth and yields components. It was increased with increasing application rates of NP up to 69/30 kg NP ha-1. The minimum value (52.4 g) was obtained from application of zero NP (P < 0.05) (Table 1).

The effect of both nitrogen and phosphorus on thousand seed weight might be attributed to the positive effect of N and P on biomass production of plants. Analysis of variance showed that application of both nitrogen and phosphorus fertilizer had significant (p < 0.05) interaction effect on total above ground dry mass (Table 1). Total above ground dry mass increased significantly when NP rates increased while combined with all rates of P but significant increase was observed when N rate increased from 0 kg ha-1 to 69 kg ha-1 with combination of P 0 to 30 kg ha-1. Similarly, significant increment of total above ground dry mass was observed as application rates of P increased and combined with all rates of N. Number of tillers per plots were not significantly to NP rates. According to this experiment, NP did not significantly affect number of tillers and infertile tillers. Maqsood et al (1999) reported that the increase in the number of fertile tillers with increasing nitrogen levels could be attributed to the well-accepted role of nitrogen in accelerating vegetative growth of plants. But, according to this experiment, the contributions of NP were also very high, it agrees with the result obtained by Prystupa et al (2004), who reported that number of productive tillers/plant was affected significantly by NP fertilizer application. The yield response of barley to NP application rates was determined for 2 years (2003–2004) in the Arjo, Gedo and Shambo highlands of Ethiopia. Results of the combined analysis of variance over 2 years showed that the mean grain yield of barley was not significantly affected by NP, nor by N×P interactions at Arjo, in which the application of 10/30 kgha-1 N/P resulted in the highest mean grain yield of barley (BARC,2006). At Gedo, the grain yield of barley was significantly affected by N application, the application of 10 kgha-1 N resulted in a better grain yield (2469 kg ha-1) than the control and other N rates. Nevertheless, the combined application of NP at a rate of 10/30 kg/ha resulted in the highest yield .At Shambo, the grain yield of barley was significantly (P<0.05) affected by P application but not by N or N×P interactions. The application of 20/30 kgha-1 N/P doubled the barley grain yield compared with the unfertilized (BARC, 2006).

Table 1. Effect of NP fertilizer on barley yield and yield components in Damot Gela, 2014.

Treatments	Plant height (cm)	Spike length (cm)	No seed/spike	No of Infertile tillers	No of fertile tillers	Yield t/ha	Biomass t/ha	Days to heading	1000seed weight	Harvest index
0 /0	61.3	6.3	36.3	34.5	253.6	0.86	2.4	65	52.4	9.3
23/10	69.6	6.3	37.8	31.5	241.1	0.106	2.6	60	53.2	13.0
46/20	74.8	6.3	39.5	12.6	300.6	1.77	3.2	52	54.1	14.2
69/30	83.1	6.6	42.6	15.8	338.0	2.02	3.9	55	54.8	15.5
LSD (0.05)	8.5	0.5	7.0	42.3	73.4	2.3	0.4	15.4	5.2	6.9
CV (%)	9.6	7.8	14.8	45.5	16.2	17.2	12.5	13.2	5.6	14.8

Table 2. Interaction effects on barely varieties in Damote Gela, 2014.

Varieties	Plant height (cm)	Spike length (cm)	No seed/spike	No of Infertile tillers	No of fertile tillers	Yield Qt/ha	Biomass t/ha	
HB-1307	71.0	6.5	37.3	17.3	275.5	22.2	2.9	NS
Shage	73.4	6.3	40.8	29.9	291.0	21.4	3.1	NS
LSD5%	5.8	0.4	4.9	29.9	51.9	0.6	0.3	

4. Conclusion

Nutrient application of plants can be varied from location to location depending on different factors such as soil and other agro-ecologies. For sustainable production of crops for a particular area, specific fertilizer recommendation is very crucial. For this reason a field experiment was conducted in Damot Gale District, Ethiopia in 2013/14. The soil was clay loam with pH of 6 (slightly acidic). This experiment was conducted to assess the effect of N and P on yield and yield components of barley. The treatments consisted of combination of four levels of phosphorus and nitrogen (0/0, 10/23, 20/46 and 30/69 kgha-1) and barely varieties (HB-1307, and Shage).The twenty four treatment combinations were replicated three times in factorial RCBD design. Many of the growth and development parameters observed responded to NP fertilization. Days to heading, number of fertile tillers and total biomass were non significantly influenced by NP application and also there is no interaction effect in verities. Other parameters, such as, plant height, number of fertile tillers, and TBM, were no significantly increased by NP rates. Therefore, N and P fertilizers are very important nutrients in limiting the growth and development of crops which has direct effect on productivity of the crops. The future studies should articulate towards and studies' involving more varies, multi-location and additional rates of N and P application, under diverse management practices, which may facilitate fine-tuning of fertilizer recommendations.

References

[1] Alemayehu Asefa, 1994. Drought stress during the belg season and selection of barley land races in the north Shewa, Ethiopia. Nile Valley and Red Sea Regional Program on cool season food legume and cereals, ICARADA/NVRSRP. Ethiopia. 209-216.

[2] Assefa Workineh Chekole. Response of Barley (Hordium vulgare L.) to Integrated Cattle Manureand Mineral Fertilizer Application in the Vertisol Areas of South Tigray, Ethiopia. Journal of Plant Sciences. Vol. 3, No. 2, 2015, pp. 71-76. doi: 10.11648/j.jps.20150302.15.

[3]   Asmare Yallew, Hailemicheal Shewayirga, Assefa Alebachew & Rahel Asrat. 1998a. Barley production practices and constraints in Meket Wereda, North Welo. pp 91–95, in: Chilot Yirga, Fekadu Alemayehu and Woldeyesus Sinebo (eds.). Barley-based Farming Systems in the Highlands of Ethiopia. Ethiopian Agricultural Research Organization, Addis Ababa, Ethiopia.

[4]   Awassa Agricultural Research Centre. 2005. Progress Report of the period 2004/2005. Awassa, Ethiopia.

[5]   BOA [Bureau of Agriculture]. 1998. Regional BasicInformation. Bureau of Agriculture Planning andProgramming Service, Awassa, Ethiopia.

[6]   Bulman, P., and D.L Smith. 1993. Grain protein response of spring barley to high rate and post anthesis application of fertilizer nitrogen. Journal of agronomy, 85(6):1109-1113.

[7]   Berhane Lakew, Hailu Gebre & Fekadu Alemayehu. 1996. Barley production and research. pp 1–8, in: Hailu Gebre and J.A.G. van Leur (eds.). Barley Research in Ethiopia: Past Work and Future Prospects. Proceedings of the 1st Barley Research Review Workshop, 16–19 October 1993, Addis Ababa. IAR/ICARDA, Addis Ababa, Ethiopia.

[8]   Chilot Yirga, Berhane Lakew & Fekadu Alemayehu. 2002. On-farm evaluation of food barley production packages in the highlands of Wolemera and Degem, Ethiopia. pp 176–187, in: Gemechu Kenini, Yohannes Gojjam, Kiflu Bedane, Chilot Yirga and Asgelil Dibabe (eds.). Towards Farmers' Participatory Research: Attempt and Achievements in the Central Highlands of Ethiopia. Proceedings of a Client-Oriented Research Evaluation Workshop, Holetta Agricultural Research Centre, Holetta, Ethiopia.

[9]   CACC (Central Agricultural Census Commission). 2002.

[10] (Hordeum vulgare L.) at Bore District, Southern Oromia Report on the Preliminary Results of Area, Production and Yield of Temporary Crops (Meher Season, Private Peasant Holdings) Part I. Addis Ababa, Ethiopia. 200 p.

[11] CLDB (Canada land development branch). 2001. Cropfertilization guide. New Nouveau Brunswick, Canada.

[12] Habtamu A., Heluf G., Bobe B., Enyew A. Fertility Status of Soils under Different Land uses at Wujiraba Watershed, North-Western Highlands of Ethiopia. Agriculture, Forestry and Fisheries. Vol. 3, No. 5, 2014, pp. 410-419. doi: 10.11648/j.aff.20140305.24

[13] Hailu Gebre and J.L. Van, 1996. Barley Research in Ethiopia: Past Work and Future Prospects. Proceeding of the first barley research review workshop IAR/ICARADA. Addis Ababa, Ethiopia.

[14] Havlin JL, JD Beaton, SL Tisdale, WL Nelson (1999). Soilfertility and fertilizers: An introduction to nutrient management. Prentice Hall, New York, 499p.

[15] Kiros Meles, 1993. Studies on barley scald (Rhynchesporium secalis (oud.) and evaluation of barley line for resistance to the disease in Ethiopia. An MSc Thesis presented to Alemaya University of Agriculture. 71p.

[16] Lakew, B., Gebre, H. and Alemayehu, F. 1996. Barley production and research in Ethiopia. In Barley Research in Ethiopia: Past Work and Future Prospects, H. Gebre and J. van leur (Eds.). Proceedings of the first barley research review workshop, 16–19 October 1993. Addis Ababa. IAR/ICARDA.

[17] MOA [Ministry of Agriculture]. 2000. Management information systems and data processing service. Ministry of Agriculture Bulletin No. 1 (August 2000).MOA, Addis Ababa, Ethiopia.

[18] M. Osundwa, J. Okalebo, W. Ngetich, J. Ochuodho, C. Othieno, B. Langat and V. Omenyo. Influence of agricultural lime on soil properties and wheat (Triticum aestivum L.) yield on acidic soils of Uasin Gishu County, Kenya. American Journal of Experimental Agriculture, 2013, 3(4): pp 806-823.

[19] Olsen SR, CV Cole, FS Watanabe, LA Dean 1954).Estimation of available phosphorus in soils by extraction with sodium bicarbonate. USA Circular. 939: 1-19.

[20] Rashid A, Khan UK, Khan DJ (2007). Comparative Effect of Varieties and Fertilizer Levels on Barley (Hordeum vulgare). ISSN Online: 1814–9596, Pakistan.

[21] Solomon Yilma, 2006. Characteristics, classification and agricultural potentials of soils of Gonde microcatchment, Arsi highlands, Ethiopia. A thesis presented to the school of graduate studies of Haramaya University. Haramaya. 74p.

[22] Wakene Tigre, Walelign Worku, Wassie Haile. Effects of Nitrogen and Phosphorus Fertilizer Levels on Growth and Development of Barley (Hordeum vulgare L.) at Bore District, Southern Oromia, Ethiopia. American Journal of Life Sciences. Vol. 2, No. 5, 2014, pp. 260-266. doi: 10.11648/j.ajls.20140205.12.

[23] Woldeyesus Sinebo & Chilot Yirga. 2002. Participatory client-orientation of research in lowinput cropping systems of Ethiopia. pp 27–43, in: Gemechu Kenini, Yohannes Gojjam, Kiflu Bedane, Chilot Yirga and Asgelil Dibabe (eds.). Towards Farmers' Participatory Research: Attempts and Achievements in the Central Highlands of Ethiopia. Proceedings of a Client-Oriented Research Evaluation Workshop. Holetta Agricultural Research Centre, Holetta, Ethiopia.

20

The Assessment of Implementing Conventional Cotton: A Regression Analysis of Meta-Data

Julian Witjaksono[*]**, Dahya, Asmin**

The Assessment Institute for Agricultural Technology, Southeast Sulawesi, Indonesia

Email address:

julian_witjaksono@yahoo.com (J. Witjaksono)
[*]Corresponding author

Abstract: This paper investigated the effects of implementing conventional cotton using meta data as the global scope from developed countries (America and Australia) and developing countries (India and China). The data base collected individual studies from more than one decade of field trials and survey. More specifically, the global effects of conventional cotton on crop yields, seed costs, pesticide costs, management and labor costs, and finally net returns were analyzed. Regression analysis was conducted to investigate and estimate the relationship between response variable and explanatory variables on these parameters. The results indicated that yield gain is the high expectation of cotton growers to optimize the net return and a strong positive correlation between yield and net return indicates that increased yield of using conventional cotton leads to higher revenue of cotton grower.

Keywords: Dependent, Economics, Independent, Indicators, Revenue

1. Introduction

Cotton is the cash crop among the farmers in the developing countries such as India, Pakistan, Indonesia, and China as well. Due to the development of Genetically Modified (GM) cotton around the world, nowadays mostly cotton growers choose GM seed for planting cotton. However, despite the higher seed cost and the uncertainty conditions such climatic conditions, conventional cotton is still needed by the farmers.

The aim of any agricultural enterprise is to maximize the profit, given limited resources or amount of inputs. The expenditure of using fertilizer, chemical matter, labor, management system and yield gain impact the net revenue of the cotton enterprise. Therefore, net income is a key measure for determining how successful a cotton grower operation has been historically, as well as an indicator of how the financial success of the farm might be in the future. What causes net returns to vary from year to year at the farm level, and more importantly, returns to vary between operations is important information for cotton producers to identify, so they can make good management decision. For instance, do

agronomic aspect (yield) has a greater effect on net return variability or do economic factors such as seed cost, pesticide cost, management and labor cost have a greater effect on net income variability? In economic analysis, inputs are the essential factors influencing yield. As a result, yield can affect net return.

At this point, more specifically, it is important to point out that the objective of this paper is to employ regression analysis to test factors influencing net return in cotton enterprise worldwide over time [1, 2, 3, 4, 5]. To determine which factors have a greater impact on net returns for cotton producers over time, historical returns were analyzed based on refereed journals, book chapters or non peer-reviewed conference proceedings through online searches from long-term studies in developed countries (USA and Australia) and developing countries (India and China). In this study, historical returns were identified from each individual study to look at variability in net returns across producers based on the input and output in economic analysis. A potential weakness of this study is that there are non-economic data evaluated in this data set (for example, variety, soil type, irrigation or non irrigation facility, rainfall data, etc.) which would help to better identify specific management styles of individual producers. Nonetheless, it is believed that results from this study can be useful for operations of all sizes as they think about what they need to focus on for long-term business survival.

2. Materials and Methods

2.1. Data Source

The data for this study were obtained from literature searched from many resources, set as the database. This study investigated the impact of conventional cotton on crop yielat the global and country level and assessed the effect of conventional cotton on farm level costs and benefits, and extends the existing literature by considering all countries and by focusing on a wide scope of literature. Four countries (USA, Australia, China and India) were considered to be chosen in terms of growing area and economic performance of conventional cotton. The database included peer-reviewed scientific articles as well as non peer-reviewed sources from grey literature. Such non peer-reviewed sources were mainly official reports from governmental organizations or agencies/institutes funded by governments, official international and national statistics as well as conference proceeding, and also from academic, governmental, civil society or from a company.

Database contained peer-reviewed and non peer-reviewed between the publication year of 1998 and 2012. A total of 129 papers were successfully collected which at least consists of one of the economic indicators (yield, net return, seed cost, pesticide cost, management and labor cost). 53 papers were successfully considered in the database then the data were tabulated and accounted for by using Microsoft Excel 2007. 16 samples (number of data tabulation) were taken based on the average data which consist of all economic indicators (yield, seed cost, pesticide cost, management and labor cost, and net return) for regression analysis. Furthermore, the data base included general information on the cotton trait (herbicide tolerance, stacked gene, Bt) from field survey and field trial.

2.2. Variable Selection

This study examined the relationship of net return with multiple variables. To simplify, net returns refer to the return to farm operator for their labor, management system, pesticide and seed, after all production expenses have been paid. Production costs refer to the expenditure of using input during the production process to produce the cotton. The question is that are net returns dependent on the yield, seed cost, pesticide cost, management and labor cost? Therefore, the technique of linear regression and correlation was used, in which case should predict the value of net returns using independent variables.

2.3. Model Establishment

Comparative statistics provide a broad overview about the agronomic and economic effects of conventional cotton. However, such statistics become less effective in separating the effects of individual changes while controlling for the effects of other variables. Individual effects of variables while controlling for the effects of others can be estimated by employing a multiple regression [6]. In this regression, net revenue is taken as the dependent variable while yield, seed cost, pesticide cost, management and labor cost are taken as the independent variables. This model is used to further explore the relationship between net return per hectare, yield and various production inputs, such as pesticide use, seed cost, management and labor cost. Based on the theoretical foundation, the regression model was established which can be written as:

$$Y = b_0 + b_1X_1 + b_2X_2 + \dots + b_iX_i + \varepsilon \quad (1)$$

Where:

b_i = partial slope coefficient (also called partial regression coefficient, metric coefficient); it represents the change in Y associated with a one-unit increase in X_i when all other independent variables are held constant. It was observed that b_0 is the sample estimate of β_0, b_i is the sample estimate of β_i, and βs are the parameters from the whole population in which the sampling was conducted. The dependent variable and the explanatory variable must be specified as:

Y = Net return
X_1 = Yield
X_2 = Seed cost
X_3 = Pesticide cost
X_4 = Management and labor cost.

We performed SPSS 16.0 to determine the intercept and regression coefficients, after that we tested them for significance by doing the Analysis of Variance (ANOVA). ANOVA determines if regression coefficients that the probable model calculates should be present in the final model as a predictor or not. A P-value or sig-value for coefficients significance test was conducted. If P-value for a coefficient was less than 0.05 (P<0.05), the coefficient is statistically significant and the related variable should be present in the model as a predictor, but if it was higher than 0.05 (P<0.05), the coefficient is not statistically significant and the related variable should not be present as a predictor [7].

Coefficient of determination or R-square (R^2) shows how the model of predictors fits the dependent or independent variables (higher R^2, higher fit of the model and higher model goodness). Moreover, significant test for intercept (b_0) is similar to regression coefficients [8]. Significance test of the coefficient and R^2 helps researchers to decide what predictor is more important and must be presented in the model. Besides this, when the number of the predictors increased, usually most of the variables are strongly correlated with each other and it is not necessary to present all of these correlated variables in the model since they can be used in place of one another [9].

3. Results

We employed a regression analysis in order to investigate the correlation between dependent variable (Y = Net Revenue) and predictor variable (X1 = Yield, X2 = Seed, X3 = Pesticide, X4 = Management and Labor). Data presented in Table 1 show that under the condition level, $\alpha = 0.05$ F = 28.448, p value = 0.000 (< 0.05). This means indicated that the goodness of fitting of equation is highly significant. Because p value of F is smaller than 0.05, therefore the overall significance is good and also indicated that there is no multicollinearity problem.

Table 1. Model summary and analysis of variance between independent an dependent variables of conventional cotton.

Model Summary[b]					
Model	R	R Square	Adjusted R Square	Std. Error of the Estimate	Durbin Watson
1	0.955a	0.912	.880	111.97310	2.100
		ANOVAb			
Model	Sum of Squares	Df	Mean Square	F	Sig.
1 Regression	1426723.719	4	356680.930	28.448	.000a
Residual	137917.719	11	12537.974		
Total	1564641.437	15			

a Predictors (Constant), Management and Labor, Seed, Yield, Pesticide
b Dependent Variable

To express the quality of fit between a regression model and the sample data, the coefficient of multiple determinations (R^2) was used ranging in value from 0.0 to 0.1. Table 1 shows the value of R^2 as 0.912 indicating that the fitting degree is high, and the linear relationship between predictors and dependent variable is significant. Higher value of R^2 indicates a better fit of the model to the sample observations. However, adding any regressor variable to this model, even an irrelevant regressor, yields a greater R^2. For this reason, R^2 by itself is not a good measure of the quality of fit. To overcome this deficiency in R^2, an adjusted value could be used. Therefore, the adjusted R^2 was used on this model which is a more reliable indicator of model quality.

We found that the value of adjusted R^2 is 0.88. As such, 88% of the variability in Net revenue in conventional cotton can be predicted from the relation of the independent variable (yield, seed, pesticide, management and labor), while the remaining can be explained by the outlier beyond the model.

In the case of one explanatory variable, the coefficient of determination is simply the square of the coefficient of correlation namely r^2. Table 2 shows the relationship between the dependent and explanatory variables. This study performed Pearson correlation matrixes focused on the strong correlation (positive or negative) between the dependent and independent variables.

Table 2. Correlation matrixes between independent variable and dependent variable of conventional cotton.

		Net Return	Yield	Seed	Pesticide	Management and Labor
Pearson Correlation	Net Return	1.000	.407	-.082	.024	-.426
	Yield	.407	1.000	.361	.443*	.577
	Seed	-.082	.361	1.000	.255	.312
	Pesticide	.024	.443	.255	1.000	.618
	Management and labor	-.426	.577*	.312	.618*	1.000
Sig. (1-tailed)	Net Return		.059	.382	.465	.050
	Yield	.059		.085	.043	.010
	Seed	.382	.085		.170	.120
	Pesticide	.465	.043	.170		.005
	Management and labor	.050	.010	.120	.005	

*. Correlation is significant at the 0.05 level (1-tailed)

Table 2 depicts that the relationship between management and labor cost and pesticide indicated a strong positive correlation (r = 0.618) with r^2 significant level < 0.05 (0.005), then yield and management and labor cost (r = 0.577) with r^2 significance level < 0.05 (0.010). The relationship between yield and pesticide cost presented a strong positive correlation (r=0.443) with r^2 significant level < 0.05 (0.043). Moreover, we found a significant negative effect between management and labor cost and net return (r=-0.426) with r^2 significance test 0.05.

Furthermore, Table 3 performed the multicollinearity test and the model test for this study. From the table 3 we represent that to independent variable yield (X1), the estimation of regression is 388.135, standard error is 45.000, t test value is 8.625, t test significance is 0.000, lower than 0.01. That is think independent variable yield is highly significant. Then, to predictors variable pesticide and management and labor, we can find that t test significance is 0.012, and 0.000 lower than 0.05, respectively. Therefore, the coefficient of independent variable is highly significant. Overall, we can say that net return variability can be significantly affected by yield, pesticide and management and labor.

Table 3. Multicollinearity test and model test of regression analysis of conventional cotton.

Model	Unstandardized Coefficient		Standardized Coefficient	t	Sig.	Collinearity Statistics	
	B	Std. Error	Beta			Tolerance	VIF
1 (Constant)	-64.890	93.569		-.694	.502		
Yield*	388.135	45.000	.979	8.625	.000	.622	1.609
Seed	-4.640	2.742	-.164	-1.692	.119	.851	1.175
Pesticide*	1.897	.633	.345	2.996	.012	.605	1.652
Manag & labor*	-1.017	.112	-1.153	-9.104	.000	.500	2.002

*. Significant at the 0.05 level

The obtained results showed that the prediction equation for net return in conventional cotton (Y) is formulated using the predictors as follows:

$$Y = -64.890 + 388.135 \, X1 - 4.640 \, X2 + 1.897 X3 - 1.017 \, X4$$

In addition, we test the multicollinearity of the model using variance inflation factor (VIF) which indicated that overall results is lower than 10. That is this model does not has the multicollinearity problem. Moreover, autocorrelation test on this model was carried out by Durbin Watson (DW) analysis which indicated that DW = 2.1. According to DW checking table, under 0.01 significant level then Du < DW < 4 – Du (n=15, K = 4) then 1.70 < 2.1 < 4 – 1.70, that is this equation has no problem with autocorrelation.

4. Discussion

Regression analysis reveals that net return mostly is affected by yield gain. That is yield gain is the main factor influencing farmers' income. The database depicts that yield gain varies from country to country, trait to trait, year to year due to the climatic conditions, site specific and geographical dependent. Moreover, the impact of yield difference on conventional cotton was dependent upon the level of pest pressure, location, year, climatic factors, and time of planting.

A question commonly asked is whether one explanatory variable is more important than the other. The effect of any given explanatory variable depends on which other variables have been included in the regression model. The question cannot be answered by simply looking at the respective values of the β coefficients, because the value of the β coefficients depends on the unit of the explanatory variable. In this case, yield gain is measured by kg/hectare and the others (seed cost, pesticide cost, management and labor cost) are measured by USD/hectare. There can be no comparison between such disparate quantities; instead we look at the t-ratios between response variable and explanatory variables, in which 8.625 was for the yield which was higher than that of any other independent variables. Therefore, the effect of the yield gain is greater than that of other explanatory variables. A strong positive correlation between yield and net return indicates that increased yield of using conventional cotton leads to higher revenue of cotton grower.

A negative t-ratio of management and labor cost showed by -9.104 indicating cotton growers with high management

and labor cost was expected to have lower net return unless they will have higher yield that can offset higher labor expenditure to optimize the return. Interestingly, this study consistent with [10, 11, 12, 13] that the implementation of GM cotton required cotton growers higher management and labor costs due to the goodness of crop management system such as consultant fee, irrigation costs, and other management costs.

Moreover, the t-ratio of pesticide cost shows a positive value (2.996), while expecting cotton growers need more chemical spray to reduce the yield losses due to the pest pressure. In other words, when farmers expect to incur large yield losses from cotton bollworm, they spray more. That is, the more they spray, the higher the expected yield. However, the higher pesticide use was due to the less resistant of conventional cotton variety againts the bollworm attack [14, 15, 16, 17]. The increased use of pesticide could also be due to the the differences in naturally occurring fluctuations in pest population especially for cotton bollworm which varied from country to country, county to county, year to year, site specific, climatic conditions and geographical dependent. In contrast, study about GM (genetically Modified) cotton that its implication rely on the chemical spray due to the secondary pest which might decrease the potential yield of GM cotton [2, 17, 18]. This means that GM cotton might face a serious problem of secondary pest infestation even its resistant to cotton bollworm.

The observed economic impacts of conventional cotton in any 'place' will depend on the yield potential of crop varieties, the pest infestation, and general and seasonal dependent climate and weather conditions, as well as government intervention [19].

5. Conclusions

Regression analysis in this paper presented the relationship between net return, yield, and production cost. The relationship is that producers expect higher yields of conventional cotton. Therefore, a significantly higher yield is needed to optimize revenue. Another correlation is due to the fact that the higher chemical spray is needed in order to optimize the yield. This is due to the fact that conventional cotton is the less resistant crop of pest infestation. Therefore, the more the chemical spray, the higher the cotton yield. Moreover, the correlation between net return and management and labor costs indicated that growing conventional cotton is time consuming for cotton harvest. Due to the less resistant of conventional cotton, consequently, the higher the chemical spray, the higher the management and labor cost which can affect the net return of cotton growers.

In this study, statistical inferences of regression analysis reveal that yield, seed cost, pesticide cost, management and labor cost effectively influence net return in conventional cotton. Other factors which determine relative economic profitability beyond those economic indicators have been ignored but should be considered and taken into account for the future research. It is a concern that this study relied on the individual studies. Thus, the data observed might not be adequately addressed to capture the effect of using conventional cotton due to the fact that these studies might use totally different methodologies to assess the economic benefit of conventional cotton. For instance, such assessment might be based on the impact different studies, using field trials or surveys, have on public research institutes or private companies which probably show presence of biases that can occur with different methodologies.

As a result of the aforementioned points, the analysis presented some interesting points that shed light on the diversity that can be observed in the literature and which helped fuel the divergent viewpoints held in the development of conventional cotton. Thus, this study is a representative of the entire economic standpoint based on the literature searched with different goals and methodologies, as well as the study's purpose.

The results presented here do support the economic analysis of growing conventional cotton and might be compared with transgenic cotton, and by adding-up individual studies through the meta-data, there is the risk of comparing aplles and oranges. Nonetheless, the analysis presented shows that conventional cotton is still being considered as the alternative crop which might contribute to poverty reduction and rural economic development, and all of these aspects should be considered in the assessment of the whole of cotton economic analyses for the future research.

References

[1]   Moucheshi Saed A, Fasihfar E, Hasheminasab H, Rahmani A, Ahmadi A, 2013. International Journal of Agronomy and Plant Production. 4(1): 127-141. Available online at http://www.ijappjournal.com

[2]   Witjaksono J, Wei X, Mao S, Gong W, Shang H, Li Y, Yuan Y, 2013. The assessment of economic indicator using GM cotton worldwide overtime. Journal of Agricultural Economic and Development. 2(27): 290-296. Available online at http://www.academeresearchjournals.org/journal/jaed

[3]   Salimi S, Moradi S, 2012. Effect the correlation, regression and path analysis in soybean genotypes (*Glycin max* L.) under moisture and normal condition. International journal of agronomy and plant production. Available online at http://www.ijappjournal.com

[4]   Pirdashti H, Ahmadpour A, Shafaati F, Hosseini Jaber S, Shahsavari A, Arab A, 2012. International Journal of Agricultural: Research and Review. 2(4): 381-388. Available online at http://www.ecisi.com

[5]   Wei Ming Liu, 2009. Analysis of High yield and efficiency technique in hybrid rice Zhongzheyou No.1. Agricultural Science and Technology. 10(2): 73-76.

[6]   Bennet, R., Ismael, Y., Morse, S., 2005. Explaning contradictory evidence regarding impacts of genetically modified crops in developing countries. Varietal performance of transgenic cotton in India. Journal of Agricultural Science, 143 (1), 35-41.

[7]   Draper NR, Smith H, 1981. Applied regression analysis. John Wiley, New York.

[8]   Kleinbaum DG, Kupper LL, Muller KE, 1998. Applied Regression Analysis and Other Multivariable Methods. PWS-Kent Publishing Co, Boston.

[9]   Manly BFJ, 2001. Statitics for environmental science and management. Chapman and hall/CRC, Boca Raton.

[10] Subramanian A, Qaim M. 2008. Village-wide effects of agricultural biotechnology: The case of Bt cotton in India. World Development 37(1): 256-267. doi: 10.1016/j.worlddev.2008.03.10.

[11] Jost P, Shurley D, Culpepper, S, Roberts, P, Nochols R, Revves, J, Anthony S. 2008 Economic comparison of transgenis and non transgenis cotton production system in Goergia. Agronomy Journal. 100: 42-51. doi: 10.2134/agronj2006.0259.

[12] Morse S, Bennet R. M, Ismael Y. 2005. Genetically Modified insect resistance in cotton:some farm level economic impacts in India. Crop Protection 24 (2005): 433-440. Available online at http://www.sciencedirect.com

[13] Pemsl D, Waibel H, Orphal J. 2004. A methodology to assess the profitability of Bt cotton: Case studiy results from the state of karnataka, India. Available online at http://www.sciencedirect.com

[14] Zhang, X. L., Liu, K. F., Chu, Z. Y., Gao, J.S., Wu, C., Shao, Y. Y., Feng, J. X., Li, X. L. 2012. Comparative analysis of the correlation about yield, Yield components of cotton hybrids and conventional varieties.

[15] Pasu, S., Nicholas, K. 2009. Understanding the adoption of cotton biotechnologies in the US: Firm level evidence. Agricultural Economics Review. 10 (1): 80-95.

[16] Holtzapffel, R., Mewett, O., Wesley, V., and Hattersley, P., 2008. Genetically modified crops: tools for insect pest and weed control in Cotton and Canola. Australian Government Bureau of Rural Sciences, Canberra.

[17] Yang, P., Iles, M., Yan, S., and Jolliffe, F., 2005. Farmers' knowledge, perceptions, and practices in transgenic Bt cotton in small producer system in Northern China. Crop Protection, 24 229-239.

[18] Lalitha N, Ramaswami B, Viswanathan P. K. 2009. India's experience with Bt cotton: case studies from Gujarat and Maharastra. Ln R. Tripp (Ed). Biotechnology and Agricultural Development: Transgenic cotton, rural institution and resource-poor farmers: 135-167. London and New York: Routledge.

[19] Finger E, Kaphengst T, Evans C, Herbert S, Lehman B, Morse S, Stupak N. 2011. A meta-analysis of farm level costs and benefits of GM crops. Sustainability., 3: 743-762. doi: 10.3390/su3050743.

21

Resource Domestication: An Introduction to Biodiversity and Wildlife in Agriculture

Benjamin E. Uchola

Faculty of Agriculture, Federal University, Dutsin-ma, Nigeria

Email address:

buchola@fudutsinma.edu.ng

Abstract: Biodiversity and Wildlife are relatively recent concepts in Agriculture. However, the meaning of each concept remains to be clearly distinguished from similar concept in Natural Resource Conservation. The concepts of Biodiversity and Wildlife in Agriculture may be better understood when explored from the perspective of Resource Domestication. Relocation of a resource from its natural habitat into human-controlled environments represents an initial phase in the process of domestication. The final phase of Resource Domestication entails selection of desired production traits in established populations through breeding programs. A more complex relationship emerges in the course of transforming a wild plant into a crop or a wild animal into a livestock. The new relationship between a resource and its domestic form served as the framework for understanding Biodiversity and Wildlife in Agriculture.

Keywords: Agriculture, Biodiversity, Domestication, Wildlife

1. Introduction

Resource Domestication is a complex interaction between humans and a valued organism. It involves the relocation of a resource from its natural habitat to artificial environments as demonstrated in the domestication of oil palm (*Elaeis guineensis*) [1-3], rubber (*Hevea brasiliensis*) [4, 5], common carp (*Cyprinus carpio*) [6, 7] and Japanese quail (*Coturnix japanicus*) [8, 9]. Domestication also entails artificial selection of preferred traits as revealed in oil palm yield improvement exercises [10-12], in the development of saponin-free quinoa seeds (*Chenopodium quinoa*) [13, 14] and selection of higher body weights in Japanese quail [15, 16]. Knowledge of Resource Domestication often provides satisfactory answers to a number of very important questions; questions such as those concerning the relationship between native Amazonian rubber trees and their plantation-grown counterpart [4, 5] or the differences in gonad maturation of wild African catfish (*Clarias gariepinus*) and their domestic form [17, 18].

The impact of Resource Domestication extends beyond its initial motivation. On the one hand, relocation of a resource into human-controlled environments was an expression of a desire to protect its dwindling supply occasioned by fluctuations in climatic conditions and expansion in local human populations [19-21]. On the other hand, artificial selection of traits explores the production potentials of a resource that is made visible through variations in individuals of the same population [10, 13, 14, 15]. The combined effect of these aspects of domestication is a transformation of primitive resources into highly productive cultivated plants or domestic animals. However, the impact of domestication also involves the evolution of new terms as a cultivated plant is referred to as a "crop" and a domestic animal a "livestock" after their partial or complete transformation [22, 23].

The use of terms such as "biodiversity" and "wildlife" is not uncommon in the field of agriculture. Wild plants are used in crop breeding programs [24-26], some wild animals are considered as livestock of the future while others have been identified as wild progenitors of modern livestock [27-37]. However, the current concept of Biodiversity or Wildlife does not sufficiently reflect the unique attribute of agriculture as a system that develops crops from wild plants and livestock from wild animals [22, 23]. More so, similar concepts in

Natural Resource Conservation accommodate undomesticated plants, flesh-eating birds and dangerous mammals which are not related to any crop or livestock [38, 39]. These broad concepts of Biodiversity and Wildlife blur the distinction that exists between conservation of natural resources and the development of crops or livestock through the process of domestication. It is, therefore, a necessity to understand the meaning of Biodiversity and Wildlife within the context of Resource Domestication in Agriculture.

2. Plants, Animals and Resources

Biotic components play important roles in processes that sustain the environment. Generally, plants serve as habitat for organisms, trap energy needed in an ecosystem but more importantly satisfy certain needs of society [22]. In the same way, animals are known to improve soil properties, aid pollination of flowers and dispersal of seeds. However, it is their utility value as adjudged by society that makes each one of them a resource [23]. But, a resource may exist only in a particular ecological zone or region. For instance, oil palm grows naturally in the rainforest of Africa, species of potato in the Andean region of South America and macadamia in Australian rainforest [1, 40, 41]. Likewise, species of wild cattle are endemic to parts of Asia, wild goat to Indian sub continent and wild sheep to Eurasia [42].

Resources seldom express their full production potential in their natural habitats. They often respond to interferences from extreme climatic conditions and biotic stress by manifesting slow growth, inconsistent fruiting pattern and low productivity [43-45]. For instance, a species of macadamia tree requires nearly two decades of growth to prepare for fruiting activities while the African bush mango grows for over a decade before the onset of maturity [46, 47]. Even more, production of fruits are often characterised by inconsistency and low yield in both cases. Like plant resources, animal resources such as wild Japanese quail and African catfish often display several attributes including late maturation and seasonal reproduction as responses to changes in environmental factors [9, 17]

The productivity of a resource is therefore generally low given the restrictions imposed by environmental factors. Estimate of fresh fruit bunch yield of wild and semi-wild oil palm trees range from 2 to 5 tonnes per hectare [1]. Similar low patterns of productivity characterise the yield of other plant resources such as rubber and macadamia [5, 41]. The quantity of products exploited from wild animals is equally low even though comprehensive data on the production performances between a resource and its domestic form are scarce. Nonetheless, the meat yield of red jungle fowl (JF) would have been less than those of its closest domestic relatives when the growth performance of JF (< 300g) is considered in relation to those of domestic chicken (>300g) within the same period [48, 49]. The differences in meat yield becomes vivid by comparing the production performance of JF (<300g) to those of improved breeds of chicken (>1200g) within the same period [48, 50].

Resource exploitation has consequences for the environment. Exploitation of a resource without appropriate management has lead to a situation in which the population of some species is declining as indicated by the status of common ostrich (*Struthio camelus*) and wild goats (*Capra aegagrus*) [51]. More still, some resources are either threatened with extinction or have gone extinct as in the case of wild cattle (auroch). For these reasons, natural populations are increasingly being protected through scientific studies and establishment of National parks [32, 34, 52].

3. Resource Domestication

Domestication is essentially an indirect approach towards the conservation of a plant or an animal resource. It involves the introduction of a resource into human-controlled environments and afterwards selection of preferred production traits [22, 23].

Relocation of a resource and selection of its traits are the major aspects in domestication projects. Seedlings of bush mango were transferred from forest to farmlands even though the preservation of the tree on farmland is still a common practice [53]. Even more, seedlings of bush mango that were propagated using vegetative methods have been used in the establishment of field banks [54, 55]. In cultivated fields, bush mango attains maturity in about half the number of years required by its wild counterpart and produces fruits with superior indices [47, 56]. The improved performances of field-grown bush mango indicate the plant is responding to selection and also suggest the likelihood of further improvement through selection of superior genotypes. Similarly, cane rat was transferred from its natural habitat into experimental farms for the purpose of studying its growth and reproductive performances. In the course of domestication, the cane rat gradually accepted feeds during daytime, manifest higher body weight and produced larger litter size [57-59]. More still, moderate to high heritability values for body weight and other production traits including their correlations suggest that production performance of cane rat could be improved through artificial selection [60, 61].

Domestication therefore facilitates better trait expression in a resource. Artificial selection improved the yield of oil palm when fresh fruit bunch yield in wild/semi-wild populations (<5 tonnes/ha) is compared to that of well managed plantations (>20 tonnes/ha) [1, 62]. Likewise, there are improvements in productivity of animals due to effects of domestication. The meat yield of Japanese quail increased from about 100g in the earliest domesticates to about 300g in modern populations [15, 16]. Similarly, a comparison of the body weights of wild jungle fowl (<300g), native chicken ecotypes (>300g) and improved meat breeds of chicken (>1200g) within the same period [48-50], suggest meat yield of jungle fowl was improved over three times through artificial selection. In the same way, selection for milk yield in cattle increased production several times given the yield estimate of local breed (<900kg) and pure breed (>2000kg) within the same period [63]. This is a trend in milk production of local cattle breed as increases in the degree of improvement leads to a corresponding increase in milk yield [64].

4. Resource Domestication, Biodiversity and Wildlife in Agriculture

The review above reveals that crops and livestock are products of Resource Domestication. Cultivated apple (*Malus domestica*) originated from a wild plant which is related to over 50 other *Malus* species [Table 1a; 65, 66]. Plantation-grown rubber, field-grown potato and cultivated rice (including their varieties/clones) are the domestic forms of wild species with each having several close relatives [4, 5, 40, 65, 67-71]. Similarly, cultured common carp (*Cyprinus carpio*) is a domesticate of a wild form known to have several *Cyprinus* species as relatives [6, 7, 72]. Domestic goat (*Capra hircus*) and sheep (*Ovis aries*) descended from different wild progenitors but each progenitor belongs to a taxonomic group with several other species [33, 35, 36, 42]. The origin of domestic chicken (*Gallus domesticus*) can be traced to red jungle fowl (*Gallus gallus*), which shares common traits with other *Gallus* species [Table 1b; 30, 31, 73].

Table 1a. *Selected Crops and their biodiversity (Wild flora & Varieties/ Cultivar/ Clones).*

Crop	Wild progenitor	Wildflora: Minimum Estimate & Selected Examples	Crop Diversity; Varieties / Clones: Selected Examples
Apple		54 species [65, 66]	Several hundred cultivars [68]
Malus domestica	*Malus sieversii*	*Malus angustifolia, M. asiatica, M. baccata, M. bracteata, M. chitralensis, M. coronaria, M. domestica, M. doumeri, M. floribunda, M. fusca, M. glabrata, M. hupehensis, M. ioensis, M. jinxianensis, M. kansuensis, M. M. lancifolia, M. melliana, M. micromalus, M. ombrophila, M. platycarpa, M. prunifolia, M. pumila, M. rockii, M. sieversii, M. spontanea, M. sylvestris, M. transitoria, M. toringoides, M. turkmenorum, M. yunnanensis, M. zumi*	*Malus domestica* Brown, Circassian, Coast, Gala, Lady, Red, Landsberger Reinette, Paide's Winter, Toko, Wealthy.
Potato		200 species [40, 65]	Several varieties [69]
Solanum tuberosum	Several species collective referred to as *Solanum brevicaule* complex.	*Solanum albornozii, S. bulbocastanum, S. bukasovii, S. burtonii, S. cardiophyllum, S. chilliasense, S. commersonii, S. demissum, S. jamesii, S. inutifoliolum, S. paucijugum, S. phureja, S. pinnatisectum, S. regularifolium, S. stoloniferum, S. stenotomum, S. ternatum, S. tuberosum*	*Solanum tuberosum* Atahualpa, Nicola, Russet Burbank, Tubira, Vitelotte.
Rice		21 species [65, 67]	Several hundred varieties [70]
Oryza glaberrima Oryza sativa	*O. breviligulata O. nivara / O. rufipogon*	*Oryza australiensis, O.barthii , O. breviligulata, O. eichingeri, O. glaberrima, O. grandiglumis, O. latifolia, O. longiglumis, O. longistaminata, O. meridionalis, O. meyeriana, O. minuta, O. neocaledonica, O. nivara, O. officinalis, O. punctata,, O. ridleyi, O. rufipogon, O. sativa, O. schlechteri*	*Oryza sativa* Kimboka, Agora, Sookha Dhan 5, NSIC 25, Nerica.
Rubber		10 species [4, 65]	Several clones [71]
Hevea brasiliensis	*Hevea brasiliensis*	*Hevea benthamiana, H. brasiliensis, H. camargoana, H. camporum, H. guianensis, H. microphylla, H. nitida, H. pauciflora, H. rigidifolia, H. spruceana*	GT l, Tjir, PB 86, PB 260, PB 312, RRII 105, RRII 430 RRIM 600, RRIM 712, RRIC 100, RRIC130.

Table 1b. *Selected Livestock and their biodiversity (Wild fauna & breeds).*

Livestock	Wild Progenitor	Wild Fauna: Estimate & Examples	Livestock Diversity*: Estimate & Selected Examples
Carp (Common)		22 (6 listed) [72]	Several [7]
Cyprinus carpio	*Cyprinus carpio* [6]	*Cyprinus acutidorsalis C. barbatus, C. carpio, C. micristius, C. rubrofuscus, C. yunnanensis*	*Cyprinus carpio* Feng, Germany mirror, Hebao red, Heyuan, Huanghe, Jian, Lotus, Molong, Songhe, Songpu, Xingguo red, Xiangyun, Ying, Yue, Scattered mirror.
Cattle		5 [42]	112 [74, 75]
Bos taurus	*B. p. primigenius +opisthonomous*	*Bos gaurus, B. javanicus, B. mutus, B. primigenius, B. sauveli*	Aberdeen Angus, Ayrshire, Braford, Brahman, Brown Swiss, Charolais, Chusco, Creole, Devon, Dexter, Galloway, Gascon, Gelbvieh, Goudali, Guersney, Hereford, Holstein, Limousin,
Bos indicus	*B.p. nomadicus* [28, 29]		Lincoln Red, Muturu, Ndama, Normande, Red Angus, Senepol, Sokoto Gudali, White fulani.
Chicken		4 [73]	101 [74, 75]

Livestock	Wild Progenitor	Wild Fauna: Estimate & Examples	Livestock Diversity*: Estimate & Selected Examples
Gallus domesticus	*Gallus gallus* [30, 31]	Sub species of *Gallus gallus bankiva, Jabouillei, murghi, spadiceus* *G. lafeyettei* *G. sonneratii* *G. varius*	Amrock, Australorp, Baladi Beheri, Bresse, Campine, Crevecoeur, Derbyshire Redcap, Dokki, Dresdener, Faverolles, Fayoumi, Gournay, Hamburgs, Hampshire, Jersey Giant, La Fleche, Minorca, NewHampshire, Orloff, Orpington, Plymouth Rock, Rhodebar, Sussex, Vorwerk, Warren, Wyandotte
Goat		8-9 [42]	40 [74, 75]
C. a. hircus	*Capra aegagrus* [35, 36]	*C. aegagrus* *C. caucasica* *C. cylindricornis* *C. falconeri* *C. ibex* *C. nubiana* *C. pyrenaica* *C. sibirica* *C. walie*	Anglo-Nubian , Angora,, Barbari, Bengal, Berber, Boer, Dutch Pied, Gaddi, Granada, Kahalari, Kamori, Karachai, Maradi, Maure, Murciana, Nigerian Dwarf, Oberhasli, Peacock Goat, Poitou, Saanen, Sahelian, Somali, Toggenburg, Tswana, Verata.
Pig			33 [74, 75]
Sus s. domesticus	*Sus scrofa* [37]	*Sus scrofa* [73]	Alentejana, American Berkshire, Berkshire, Chester White, Dalland, Duroc, Ghori, Haitian, Jersey Red, Lacombe, Large Black, Large White, Mangalitsa, Meishan, North Caucasus, Pelon, Pietrain, Saddleback, Seghers, Siska, Spotted, Tamworth, Turopolje, Welsh, Wessex Saddleback.
Sheep		6 [42]	100 [74, 75]
Ovis aries	*Ovis orientalis* [33]	*O. ammon* *O. canadensis* *O. dalli* *O. orientalis* *O. nivicola* *O. vignei*	Australian Merino, Awassi, Blue Texel, Bond , British Milksheep, Chios, Coopworth, Corriedale, Devon Longwool, Dormer, Dorper, Dorset, Dorset Down, Drysdale, Finnsheep, North Ronaldsay, Quessant, Pool Merino, Polwarth, Polypay, Portland, Santa Cruz, Texel, Van Rooy, West African Dwarf, Zwartbles.

*Trans-boundary Breeds/Hybrids

The development of a crop or livestock establishes new relationships. Cultivated rice as a special grain-producing plant plays the role of a food crop while other species of rice remain mere plants of an ecosystem. More still, cultivated rice responds to series of further selection for disease resistance and other traits resulting in the development of several varieties (Table 1a). The movement of rice from a mere plant or plant resource to a crop with varieties represents a process of rice development [22]. This development process establishes a new and complex relationship which consists of the wild progenitor of rice as well as other *oryza* species, earliest cultivated rice and its varieties. Put differently, cultivated rice and its varieties are the most advanced forms of rice while other *Oryza* species are wild flora, wild relatives or less developed forms. Cultivated rice, its varieties and wild flora when summed up represents the biodiversity of rice (Rice Biodiversity). In the same way, the biodiversity of other crops would be the sum of their earliest cultivated forms, varieties/clones and wild flora (Table 1a). Like crops, domestic chicken is an animal kept for a purpose while other *Gallus* species are mere birds or bird resource of an ecosystem. The development of domestic chicken establishes its wild fauna as less developed forms while domestic chicken and its breeds are the most advanced in terms of productivity (Table 1b). Therefore, the constituents of chicken biodiversity are the earliest form of domestic chicken, its breeds and wild fauna. Similarly, the biodiversity of other livestock such as cattle, sheep and pig would be the sum of their individual wild fauna, earliest domestic form and breeds.

From Resource Domestication perspective, biodiversity is a specific concept that excludes all botanical entities that are not related to any particular crop or zoological entities that are unrelated to a known livestock. Accordingly, the biodiversity of rice excludes other members of the grass family (*Poaceae or Gramineae*). As a result, grasses such as spear grass (*Imperata cylindrica*) and other similar plants that are not directly related to any crop may not be categorised as part of Biodiversity in Agriculture. Likewise, the biodiversity of cattle excludes all members of the family (*Bovinae*) except its wild fauna, domestic form and breeds. In other words, cattle-like animals like African buffalo (*Syncerus caffer*) which fatally attack humans or American bison (*Bison bison*) whose domestication has largely been unsuccessful, do not belong to Cattle Biodiversity. These organisms and others like African grey parrot (*Psittacus erithacus*) and mountain gorilla (*Gorilla beringei*) are considered part of biodiversity by Natural

Resource Conservation-based Organisations [38, 39]. Interestingly, a Resource Domestication-based concept of biodiversity alters the concept of Wildlife to that which revolves around wild flora of crops and wild fauna of livestock.

5. Conclusion

The use, of basic terms like crop or livestock, is an implicit acknowledgement of Resource Domestication in Agriculture. Development of crops and livestock from wild resources reveals that Biodiversity and Wildlife have their roots in Resource Domestication. However, Biodiversity and Wildlife are distinct concepts in Agriculture. Biodiversity expresses the sum of different development levels of a particular crop or livestock while Wildlife is a collective term for either wild flora of crops, wild fauna of livestock or wild forms of both crops and livestock.

References

[1] A. C. Zeven, The semi-wild oil palm and its industry in Africa. Agricultural Research Report, No. 687,178 p. 1967.

[2] A. C. Zeven, "The partial and complete domestication of oil palm (Elaeis guineensis)". Economic Botany Vol. 26, pp. 274–279. 1972.

[3] R. H. V. Corley, B. S. Gary and S. K. Ng, "Productivity of the oil palm (Elaeis guineensis Jacq.) in Malaysia". Experimental Agriculture Vol. 7, pp. 129-136, 1971a.

[4] R. E. Schultes, "A brief taxonomic view of the genus Hevea. Malaysian Rubber" Research and Development Board, Kuala Lumpur. Monograph no. 14, 1990.

[5] R. E. Schultes, "The domestication of the rubber tree: economic and sociological implications". Amer J Econ Soc Vol. 52 (4), pp. 479-485. 1993.

[6] E. K. Balon, "Origin and domestication of the wild carp, Cyprinus carpio: from Roman gourmets to the swimming flowers". Aquaculture Vol.129, pp. 3–48. 1995.

[7] Z. Jeney and Z. Jian, "Use and exchange of aquatic resources relevant for food and aquaculture: common carp (Cyprinus carpio L.)" Reviews in Aquaculture Vol.1, pp.163–173. 2009.

[8] Kerr H. W. Quailology: The domestication, propagation, care and treatment of wild quail in confinement. Little Sioux, Iowa, U. S. A: The Taxiderm Company. 1903.

[9] G. B. Chang, X. P. Liu, H. Chang, G. H. Chen, W. M. Zhao, D. J. Ji, R. Chen, Y. R. Qin, X. K. Shi and G. S. Hu "Behavior differentiation between wild japanese quail and domestic quail". Poultry Science Vol.88, pp. 1137–1142. 2009.

[10] R. H. V. Corley, J. J. Hardon and G. Y. Tan, "Analysis of growth of the oil palm (Elaeis guineesis Jacq) I. Estimation of growth parameters and applicaton in breeding". Euphytica Vol. 20, pp. 304-315. 1971b.

[11] J. J Hardon; R. H. V Corley and C. H. Lee. Breeding and selecting the oil palm. In: Abott, A. J and Atkin, R. K (Eds). Improving Vegetatively Propagated Crops. Academic Press Ltd, London. Pp 63-68. 1987.

[12] H. Limburg and H. D. "Mastebroek Breeding high yielding lines of Chenopodium quinoa Willd. with saponin free seed". Proceedings of COST-Workshop., 22–24/2 1996, European Commission EUR 17473/KVL, Copenhagen Copenhagen: KVL, pp. 103–114. 1996.

[13] H. D. Mastebroek, E. N. Van Loo and O. Dolstra, "Combining ability for seed yield traits of Chenopodium quinoa breeding lines". Euphytica Vol. 125(3), pp. 427-432, 2002.

[14] K. E. Nestor, W. L. Bacon and A. L. Lambio, "Divergent selection for body weight and yolk precursor in Coturnix coturnix japonica. 1. Selection response". Poultry Science Vol. 61, pp. 12-17. 1982.

[15] H. L. Marks. Long-term selection for body weight in japanese quail under different environments. Poultry Science Vol. 75, pp. 1198-1203. 1996.

[16] N. B. Anthony, K. E. Nestor and H. L. Marks. Short-term selection for four-week body weight in japanese quail. Poultry Science Vol.75, pp. 1192-1197. 1996.

[17] D. O. Owiti and S. Dadzie, "Maturity, fecundity and the effect of reduced rainfall on the spawning rhythm of a siluroidcatfish, Clarias mossambicus (Peters)". Aquaculture and Fisheries Management Vol. 20, pp. 355-368. 1989.

[18] C. J. J. Richter, W. J. A. R. Viveen, E. H. Eding, M. Sukkel, A. J. Rothuis, M. F. P. M. Van Hoof, F. C. J. Van Den Berg et al, The significance of photoperiodicity, water temperature and an inherent endogenous rhythm for the production of viable eggs by the African catfish, Clarias gariepinus, kept in subtropical ponds in Israel and under Israeli and Dutch hatchery conditions. Aquaculture Vol. 63, pp. 169-185. 1987.

[19] M. A. Blumler, "Independent inventionism and recent genetic evidence on plant domestication". Econ. Bot. Vol.46, pp. 98-111. 1992.

[20] J. Diamond, "Evolution, consequences and future of plant and animal domestication". Nature. Vol.418, pp. 700–707. 2002.

[21] F. Salamini, H. Özkan, A. Brandolini, R. Schäfer-Pregl and W. Martin, "Genetics and geography of wild cereal domestication in the near east. Nature Review Genetics Vol. 3, pp. 429–441. 2002.

[22] B. E. Uchola, "Agriculture: From a development perspective to Plant Resource Domestication". Amer. J. Agric. Forest. Vol. 3(4), pp. 127-134. 2015a.

[23] B. E. Uchola, "Agriculture: From a development perspective to Animal Resource Domestication. J. Res. Agric. Anim. Sci. Vol. 3(2), pp. 05-12. 2015b.

[24] J. G. Hawkes, "Significance of wild species and primitive forms for potato breeding". Euphytica. Vol.7, pp. 257–270. 1958.

[25] D. S. Brar, R. Dalmacio, R. Elloran, R. Aggarwal, R. Angeles and G. S. Khush, "Gene transfer and molecular characterization of introgression from wild Oryza species into rice". Khush G. S (Ed.) Rice Genetics Ill. Proceedings of the Third International Rice Genetics Symposium. Manila, Philipines. International Rice Research Institute, Manila-Philippines. pp. 477-486. 1996.

[26] R. Hajjar and T. Hodgkin, "The use of wild relatives in crop improvement: a survey of developments over the last 20 years". Euphytica, Vol.156: pp. 1–13. 2007.

[27] NRC. Microlivestock: Little-Known Small Animals with a Promising Economic Future. (Washington, DC. National Academy Press 1991). 448p. 1991.

[28] R. T. Loftus, D. E. MacHugh, D. G. Bradley, P. M. Sharp and P. Cunningham. Evidence for two independent domestication of cattle. Proceedings of the National Academy of Sciences USA, Vol. 91(7), pp. 2757–2761. 1994.

[29] D. G. Bradley, D. E. MacHugh, P. Cunningham and R. T. Loftus, "Mitochondrial DNA diversity and the origins of African and European cattle". Proceedings of the National Academy of Sciences USA, 93(10), pp. 5131–5135. 1996.

[30] A. Fumihito, T. Miyake, M. Takada, R. Shingu, T. Endo, T. Gojobori, N. Kondo and S. Ohno, "Monophyletic origin and unique dispersal patterns of domestic fowls". Proceedings of the National Academy of Sciences USA. Vol. 93(13), pp. 6792–6795. 1996.

[31] H. Sawai, H. L. Kim, K. Kuno, S. Suzuki, H. Gotoh, M. Takada, N. Takahata, Y. Satta, and F. Akishinonomiya, "The origin and genetic variation of domestic chickens with special reference to Jungle fowls *Gallus g. gallus* and *G. varius*". *PLoS ONE* 5(5): 2010.

[32] R. A. Fuller, J. P. Carroll and P. J. K. McGowan (eds.). Partridges, Quails, Francolins, Snowcocks, Guineafowl, and Turkeys. Status Survey and Conservation Action Plan 2000–2004. (Gland, Switzerland; Cambridge, UK: IUCN, and Reading, UK: the World Pheasant Association, 2000). vii + 63 pp.

[33] S. Hiendleder, K. Mainz, Y. Plante and H. Lewalski, "Analysis of mitochondrial DNA indicates that the domestic sheep are derived from two different ancestral maternal sources: no evidences for the contribution from urial and argali sheep". J. Hered., Vol. 89, pp. 113–120. 1998.

[34] Shackleton, D. M (Ed). Status and distribution of Caprinae by region. In: Wild Sheep and Goats and their Relatives. Status Survey and Conservation Action Plan for Caprinae. Shackleton, D. M. (ed.) and the IUCN/SSC Caprinae Specialist Group. IUCN, Gland, Switzerland and Cambridge, UK. 390 + vii pp. 1997.

[35] G. L. Luikart, L. Gielly, L. Excoffier, J-D. Vigne, J. Bouvet and P. Taberlet, "Multiple maternal origins and weak phylogeographic structure in domestic goats". Proceedings of the National Academy of Sciences USA, Vol. 98(10). pp. 5927–5930. 2001.

[36] M. B. Joshi, P. K. Rout, A. K. Mandal, C. Tyler-Smith, L. Singh and K. Thangaray, "Phylogeography and origins of Indian domestic goats". Mol. Biol. Evol., Vol. 21(3), 454–462. 2004.

[37] E. Guiffra, J. M. H. Kijas, V. Amarger, Ö. Calborg, J. T. Jeon and L. Andersson, "The origin of the domestic pigs: independent domestication and subsequent introgression". Genetics, Vol. 154(4), pp. 1785–1791. 2000.

[38] IUCN. The IUCN Red List of Threatened Species. International Union for the Conservation of Nature. www.iucnredlist.org.

[39] WWF. Endangered Species Conservation. World Wildlife Fund. www.worldwildlife.org.

[40] R. J. Hijmans and D. M. Spooner, "Geographic distribution of wild potato species". Amer. J Bot. Vol. 88(11), pp. 2101–2112, 2001.

[41] J. M. Neal, C. M. Hardner and C. L. Gross, "Population demography and fecundity do not decline with habitat fragmentation in the rainforest tree Macadamia integrifolia (Proteaceae)". Biol. Cons., Vol.143, pp. 2591–2600. 2010.

[42] The IUCN Red List of Threatened Species. *Bos, Capra and Ovis.* www.iucnredlist.org. Downloaded on 23 February 2016.

[43] C. M. Herrera, P. Jordano, J. Guitian and A. Traveset, "Annual variability in seed production by woody plants and the masting concept: Reassessment of principles and relationship to pollination and seed dispersal". Amer. Naturalist Vol. 154, pp. 576– 594. 1988.

[44] D. Kelly and V. L. Sork, "Mast seeding in perennial plants: Why, how, where". Annual Rev. Ecol. Syst. Vol. 33, pp. 427–447, 2002.

[45] E. E. Goldschmidt, "The evolution of fruit tree productivity: A review". Econ Bot. Vol. 67(1), pp. 51–62, 2013.

[46] C. M. Hardner, C. Peace, A. J. Lowe, J. Neal, P. Pisanu, M. Powell, A. Schmidt, C. Spain and K. Williams, "Genetic Resource and Domestication of Macadamia". Horticultural Reviews Vol. 35, pp. 1–125, 2009.

[47] D. O. Ladipo, J. M. Fondoun and N. Ganga, "Domestication of the bush mango (Irvingia spp.): some exploitable intraspecific variations in west and central Africa", in Domestication and commercialization of non-timber tree products for Agro-forestry. FAO Tech Paper, No. 9. FAO, Rome. Pp. 193–205, 1996.

[48] I. Zulkifli, H. S. Iman Rahayi, A. R. Alimon, M. K. Vidyadaran and S. A. Babjee, "Responses of choice-fed red jungle fowl and commercial broiler chickens offered a completed diet, corn and soybean". Asian-Australasian J. An. Sci. Vol. 14(12), pp. 1758-17562. 2001.

[49] J. A Oluyemi, D. F. Adene and G. O. laboye, "Comparison of Nigeria indigenous fowl with White Rock under conditions of disease and nutritional stress". Trop. Anim. Hlth Prod. Vol. 11, pp. 199-202. 1979.

[50] B. D. Binda, I. A. Yousif, K. M. Elamin and H. E. Eltayeb, "A comparison of performance among exotic meat strains and local chicken ecotypes under Sudan conditions". Int. J. Poul. Sci. Vol. 11(8), pp. 500-504. 2012.

[51] The IUCN Red List of Threatened Species. *Struthio camelus, Capra aegagrus.* www.iucnredlist.org. Downloaded on 23 February 2016.

[52] FAO, DFSC, IPGRI. Forest genetic resources conservation and management. Vol. 2: In managed natural forests and protected areas (*in situ*). International Plant Genetic Resources Institute, Rome, Italy. 98pp +. 2001.

[53] E. T. Ayuk, B. Duguma, S. Franzel, J. Kengue, M. Mollet, T. Tiki-Manga and P. Zenkeng "Uses, management and economic potential of Irvingia gabonensis in the humid lowlands of Cameroon". Forest Ecol Mgt, Vol. 113, pp. 1–9. 1999.

[54] P. N. Shiembo, A. C. Newton and R. R. B. Leakey, "Vegetative propagation of Irvingia gabonensis, a West African fruit tree". Forest Ecol Mgt Vol. 87, pp. 185-192. 1996.

[55] Z. Tchoundjeu, B. Duguma, J-M. Fondoun and J. Kengue, "Strategy for the domestication of indigenous fruit trees of West Africa: case of Irvingia gabonensis in southern Cameroon". Cameroon J Biol Biochem Sci Vol. 4, pp. 21-28, 1998.

[56] A. R. Atangana, Z. Tchoundjeu, J-M. Fondoun, E. Asaah, M. Ndoumbe and R. R. B. Leakey, "Domestication of Irvingia gabonensis: I. Phenotypic variation in fruit and kernel traits in two populations from the humid lowlands of Cameroon". Agroforestry Systems. Vol. 53, pp. 55-64, 2001.

[57] S. S. Ajayi and O. O. Tewe "Food preference and carcass composition of the grasscutter (*Thryonomys swinderianus*) in captivity". Afr. J. Ecol. Vol. 18 (2-3), pp. 133–140. 1980.

[58] C. H. Steir, G. A. Mensah and C. F. Gall, "Breeding of cane rats (*Thrynomys swinderianus*) for the production of meat". World Anim. Rev. Vol. 69, pp. 44-49. 1991.

[59] S. A. Onadeko and F. O. Amubode, "Reproductive indices and performance of captive reared grasscutters *(Thryonomys swinderianus Temminck)*" Nig. J. Anim. Prod. Vol. 29(1), pp. 142-149. 2002.

[60] S. Y. Annor, B. K. Ahunu, G. S. Aboagye, K. Boa-Amponsem and J. P. Cassady, "Phenotypic and genetic estimates of grasscutter production traits. 1. (Co) variance components and heritability". Glo. Adv. Res. J. Agric. Sc. Vol. 1(6), pp. 148-155. 2012a.

[61] S. Y. Annor, B. K. Ahunu, G. S. Aboagye, K. Boa-Amponsem and J. P. Cassady, "Phenotypic and genetic estimates of grasscutter production traits.2. Genetic and phenotypic correlations. Glo. Adv. Res. J. Agric. Sc. Vol. 1(6), pp. 156-162. 2012b.

[62] C. D. Ataga and H. A. M. Van Der Vossen "Elaeis guineensis Jacq". In: Van Der Vossen, HAM & Mkamilo GS (eds). PROTA 14: Vegetable oils/Oléagineux. [CD-Rom]. Wageningen, Netherlands. 2007.

[63] R. I. Ogundipe and A. A. Adeoye, "Evaluation of the dairy potential of Friesian, Wadara and their crossbreds in Bauchi State". Scholarly J Agric Sc. Vol. 3(6), pp. 223-225. 2013.

[64] V. Buvanendran, M. B. Olayiwole, K. I. Protrowskiu and B. A. Oyejola, "A comparison of milk production traits in Friesian x White Fulani crossbred cattle". Anim prod Vol. 32, pp. 165-170. 1981.

[65] The plant list 2013. www.theplantlist.org.

[66] J. B. Phipps, K. R. Robertson, P. G. Smith, J. R. Rohrer, "A checklist of the subfamily Maloideae (Rosaceae)". Can. J. Bot. Vol. 68, pp. 2209–2269. 1990.

[67] G. S. Khush, "Origin, dispersal, cultivation and variation of rice". Plant Molecular Biology Vol. 35, pp. 25–34, 1997.

[68] Germplasm Resources Information Network, United States Department of Agriculture. http://www.ars-grin.gov Retrieved 2015-11-01.

[69] FAO International Year of Potato, 2008. Potato varieties. http://www.fao.org/potato-2008/en/potato/varieties.html

[70] IRRI Rice varieties. Global release: 2014. irri.org

[71] Rubber Board 2002. Rubber clones. http://rubberboard.org.

[72] Froese R. and Pauly D., eds. Species of *Cyprinus* in FishBase. August 2011 version.

[73] The IUCN Red List of Threatened Species: *Gallus and Sus*. www.iucnredlist.org. Downloaded on 23 February 2016.

[74] FAO. The State of the World's Animal Genetic Resources for Food and Agriculture, Barbara Rischkowsky & Dafydd Pilling (Ed). FAO, Rome. Pp. 31-36. 2007.

[75] FAO Domestic Animal Diversity Information System. dad.fao.org Retrieved 2016-01-28.

22

Economics of Rural Livelihoods: A Case Study of Bitter Kola Marketing in Akwa Ibom State, Nigeria

Asa Ubong Andem, Daniel Enwongo Aniedi, Ebong Effiong Okon

Department of Agricultural Economics and Extension, University of Uyo, Uyo, Akwa Ibom State, Nigeria

Email address:

keana0772@yahoo.com (A. U. Andem)

Abstract: The study examined the economics of bitter kola marketing in rural areas of Akwa Ibom State, Nigeria. A sample size of 120 bitter kola marketers was selected for the study using a two-stage sampling procedure. Data obtained were analysed using descriptive statistics and budgeting technique. Findings reveal that 61.7% of the respondents were females with an average age of 37 years; 58.3% were married and 85% had formal education. The average household size of the respondents comprised of six persons and their average years of bitter kola marketing experience was also six years. Budgetary analysis indicated the bitter kola marketing in the study area is profitable with a marketing efficiency of 135.2%. Transportation costs, poor marketing channels, price fluctuation, perishability and seasonality of the product were the most severe constraints to bitter kola marketing faced by the respondents. -Findings recommend, among others, that the Government (State and Local Government) endeavour to provide basic infrastructure such as good road networks in the rural areas of the state since high cost of transportation was the most severe constraint to bitter kola marketing in the study area.

Keywords: Bitter Kola, Marketing, Rural Areas, Akwa Ibom State

1. Introduction

Garcinia kola commonly known as 'bitter kola' is an economic and highly valued nut-bearing tropical tree available in large quantity in West Africa [1]. The tree is commonly found in timid lowland forests of Nigeria, Cameroon, Ghana and the Benin Republic [2]. The tree produces edible and medicinal seeds which are widely consumed [3]. The nuts have a bitter taste, followed by slight sweetness hence the name 'bitter kola'. Despite its bitter taste, *Garcinia kola* nuts are commonly eaten as snacks and used for their stimulant effects due to high caffeine content [4].

Garcinia kola has great economic value across West Africa; and the seeds are of particular importance in the social-cultural lives of the people in the tropics[5]. The nuts of bitter kola are highly valued for their perceived medicinal attributes, and consumption of large quantities of them does not cause indigestion[6]. [7] reported that the seeds, nuts and bark of *Garcinia kola* plants have been used extensively in African traditional medicine for the treatment of various diseases. Currently, *Garcinia kola* is being harnessed as a cure for Ebola virus infection and flu[8]. Its economic contribution to both domestic and national markets, according to [9], raises the standard of living of those involved in its trading activities both in the rural and urban centers. Trading in bitter kola is more profitable than trading in non-timber forest products in most developing countries [6]. This is because of the high amenability of bitter kola (both in fresh and dried forms) to storage. In Nigeria, where employment opportunities for traditional industries are declining, workers looking for alternative sources of income often turn to the collection of non-timber forest products, such as bitter kola, from nearby forests[10].

Inspite of the importance of bitter kola, there has been a dearth of information on the profitability of bitter kola marketing in the rural areas of Akwa Ibom State, Nigeria[11]. This study, therefore, aimed at filling this research gap by ascertaining the economics of bitter kola marketing in rural areas of Akwa Ibom State, Nigeria. Specifically, the objectives were to examine the socio-economic characteristics of bitter kola marketers in the rural areas of Akwa Ibom State, determine the profitability of bitter kola marketing in the study area and ascertain the constraints to bitter kola marketing in the study area.

2. Methodology

The study was -conducted in AkwaIbom State - that lies within the South-Eastern axis of Nigeria, wedged between Cross River, Abia and Rivers States. On the southern margins of the State is the Atlantic Ocean which stretches from Ikot Abasi Local Government Area to Oron Local Government Area. The Administrative Capital of the State is Uyo Local Government Area. The State lies between 4^0 3" and 5^0 3" North latitudes and 7^0 35" and 8^0 25" East longitudes. It has an estimated population of 3,920,208 -out of the total land area of 7,245, 935km², [12]. A two–stage sampling procedure was employed for this study. The first stage involved the simple random selection of three out of the six Agricultural Development Project (ADP) zones in Akwa Ibom State. The selected ADP zones were Abak, Ikot Ekpene and Uyo and, the second stage of the sampling procedure involved the purposive selection of forty bitter kola marketers from rural areas in each of the three selected ADP zones. Purposive sampling was employed to ensure that only bitter kola marketers in the rural areas of the zones were selected for the study. -Selection resulted in a sample size of 120. Data collection was done through primary sources using a validated questionnaire. -The study wascarried out between January 2014 and October 2014.

Analysis of data was done using descriptive statistics and budgetary analysis. Frequency counts, percentages, means and ranks were used to analyse the socio-economic characteristics of bitter kola marketers in rural areas of AkwaIbom State. Budgetary analysis was employed to estimate the cost, revenue, gross margin and net profit accruable to the marketers. The equation used in estimating the gross margin is defined below:

$$GM = TR - TVC \qquad (1)$$

Where:
GM= Gross Margin, TR= Total Revenue and TVC = Total Variable Cost

The equation used in ascertaining the net profit is as follows:

$$Л = TR - TC \qquad (2)$$

Where:
Л = Profit
TR = Total Revenue
TC = Total marketing Cost

The efficiency of marketing was measured by Shepherd-Futrel model for accurate measurement of efficiency in the productivity of resources invested in the marketing process in quantitative terms[13]. The model is given as:

$$ME = \frac{TR}{TC} \times 100 \qquad (3)$$

Where:
ME = Marketing Efficiency
TC = Total marketing Cost

TR = Total Revenue

In order to ascertain the constraints to bitter kola marketing faced by the respondents, twelve constraints were identified through Focus Group Discussions (FGDs), interviews and literature; and the respondents were requested to indicate the severity in each of the constraint item. This was done with the aid of a 3-point rating scale, with nominal values assigned to the points in the scale, as follows: Not a constraint = 1, Mild constraint =2 and Severe constraint = 3. A mean score was computed for each constraint item, and the mean score was used to rank the constraints in order of severity.

3. Results and Discussion

Table 1. *Socio-economic characteristics of bitter kola marketers in Akwa Ibom State.*

Variables	Categories	Frequency (n = 120)	Percentage	Mean
Age	21-30 years	42	35.0	37
	31-40 years	30	25.0	
	41-50 years	34	28.3	
	51-60 years	14	11.7	
Sex	Male	46	38.3	
	Female	74	61.7	
Marital status	Single	32	26.7	
	Married	88	73.3	
Educational status	No formal education	18	15.0	
	Primary education	47	39.2	
	Secondary education	55	45.8	
Religion	Christianity	108	90.0	
	Non-Christianity	12	10.0	
Household size	1-3 person(s)	19	15.8	6
	4-6 persons	61	50.8	
	7-9 persons	40	33.3	
Marketing experience	1-7 year(s)	92	76.7	6
	8-14 years	28	23.3	
Access to credit	Yes	19	15.8	
	No	101	84.2	
Access to extension	Yes	27	22.5	
	No	93	77.5	

Source: Field Survey data, 2014

Socio-economic characteristics of bitter kola marketers: Table 1 shows the socio-economic characteristics of bitter kola marketers in Akwa Ibom State. The Table reveals that the mean age of bitter kola marketers in the study area was 37 years which agrees with [14] who reported that majority of the bitter kola marketers in Nigeria were within the age range of 20-50 years. About 61.7% of the respondents were females and majority of them (58.3%) were married. Marriage is a highly cherished value among people in rural areas of Akwa Ibom State, Nigeria[15]. From the table, majority of the respondents had formal education (85.0%) and - 90.0% were Christians. [16] reported a high level of literacy among fruit marketers in Akwa Ibom State. The average household size of the respondents comprised of six persons. The finding agrees with [17] who reported that the average household size of rural dwellers in Akwa Ibom is low. The average years of experience in bitter kola marketing of the respondents was six years, which is relatively low. Over eighty four percent of the respondents (84.2%) had no access to credit facilities and 77.5% of them had no access to extension services.

Profitability of bitter kola marketing in the study area: Table 2 shows the profitability of bitter kola marketing ascertained using budgetary analysis. - The total cost of marketing bitter kola during the period of the studywas ₦2, 524, 590.00 and the total revenue of ₦3, 413, 591.00 was realized, hence a net profit of ₦ 889, 001.00 was realized from sales of bitter kola by the marketers between January, 2014 to October, 2014. The net profit shows that bitter kola marketing in the study area was profitable. The finding agrees with [11] who reported that bitter kola marketing is a profitable enterprise in Akwa Ibom State. [6] also reported that bitter kola trading is a significant economic livelihood of people in the rain forest and derived savanna ecological zones of Nigeria. The marketing efficiency of bitter kola marketing in the study area was 135.2 as shown in Table 2. This implies that bitter kola market in the study area is efficient.

Table 2. *Budgetary analysis of bitter kola marketing in rural areas of Akwa Ibom State, Nigeria.*

Item	Total cost for all respondents (₦)	Percentage of TC
A. Fixed cost		
Basin	57, 000.00	2.26
Bags	4, 600.00	0.18
Rent	235, 970.00	9.35
Tax	89, 150.00	3.53
Equipment/machinery	180, 450.00	7.15
Total Fixed Cost (TFC)	567, 170.00	22.47
B. Variable cost		
Supply Price	1, 385, 000.00	54.86
Transportation	218, 000.00	8.64
Labour	209, 930.00	8.32
Processing cost	40, 022.00	1.59
Wheel barrow hire	72, 250.00	2.86
Storage	13, 408.00	0.53
Water	18, 810.00	0.75
Total Variable Cost (TVC)	1,957,420.00	77.53
Total Cost = TFC + TVC	2,524,590.00	100
C. Total Revenue (TR)	3,413,591.00	
D. Net Profit, Л	889, 001.00	
E. Gross Margin (GM)	1,456,171.00	
E. Marketing efficiency (ME)	135.2	

Note: Naira (₦) is the Nigerian currency and 1.00 US Dollar is equal to 199.20 Nigerian Naira as at November 7, 2014
Source: Computed from Field Survey data, 2014

Constraints to bitter kola marketing in rural areas of Akwa Ibom State: Table 3 shows the constraints to bitter kola

marketing faced by_the respondents in the study area. Transportation cost (x =0.99), poor marketing channels (x = 0.98), price fluctuation (x =0.88), perishability (x =083) and seasonality of the product (x =0.81) were the most severe constraints to bitter kola marketing in rural areas of Akwa Ibom State. The findings corroborate [18] who reported that transportation, irregular supply of product, poor marketing and price fluctuation were major challenges facing the marketing of non-timber forest products such as bitter kola in Nigeria. [19] opined that price fluctuation as a major constraint to bitter kola marketing is due to the fact that the forest fruits are not always available throughout the year due to their seasonal nature and perishability which makes them scarce leading to unwarranted changes in prices of the fruits. The least severe constraints to bitter kola marketing in the study area were: lack of relevant extension services (x =0.50) and high cost of registration/selling permit (x =0.48).

Table 3. *Constraints to bitter kola marketing in rural areas of Akwa Ibom State.*

Constraints	Mean	Rank
1. Transportation cost	0.99	1
2. Lack of storage facilities	0.66	7
3. Price fluctuation	0.88	3
4. Deforestation	0.59	10
5. Lack of modern processing technologies	0.63	8
6. Seasonality of the product	0.81	5
7. Lack of capital/finance	0.69	6
8. Poor marketing channels	0.98	2
9. Perishability	0.83	4
10. Lack of relevant extension services	0.50	11
11. Poor patronage	0.62	9
12. High cost of registration/selling permit	0.48	12

Note: * = Rank 1 is considered the most severe constraint while rank 12 is the least severe constraint
Source: Field Survey data, 2014

4. Conclusions and Recommendations

The study has shown that bitter kola marketing in the rural areas of Akwa Ibom State is profitable as evidenced from the result of budgetary analysis. The study also revealed that the most severe constraints to bitter kola marketing in the study area were: transportation cost, poor marketing channels, price fluctuation, perishability and seasonability of bitter kola nuts.

Based on the findings of the study, the following recommendations are made:

a. The Government (State and Local) should endeavour to provide basic infrastructure such as good road networks in the rural areas of the state since high cost of transportation was the most severe constraint to bitter kola marketing in the study area.

b. The officials of the Ministry of Agriculture in Akwa Ibom State, Non-Governmental Organisations and Community-based Organisations organize training programmes on effective marketing for rural-based bitter kola marketers in the State since poor marketing channel was another major constraint to bitter kola marketing faced by the respondents.

c. Subsidized storage facilities /modern processing equipment should be made available to bitter kola marketers by both Governmental and Non-Governmental agencies in the State.

d. Bitter kola marketers in the study area should endeavour to form co-operatives to enhance their ability to access storage and transportation facilities thereby overcoming some of the major constraints to their marketing activities.

References

[1] Ikpesu, T. O., Tongo, I. and Ariyo A. (2015) Restorative Prospective of Powdered seeds extract of G. Kola in *Chrysihthysfurcatus* induced with Calyphosate Formulation, *Nature and Science*, 13 (11): 91-100.

[2] Unaeze, H. C., Oladele A. T. and Agu, L. O. (2013) Collection and marketing of bitter cola (*Garcinia kola*) in Nkwerre local Government Area, Imo state, Nigeria, *Egyptian Journal of Biology*, (1) 5: 37-43.

[3] Okigbo, B. N. (1977) Neglected plants of importance in traditional faming system of Tropical Africa, *Acta. Horticulture*, 5(3): 131-150.

[4] Ayensu, E. S. (1978) *Medicinal Plants of West Africa*, Reference Publ.Inc. Algonac, ML., U.S.A, pp 162-163.

[5] Adebayo, S. A. and Oladele, O. I. (2012) Medicinal values of kolanut in Nigeria: implications for extension service delivery, *Life Science Journal*, 9 (2): 887-891.

[6] Adebesi A. A. (2004). A case study of Garcinia kolanut production to consumption system in J4 area ofomo forest reserve south west Nigeria. In: sunder land T. and Ndoye O. (eds) forest products livelihoods and conservation case studies of non-timber forest products systems. Vol. 2. Africa. CIFOR ISBN 979-3361-25-5. pp 115-132.

[7] Okoli, U. J., (1991). An Investigation into the Hypoglycemic Activity of GBI Biflavonoids of Garcinia kola, B. Pharm. Project, University of Nigeria, Nsukka.

[8] Adaramonye, O. A., Nwaneri, V. O. Anyanwu, K. C., Farombi, E. O. and Emerole, G. O. (2009) Possible anti-antherogenic effect of kolaviron (a Garcinia kola seed extract) in hypercholesterolaemicrats, *African Journal of Biotechnology*, 32(1-2): 40-46.

[9] Yakubu, F. B., Adejoh, O. P.,Ogunade, J. O. and Igboanugo, A B. I. (2014) Vegetative propagation of Garcinia kola (Hcckel), *World Journal of Agricultural Science*, 10 (3): 85-90.

[10] Adepoju, A. A. and Salau, A. S. (2007) Economic valuation of Non-Timber forest products MPRA. Paper No. 2689.Available online at http://mpra.ub.uni-muenchen.de/2689/. Accessed on August 08, 2007.

[11] Akpan, V. U. (2015) Economics of the bitter kola marketing in Uyo Metropolis of Akwa Ibom State, B. Agric, Department of Agricultural Economics and Extension, University of Uyo, Uyo. 65pp.

[12] National Population Commission (2006) Population and Housing Census of the Federal Republic of Nigeria. Analytical Report at the National Population Commission, Abuja, Nigeria.

[13] Omofonwam, E. I., Ashaolu O. F., Ayinde I. A. and Fakoya E. O. (2013) Assessment of palm wine market in Edo State, *Journal of Science and Multidisciplinary Research*, 5(2): 141 – 151.

[14] Agbelade, A. D. and Onyekwelu, J. C. (2013) *Poverty alleviation through optimizing the marketing of Garcinia kola and Irvingia Gabonensis in Ondo State, Nigeria,* Hindawi Publishing Corporation, Nigeria, pp 1-5.

[15] Ekong, E. E (2003) *An Introduction to Rural Sociology* (2nd Edition), Dove Educational Publishers, Uyo, Nigeria. pp 341-395.

[16] Ekerete, B. A. and Asa, U. A. (2014) Constraints to watermelon marketing in Uyo Metropolis of Akwa Ibom State, Nigeria, *Journal of Agricultural and Environmental Sciences*, 3 (4): 63-69.

[17] Asa, U. A. and Eyo, E. J. (2015) Constraints to palm wine marketing in rural areas of Akwa Ibom State, Nigeria, *British Journal of Science*, 13 (1): 45-52.

[18] Yusuff, A. O., Adams, B. A., Adewole, A. T. and Olatoke, T. I. (2014) NTFPs collection as an alternative source of income for poverty alleviation among rural farmers in Egbeda Local Government Oyo State, Academic Journal of Interdisciplinary Studies, 3(6): 467-474.

[19] Famuyide, O. O., Adebayo, O. Arabomen, O. and Jasper, A. A. (2012) Economic assessment of marketing of non-wood forest products in Ibadan Metropolis, *Elixir International Journal*, 52 (2012): 11645-11649.

Floristic Composition and Vegetation Structure of Woody Species in Lammo Natural Forest in Tembaro Woreda, Kambata-Tambaro Zone, Southern Ethiopia

Melese Bekele Hemade[1], Wendawek Abebe[2]

[1]Ethiopian Biodiversity Institute, Forest and Rangeland plants Biodiversity Directorate, Addis Ababa, Ethiopia
[2]Hawassa University, Hawassa, Ethiopia

Email address:
melesebekeles@gmail.com (M. B. Hemade)

Abstract: This study was conducted to investigate the floristic composition of woody species in Lammo natural Forest, Tembaro woreda, Kembata-Tambaro zone, Southern Ethiopia. 52 quadrants, each with 20 m x 20 m (400 m^2) were systematically laid to collect vegetation data along line transects at the distance of 60m from each other. Vegetation structures such as DBH, basal area, height, frequency, IVI and Species population structure were computed. Specimens were collected, pressed, dried and taken to Ethiopian National Herbarium for identification. Data analysis was carried out using Shannon-Wiener Diversity index, Microsoft Excel, R-package, past and Sorensen's similarity index. 54 woody species belonging to 46 genera from 29 families were identified. The dominant family was Myrtaceae followed by Euphorbiaceae. The vegetation cluster classification resulted in three plant communities. Most of the species in the study area fall under middle diameter and height classes. The results of population structure for the forest, revealed the signs of some disturbances and hence some management and conservation practices need to be in place.

Keywords: Basal Area, Community Classification, DBH, Floristic Composition

1. Introduction

Tropical forests constitute the most diverse plant communities on earth [1]. Ethiopia is one of the tropical countries with diverse flora and fauna. However, the forest cover in Ethiopia has been declining rapidly due to anthropogenic impacts [2]. A reduction of biological diversity will negatively affect vital ecosystem functions that regulate the Earth system upon which humans ultimately depend [3]. An obvious approach to conserve plant biodiversity is to map distributional patterns and look for concentrations of diversity and endemism [4]. Further, management of forest requires understanding of its composition in relation to other forests, the effects of past impacts on the present status and the present relationship of the forest with surrounding land uses [5].

Most of the remaining forests of Ethiopia are confined to the south and southwest parts of the country, which are less accessible, and/or less populated [6]. Nowadays, even the remnant natural forests in these areas are continuously threatened by human activities. Lammo Natural Forest is also one of the remaining forests in south region and so far, no studies have been reported on the forest. Therefore, this study was undertaken to describe and provide valuable information on floristic composition and vegetation structure of woody species in the forest.

2. Materials and Methods

2.1. Description of the Study Area

The study area is located within the geographic co-ordinates of 7° 17' 030"N to 7° 19' 55N latitudes and 37° 33' 13"E to 37° 55' 40"E longitudes. It is situated at a distance of about 400kms South of Addis Ababa, between the altitudinal ranges of 2010 and 2484-m.as.l. The natural vegetation of the area is a broad–leaved and evergreen with the most dominant tree species including *Syzygium guineense, Croton macrostachyus, Ficus sycomorus,* and *Ekebergia capensis.*

2.2. Data Collection and Identification

The woody species specimens were collected by laying a quadrates of 20×20 m (400m^2) at a distance of 60m along the transects. 52 plots of the forest were sampled and the specimens of these plots were counted, numbered and pressed; height and Diameter at Breast Height (DBH) of each species with height greater than or equal to 2.5 m and DBH greater than or equal to 3 cm in each plot were measured. The collections were first named using the folk taxonomy as field identification and formal taxonomic identification to species level was made later using the voucher specimens at the National Herbarium, Addis Ababa University.

3. Methods of Data Analysis

3.1. Vegetation Classification

Cluster analysis was used for the purpose of vegetation classification into different community types using the statistical software R-package for windows version 2.15.0 [7]. The Indicator Species Analysis was made to compare the species present in each community.

3.2. Diversity Index

Species diversity and evenness are often calculated using Shannon-Wiener diversity index [8].

$$H' = -\sum_{i=1}^{s} pi \ln pi$$

Where: H'= Shannon Diversity Index

s = the number of species
Pi = the proportion of individuals or the abundance of the ith species expressed as a proportion of total cover and
ln = natural logarithm.
And that of Sorensen's similarity was calculated by using the formula Ss = 2 a /2a+b+c
Where: Ss = Sorenson's similarity coefficient
a = Number of species common to both communities
b = number of species in community1
c = number of species in community 2

3.3. Vegetation Structure

To describe the vegetation structure of the forest, density, frequency, height, Diameter at Breast Height (DBH), composition, family importance value (FIV), species importance value (SIV) and basal area were calculated following [9].
BA = πd^2/4, where BA=Basal Area in m^2 per hectare
d = diameter at breast height (m)
π = 3.14

3.4. Species Population Structure

Woody species in the forest with a diameter at breast height (DBH) greater than 3 cm, and height greater than 2.5 m were measured to analyze the DBH class distribution by classifying the DBH values of the species in to seven class intervals (3-20cm, 21-40cm, 41-60cm, 61-80cm, 81-100cm, 101-120cm and >120cm. Individuals with DBH less than 3 cm, and height less than 2.5 m were counted. Similarly, their height was measured and classified into eight class intervals. Thus, height classes were ranked as <5m, 5.1-10m, 10.1-15m, 15.1-20m, 20.1-25m, 25.1-30m, 30.1-35m and >35m.

4. Results and Discussion

4.1. Floristic Composition

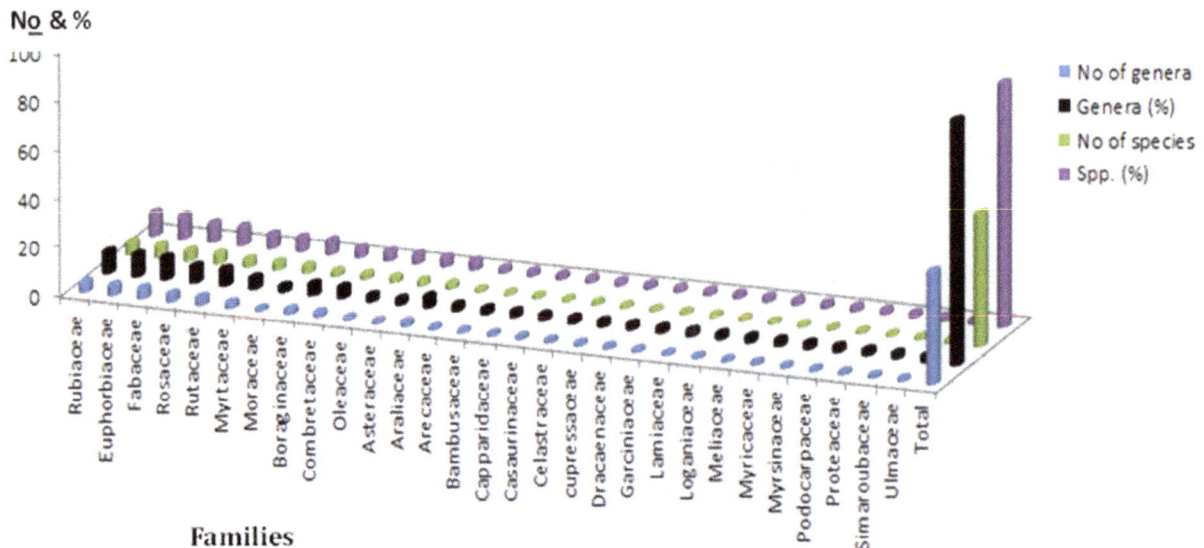

Fig. 1. *Showing families with their representative genera, species and their percentage.*

In the study, 54 woody species, belonging to 46 genera and 29 plant families were recorded. Rubiacea, Euphorbiaceae and Fabaceae each represented by 8.7% of the genera followed by Rutaceae and Rosaceae each representing 6.5% of the genera and 4.35% of the genera was represented by each of the other four families (Myrtaceae, Boraginaceae, Combrtaceae and Araliaceae). The remaining 20 families were represented by only a single genus. With respect to species composition, Rubiacea and Euphorbiaceae were found to be represented by five species each (9.26%) followed by Fabaceae and Rosaceae each having four species (7.4%). Rutaceae, Myrtaceae and Moraceae each represented by three species; Boraginaceae, Combretaceae, Oleaceae, Asteraceae and Araliaceae each represented by two species

and the remaining 17 families were found to be represented by only a single species and it is shown in figure 1.

4.2. Vegetation Classification

From cluster analysis of the forest, three plant communities were identified as shown in figure 2. Each community was named by the species having higher indicator value. Accordingly, the three identified communities were described as Myrica salicifolia- Maytenus ovata (Community type1), Millettia ferrugina-Syzygium guineense (Communitytype 2), and Macaranga capensis- Galiniera saxifraga (Community type 3).

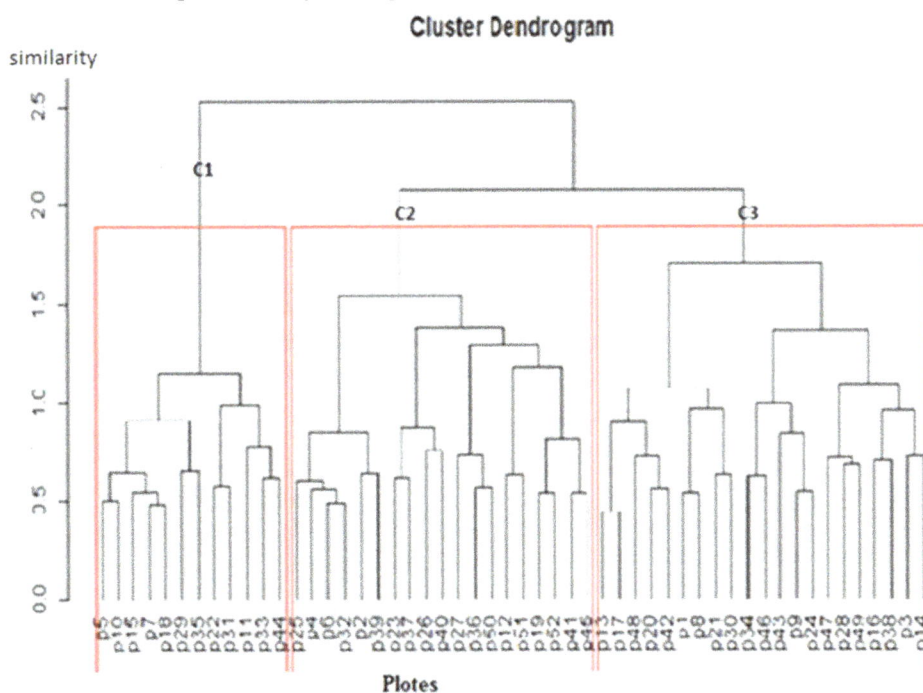

Fig. 2. *Dendrogram showing the three community types obtained from cluster analysis of the Forest.*

4.2.1. Myrica salicifolia- Maytenus Ovata Community Type

This community comprised of 12 quadrates (about 0.48 hectar) and two species were found to be indicator species (Myrica salicifolia and Maytenus ovata) with significant indicator values in the community (Annex1)

4.2.2. Millettia Ferrugina- Syzygium Guineense Community Type

This community type was represented by 19 quadrates (0.76ha) and two species (*Millettia ferrugina* and *Syzygium guineense* were found to be indicator species of the community as they displayed the highest indicator values.

4.2.3. Macaranga Capensis- Galiniera Saxifraga Community Type

The two indicator species of this community were *Macaranga capensis and Galiniera saxifraga* and it comprised of 21 quadrates (0.84ha).

4.3. Species Diversity, Evenness and Richness in the Communities

Shannon Wiener diversity index revealed that *Macaranga capensis- Galiniera saxifraga* community type had the highest species diversity and richness compared to the other two community types. The possible reason for high diversity and richness could be its being situated at the medium average altitude interval (2230 m.a.s.l) which is relatively more favorable for growth and reproduction of a variety of species in the area. In contrast, *Myrica salicifolia- Maytenus ovata* community type had the least species diversity and richness. This could be associated to growth at a relatively higher altitude in which only more adaptive species potentially grow better than the others. The third community type, which had lowest average altitude range of the study site, was characterized by the intermediate species diversity and richness (Table 1).

Table 1. Shanon-Wiener Diversity Index of the three communities in Lammo natural forest.

Communities	Average altitude (ma.s.l)	Species Richness	Diversity Index (H')	H'max	Evenness (H'/H'max)
1	2280	43	3.751	3.964278	0.9462
2	2190	45	3.774	3.970542	0.9505
3	2230	49	3.789	3.93213	0.9636

4.4. Similarity of Species Composition Among the Three Community Types

The result of Sorensen's similarity coefficient of the three communities displayed, community 1 and 3 share more species in common (90.3%) followed by 2 and 3 (87.9%) and that of community 1 and 2 was relatively low (82%).

Table 2. The similarity index of plant species composition in three communities.

Community types	1	2	3
1	-	0.82	0.903
2	-	-	0.879
3	-	-	-

4.5. Density

Of all the collected and identified families of woody species in the forest, Myrtaceae family was found to have highest number of tree individuals per hectare (12.9%) followed by Euphorbiaceae (12. 87%). On the other hand, Garciniaceae (0.37%) was found to have the least number of individuals per hectare.

4.6. Dominant Species of the Forest

The species having relatively higher species importance value (SIV) and basal area (*BA)* were considered as dominant species. Accordingly, *Croton macrostachyus with SIV (31), Syzygium guineense (18.97), Ficus sycomorus (17.6), Eucalpytus globulus (12.3), Galiniera coffeoides (12.2), Millettia ferrugina (9.8), Schefflera abyssinica (9.6), Ekebergia capensis (9.2), Myrica salicifolia (7.2), Eucalyptus camaldulensis (6.7)* were some of the dominating species of

the forest. The dominance of these species was because of their abundance in distribution and high basal area within the forest.

4.7. Commercially Important Tree Species in the Forest

The forest contains some of the major commercially important indigenous tree species reported by Ethiopian Forestry Action Plan [10]. These tree species include *Albizia gummifera, Celtis africana, Croton macrostachyus, Ekebergia capensis, Hagenia abyssinica, Olea welwitschii, Prunus africana* and *Syzygium guineense*. In addition, some endemic tree species of Ethiopia, such as *Millettia ferruginea, Vepris dainellii* and *Erythrina brucei* were found in the forest.

4.8. Vegetation Structure of the Forest

4.8.1. Diameter at Breast Height (DBH)

In contrast to the similar study conducted in Sese forest that showed the inverted J-shape pattern of distribution [11], the distribution of individual woody species across the DBH classes displayed irregular shaped pattern of distribution. The middle DBH classes were found to have relatively higher number of individuals than that of the lower and the top classes. About 70% of the individuals were found in the middle four classes (21-40cm, 41-60cm, 61-80cm and 81-100cm). Less percentage of individuals was recorded for both the lowest and the highest DBH classes. This indicates that small sized and very large sized individuals were not common in Lammo natural forest. Hence, this distribution depicts that the forest is on the status of low regeneration and there seems to be selective removal of bigger individuals for different purposes.

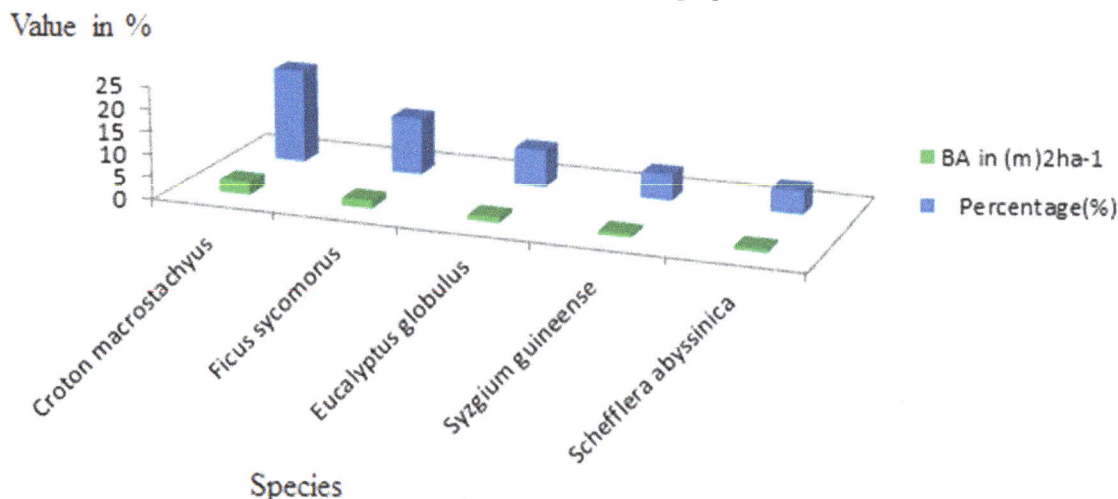

Fig. 3. BA of the five most important tree species in the forest.

4.8.2. Basal Area of the Species in the Forest

The total basal area (BA) of the recorded species in the forest was found to be 13.35m²/ha. *Croton macrostachyus, Ficus sycomorus, Eucalyptus globules, Syzgium guineense* and *Schefflera abyssinica* covered about 52% of the total basal area. According to the standard set by Dawkins, (1959) cited in [12], the normal value of basal area (BA) for virgin tropical rain forest in Africa is expected to be between 23 and 37-m2 ha⁻¹. In this regard, Lammo natural forest displayed lower basal area that might indicate the presence of different factors that could potentially affect the composition in the area. These factors include –the destruction of mature trees for logging, timber production and fuel wood collection. Compared to similar studies in Ethiopia, the BA of Lammo natural forest is relatively greater than that of Beschillo and Abay Riverine Vegetation [12] with total BA (12.6) and far less than that of Komto Afromontane Rainforest [13] with total BA (50.72m²/ha).

4.8.3. Height Class

Most of the individuals in the forest belong to the third height class (10.1-15m) followed by the second height class (5.1-10m). The least number of individuals corresponds to the biggest height class (>35m). This indicates that most of the recorded species of the forest (about 60%) were short (<15m in height). Few in number or the absence of large individuals in the forest might be associated with the selective cutting of species for various purposes.

4.8.4. Family Importance Value (FIV)

Six of the families (Euphorbiaceous, Myrtaceae, Moraceae, Rubiaceae, Fabaceae and Rutaceae) together contributed about 50% of the total FIV. Euphorbiaceae displayed the highest FIV and is represented by five species. Families Myrtaceae, Moraceae, Rubiaceae, Fabaceae and Rutaceae were respectively ranked to be with the second, third, fourth, fifth and sixth highest FIVs. Loganiaceae family, which was represented by a single species, displayed the least FIV.

4.8.5. Species Importance Value (SIV)

About 35% of the total SIV was contributed by the six most dominant species in the forest. Accordingly, these species include Croton macrostachyus, Syzgium guineense ssp.afromontanum, Ficus sycomorus, Eucalyptus globules, Galiniera coffeoides and Millettia ferrugina. In contrast to this, most of the species displayed very low SIV. The possible reason for this could be either the selective cutting of these species by the local people or unfavorable conditions. It indicates that the requirement of conservation and management of the forest as a whole.

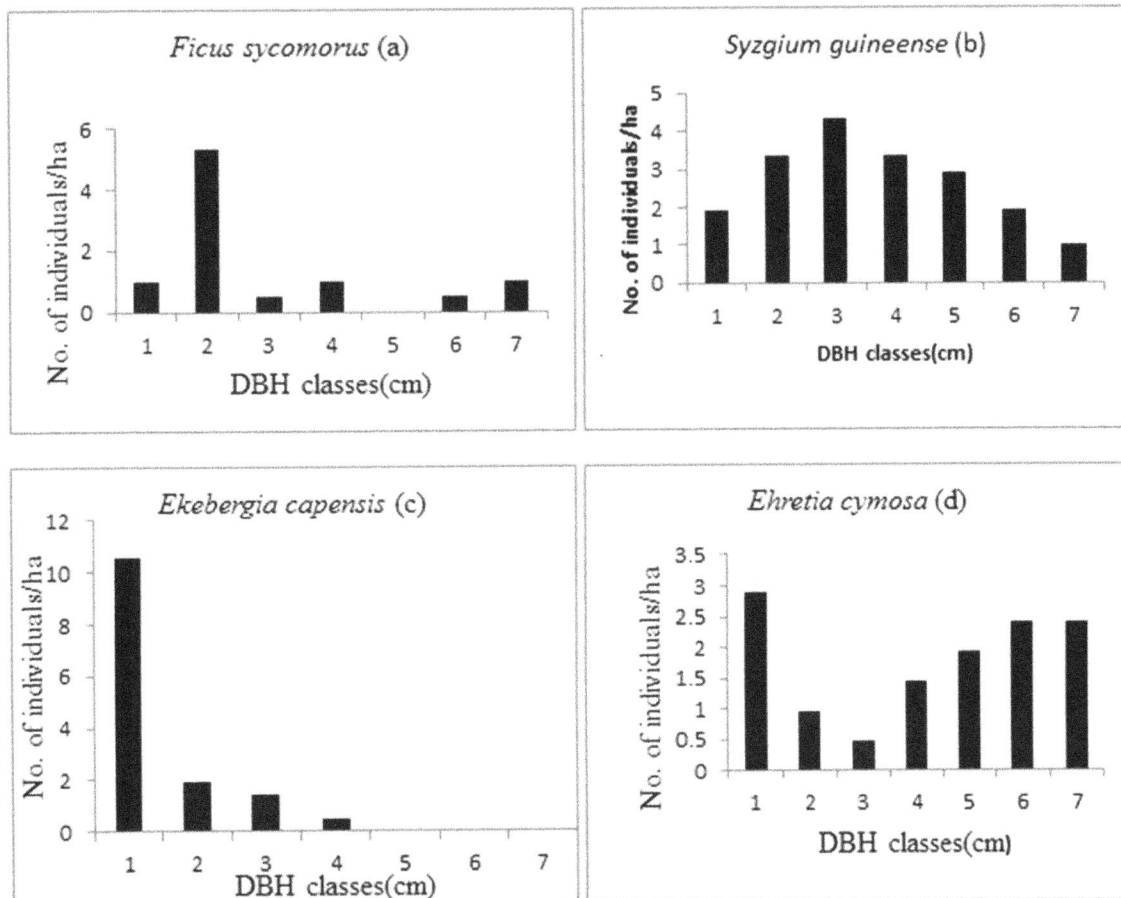

Fig. 4. *Four patterns of Species population structure in Lammo natural forest.*

4.8.6. Species Population Structure

Four different patterns of distribution were recognized in the forest. The first pattern is described as irregular where no defined pattern will be observed when one goes across the DBH classes. Most of the species in the forest displayed such pattern of population structure. This type of pattern could be the result of selective cutting of individual species by the local people. *Ficus sycomorus* was found to be the ideal representative example for this pattern of distribution.

The second pattern showed a bell-shaped distribution pattern. The number of individuals increases with increasing DBH up to certain point then decreases with increase in DBH. This pattern of distribution suggests that the lower regeneration and recruitment capacity of these species and indicates the management and conservation problems (Example, Syzgium *guineense).*

The third pattern of population structure exhibited the inverted J-shaped distribution. This pattern of distribution suggests a good reproduction and recruitment capacity of the species. It shows that a higher number of individuals in the lower DBH classes and decreases with increasing in DBH. *Ekebergia capensis* was among the representative examples of this pattern of distribution.

The fourth pattern was found to be a U-shaped pattern of distribution. This pattern is characterized by having large number of individuals in the lower DBH class and lower or the total absence in the next two middle classes and gradually increasing to the higher DBH class. *Ehretia cymosa* was in this category of distribution patterns.

5. Conclusion

In the study from 52 quadrates (2.08ha), 54 plant species belonging to 46 genera and 29 families were identified. Myrtaceae was found to be the most abundant and dominant family followed by Euphorbiaceae. The vegetation was classified into three plant community types. The variation in species composition and diversity among these communities could be because of different environmental factors (anthropogenic and natural). The forest contains some of commercially important and endemic plant species of Ethiopia. For structural analysis, DBH and height records were classified in to class intervals and more number of individuals was recorded for the intermediate classes. The number of matured individuals (top class) as well as seedlings and saplings were found to be low and in some species, no single seedling and/or sapling were recorded. The general distribution pattern of species population structure in the forest based on the DBH value revealed four structural patterns. This population dynamics within the forest revealed the signs of some disturbances and requires for management and conservation practices as soon as possible.

Annex 1

Annex 1. The indicator species and their P*.value in three communities of Lammo natural forest.

species	Com.1	Com.2	Com.3	P*.value
Acacia tortilis	0.000000000	0.097435897	0.080971660	0.602
Afrocurps falcatus	0.077896787	0.049367089	0.000000000	0.426
Albizia gummfera	0.019180295	0.048622047	0.067343556	0.848
Brucea antidysenterica	0.017603595	0.022312556	0.115374253	0.411
Buddleja polystachya	0.000000000	0.111038961	0.013670540	0.233
Canthium lactescens	0.022890187	0.009671104	0.080369287	0.63
Canthium oligocerpum	0.115841886	0.065257596	0.018076896	0.507
Casurina equstifolia	0.000000000	0.016101695	0.071364853	0.391
Celtis africana	0.011173841	0.151070336	0.005230967	0.049
Clausena anisata	0.011009706	0.093032015	0.061849539	0.703
Coffea arabica	0.033444816	0.056521739	0.000000000	0.775
Combretum molle	0.137953290	0.006476141	0.014351558	0.224
Cordia africana	0.000000000	0.117525773	0.065111232	0.342
Croton macrostachyus	0.075276406	0.107339450	0.125543216	0.902
Dracaena steudneri	0.000000000	0.170149254	0.007855460	0.076
Ehretia cymosa	0.038036134	0.008035133	0.155805910	0.228
Ekebergia capensis	0.120263736	0.070571429	0.086015038	0.872
Erythrina brucei	0.032954644	0.013923337	0.015427520	1
Eucalyptus camaldulensis	0.000000000	0.173780488	0.032092426	0.127
Eucalyptus globulus	0.098623026	0.034723524	0.071819662	0.78
Euphorbia ampliphylla	0.199731939	0.004219337	0.042076493	0.108
Euphorbia candelabrum	0.120677488	0.000000000	0.050217286	0.214
Ficus sur	0.106487320	0.007498482	0.008308567	0.347
Ficus sycomorus	0.053644282	0.063461186	0.070317103	1
Ficus vasta	0.028742153	0.012143559	0.040366403	0.918
Galiniera saxifraga	0.010798215	0.004562246	**0.242645772**	0.03
Galiniera coffeoides	0.116183726	0.044625113	0.147102173	0.666
Garcinia buchanani	0.091968860	0.006476141	0.028703116	0.462
Gardenia ternifolia	0.020968988	0.035437590	0.039266027	1
Grevillea robusta	0.037717122	0.132795699	0.035314092	0.463

species	Com.1	Com.2	Com.3	P*.value
Hagenia abyssinica	0.017812778	0.045155393	0.033355784	0.964
Juniprus procera	0.004347229	0.110202261	0.185196781	0.293
Macaranga capensis	0.012651274	0.016035490	**0.248749986**	0.048
Maytenus ovatus	**0.213818489**	0.033876867	0.004170744	0.049
Millettia ferrugina	0.016641485	**0.221477370**	0.068167858	0.046
Myrica salicifolia	**0.270883287**	0.041617523	0.004611360	0.019
Olea europea	0.110166216	0.062060302	0.005730406	0.454
Olea welwitschii	0.100334448	0.141304348	0.000000000	0.253
Oxytenanthera abyssinica	0.081422756	0.022934076	0.025411719	0.582
Phoenix abyssinica	0.106681640	0.045072993	0.037456781	0.614
Pilostigma thonningii	0.121401441	0.005699123	0.037888908	0.352
Polyscias fuluva	0.000000000	0.073076923	0.053981107	0.654
Premna schimperi	0.077035702	0.054245974	0.064113431	0.978
Prunus africana	0.011118589	0.164416128	0.020820404	0.196
Ricinus commnus	0.076570450	0.008087754	0.035846000	0.833
Ritichea albersii	0.066151670	0.069872702	0.023226383	0.903
Rubus apetalus	0.066395842	0.042078365	0.007770704	0.718
Schefflera abyssinica	0.061831347	0.078371232	0.032564224	0.892
Syzygium guineense	0.122509511	**0.215004191**	0.079410597	0.048
Teclea nobilis	0.098503013	0.020808762	0.076855998	0.747
Terminalia brownii	0.000000000	0.027536232	0.152555301	0.177
Vepris dainellii	0.106261506	0.022447743	0.049745691	0.608
Vernonia amygdalina	0.184946341	0.023441949	0.017316306	0.114
Vernonia auriculifera	0.000000000	0.027536232	0.152555301	0.156

Com.1, 2, 3= community 1, 2, 3; and the bolded are the value of species with higher IVI

References

[1] Supriya L. D. and Yadava P. S., (2006): Floristic Diversity Assessment and Vegetation Analysis of Tropical Semievergreen Forest of Manipur, North East India.

[2] Feyera Senbeta, Tadesse Woldemariam, Sebsebe Demissew and Denich, M. (2007). Floristic diversity and composition of Sheko Forest, Southwest Ethiopia. *Ethiop. J. Biol.*

[3] Kelvin Seh-Hwi Peh(2009): The Relationship between Species Diversity and Ecosystem Function in Low- and High-diversity Tropical African Forests.

[4] Greig-Smith, P. (1979). Pattern in vegetation. Journal of Ecology 67(3): 755–779.

[5] Geldenhuys, C. J. & B. Murray. (1993). Floristic and structural composition of Hanglip forest in the South Pansberg, Northern Transvaal. *South African Forestry Journal* 165: 9-20.

[6] Kumelachew Yeshitela and Tamrat Bekele (2002). Plant community analysis and ecology of Afromontane and transitional rainforest vegetation of southwest Ethiopia. *SINET: Ethiop. J. Sci.*, 25: 155-175.

[7] Venables W. N and D. M. Smith, (2012). Introduction to the R packaging system.

[8] Kent, M. and Coker, P. (1992). Vegetation Description and Analysis. A practical approach. John Wiley and Sons, New York, 363p.

[9] Muller-Dombois, D. and Ellenberg, H. (1974). *Aims and Methods of Vegetation Ecology.*

[10] EFAP (1994). The Challenge for Development. Ministry of Natural Resources, Addis Ababa. Volume III.

[11] Shiferaw Belachew (2010). Floristic composition, structure and regeneration status of woody plant species of Sese forest; Oromia National Reginal State, Southwest Ethiopia.

[12] Getaneh Belachew, (2006): Floristic composition and structure in Beschillo and Abay (Blue-Nile) Riverine Vegetation.

[13] Fekadu Gurmessa, (2010): Floristic Composition and Structural Analysis of Komto Afromontane Rainforest, East Wollega Zone of Oromia Region, West Ethiopia.

24

Inhibitory Effects of Oligochitosan on Pathogenic Fungi Isolated from *Zanthoxylum bungeanum*

Peiqin Li[1, *], Zhou Wu[1], Tao Liu[1], Yanan Wang[2]

[1]Department of Forestry Pathology, College of Forestry, Northwest A&F University, Yangling, Shaanxi, China

[2]Department of Landscape Architecture, College of Landscape Architecture and Arts, Northwest A&F University, Yangling, Shaanxi, China

Email address:

lipq@nwsuaf.edu.cn (Peiqin Li), lipq110@163.com (Peiqin Li), 15229371653@163.com (Zhou Wu), 937215187@qq.com (Tao Liu), 2371754065@qq.com (Yanan Wang)

[*]Corresponding author

Abstract: To explore nontoxic degradable natural substances which could be used to control *Zanthoxylum bungeanum* diseases, the effects of oligochitosans, i.e., OCHA and OCHB, on pathogenic fungi *Pseudocercospora zanthoxyli*, *Fusarium sambucinum* and *Phytophthora boehmeriae* were investigated. Excellent inhibitory effects of OCHA and OCHB on the growth of all tested pathogens were observed, which were calculated by RGI and BGI. The highest inhibitions for *P. zanthoxyli* and *F. sambucinum were induced by* 1.0 mg/mL OCHB with the corresponding RGI values as 51.25% and 95.69%, and BGI values as 44.76% and 92.34%. For *P. boehmeriae*, the maximum values of RGI and BGI were induced by 1.0 mg/mL OCHA with the corresponding values as 82.35% and 53.24%. Desirable results obtained from the present research might establish the foundation for the utilization of oligochitosan for the nuisanceless control of *Z. bungeanum* diseases.

Keywords: Oligochitosan, Pathogenic Fungi, Growth Inhibition, *Zanthoxylum bungeanum*

1. Introduction

These years, the control of plant diseases and pests has been facing a huge challenge. Although utilization of chemical pesticide to control plant disease could bring great advantages, the excessive use has taken its toll on human health and natural environment [1-4]. Furthermore, the registration procedures of broad-spectrum pesticides but low-toxic to human are extremely complicated and time-consuming, while the resistance of plant pathogen to pesticide is also one main problem faced by growers [5, 6]. Hence, there is a worldwide trend to explore new alternatives for synthetic pesticides, which can not only protect plant against the infection of pathogen but also avoid negative and side effects on human health because of abuse of chemical pesticides.

Oligochitosan (OCH), which is the degraded product of chitosan or chitin and abundant in nature, has aroused an increasing attention for its various biological properties, such as anti-tumor and antioxidant activities, antimicrobial activities, immuno-modulating effects, and so on [7]. Of them, the antimicrobial activity of OCH has been especially concerned by many plant pathologists because of its nontoxic and degradable characteristics [8]. The inhibitory effects of OCH on the growth of plant pathogen have been frequently reported in previous researches, such as *Fusarium solani* [9], *Puccinia arachidis* [10], *Alternaria alternate* [11], *Aspergillus niger* [12], *Botrytis cinerea* [13, 14], and so on. However, up till now, there are no related researches on the inhibitory effects of OCH on the pathogenic fungi isolated from *Zanthoxulum bungeanum*.

*Z. bungeanum*is an aromatic plant of Rutaceae *Zanthoxylum* as shrubs or small trees and native to southwestern China, which has been considered as an important economic crop for its special seasoning function [15, 16]. However, *Z. bungeanum*is frequently infected by different kinds of plant pathogenic fungi during its growth process, which has generated serious effects on the yield and quality of *Z. bungeanum* [17]. The most reported pathogenic fungi isolated from *Z. bungeanum* were *Pseudocercospora* *zanthoxyli*, *Fusarium sambucinum* and *Phytophthora boehmeriae* in Shaanxi and Gansu districts, which respectively resulted in prickly-ash leaf mold, stem blight and dry rot [18-20]. Up until now, the main method of controlling the above mentioned pathogenic fungi is still utilization of chemical pesticide [21]. Although the application of chemical pesticides has brought great advantages, the side effects of synthetic pesticides on crops should not be neglected [6, 22]. Hence, it is very necessary to explore new nontoxic and efficient alternatives to synthetic pesticides to control plant pathogenic fungi.

The main aim of this research is to investigate the effects of OCH on the growth of pathogenic fungi of *Z. bungeanum*, i.e., *P. zanthoxyli*, *F. sambucinum* and *P. boehmeriae*, which would establish the basement for the nuisanceless prevention and management of *Z. bungeanum* diseases.

2. Materials and Methods

2.1. Chemicals

OligochitosanA (OCHA) and oligochitosan B (OCHB) were purchased from Qindao BZ-Oligo Biotech Co. Ltd (Qiandao, China). OCHA was the hydrolyzate mixture of chitosan by acid, but OCHB was the hydrolyzate mixture of chitosan by enzyme. All the other chemicals were purchased from JieCheng Chemical and Glass Company (Yangling, Shaanxi, China). The preliminary structural analyses of OCHA and OCHB were demonstrated in Section of 3.1.

2.2. Plant Pathogenic Fungi

The tested pathogenic fungi of *Pseudocercospora zanthoxyli*, *Fusarium sambucinum* and *Phytophthora boehmeriae*were isolated and preserved in our lab by previous study [18-20], all of which were cultured and preserved on Potato Dextrose Agar (PDA) medium.

2.3. Determination of the Effects of OCHA and OCHB on the Radial Growth of Tested Pathogenic Fungi

The inhibitory effects of OCHA and OCHB on the tested pathogenic fungi were firstly conducted by checking the growth of radial colony on PDA. Both of OCHA and OCHB were separately dissolved in sterile distilled water, and then filtered through a sterile filter membrane (pore size, 0.45 um). For the radial colony growth determination, the sterile oligochitosan solution was added into PDA at 60°C with the final concentrations of 0.2, 0.4, 0.6, 0.8 and 1.0 mg/mL, respectively, which were then mixed rapidly and poured into Petri dishes (diameter, 9 cm). The same volume of sterile distilled water was added into PDA as the control. After the PDA plate cooled, a 5-mm-diameter mycelial plug of *tested pathogen* was inoculated on the center of the PDA plate and incubated at 25°C in dark. When the tested pathogenic fungi were cultured for seven days, the diameter (R) of each colony was measured respectively [23]. Each treatment was carried out for three duplicates. Radial growth inhibition (RGI) was calculated according to the following formula:

$$RGI\ (\%) = (R_0\text{-}R) \times 100\%/R_0 \qquad (1)$$

where R_0 is the diameter of colony of control, R is the diameter of colony of treatment.

2.4. Evaluation of the Effects of OCHA and OCHB on Mycelial Biomass Growth of Tested Pathogenic Fungi

To determinate the effects of oligochitosan on mycelial biomass of *tested pathogenic fungi*, the submerged culture was carried out in PDB liquid medium. Each 1000-mL flask was filled with 300 mL PDB medium, and then sterile oliogochitosan solution was added into PDB. The final concentrations of both OCHA and OCHB in PDB were separately 0.2, 0.4, 0.6, 0.8 and 1.0 mg/mL. And then, 0.5 mL seven-day-old suspension culture of *tested pathogen* in PDB was taken as the inoculum and injected into each flask containing PDB liquid medium with oligochitosan supplementation. All the flasks were maintained on a rotary shaker at 150 rpm at 25°C. Seven days after inoculation, the mycelial biomass of each tested pathogen in each flask was respectively by filtrating under vacuum to obtain the mycelia, which were further lyophilized to a constant dry weight (DW) and expressed as gram per liter [24]. The effect of oligochitosan on mycelial biomass of each tested pathogenic fungus was calculated by the biomass growth inhibition (BGI) according to the following formula:

$$BGI\ (\%) = (DW_0\text{-}DW) \times 100\%/\ DW_0 \qquad (2)$$

where DW_0 is the mycelial dry weight of control, DW is the mycelial dry weight of treatment.

2.5. Monitoring of Time Dynamics of RGI and BGI Respectively for the Three Tested Pathogenic Fungi

To investigate the effects of OCHA and OCHB on the radial colony and biomass growth of *P. zanthoxyli*, *F. sambucinum* and *P. boehmeriae* as culture time varied under certain concentration, time dynamics of RGI and BGI were calculated. The selected concentrations were respectively 0.5 and 1.0 mg/mL. And the checking time were designated on the culture day of 2, 4, 6, 8, 10, 12 and 14, separately. The evaluations of RGI and BGI were the same as described above.

3. Results and Discussion

3.1. ESI-MS and IR Analyses of OCHA and OCHB

The ESI-MS spectra of OCHA and OCHB were presented in Figure1. As shown in Figure 1, the molecular weight of OCHA or OCHB was below 1000 Da, and some fragment ion peaks of oligochitosan were observed. Figure 1A showed the oligosaccharide nature of OCHA with one oligochitosan residue molecular weight ($C_6H_{11}O_4N$, 161 Da) differences among the ion peaks with the m/z respectively at 985.46,

824.28, 663.26, 502.22 and 341.19, or 806.57, 645.19, 484.21, 323.20 and 162.05. For OCHB, one oligochitosan residue molecular weight ($C_6H_{11}O_4N$, 161 Da) difference was also observed among the ion peaks with the m/z respectively at 824.39, 663.26 and 502.23, or 789.46, 645.27, 484.19, 323.18 and 162.06 (Figure 1B).

The FT-IR of OCHA and OCHB were shown in Figure 2, which demonstrated the characteristic absorption peaks of oligochitosan, including those peaks respectively at 3448 cm^{-1} (the absorption of stretching vibration of associated –OH groups), 2932 cm^{-1} (the absorption of stretching vibration of C–H bond), 1632 cm^{-1} (the absorption of flexural vibration of N–H bond), 1410 cm^{-1} (the absorption of flexural vibration of O–H bond), 1340 cm^{-1} (the absorption of stretching vibration of C–N bond), 1070 cm^{-1} (the absorption of stretching vibration of C–O–C bond) and 1020 cm^{-1} (the absorption of stretching vibration of C–OH bond). Besides, the FT-IR spectrum of OCHA was extremely similar to that of OCHB, which might indicate the consistency of chemical components of OCHA and OCHB.

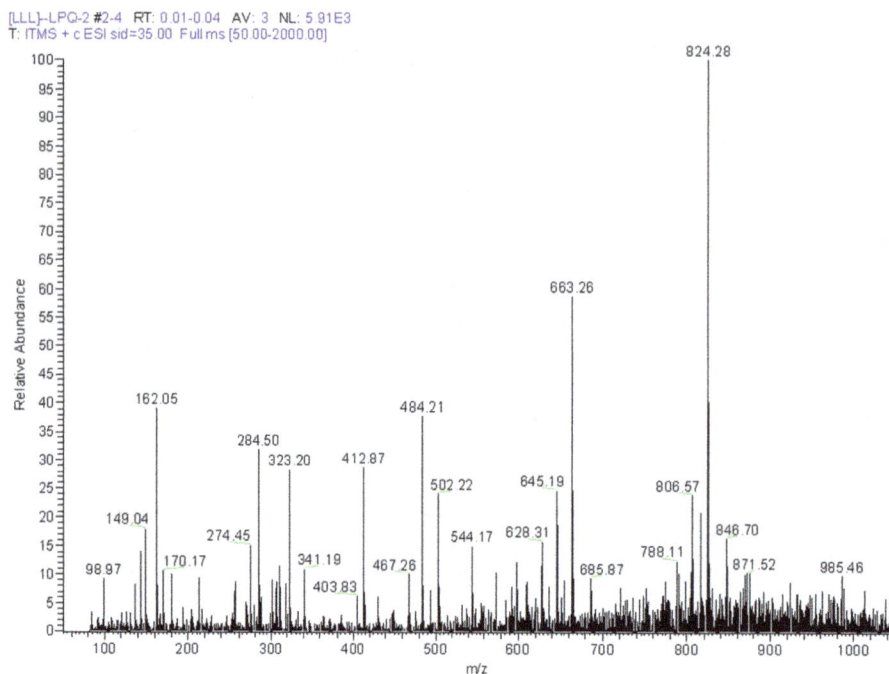

A

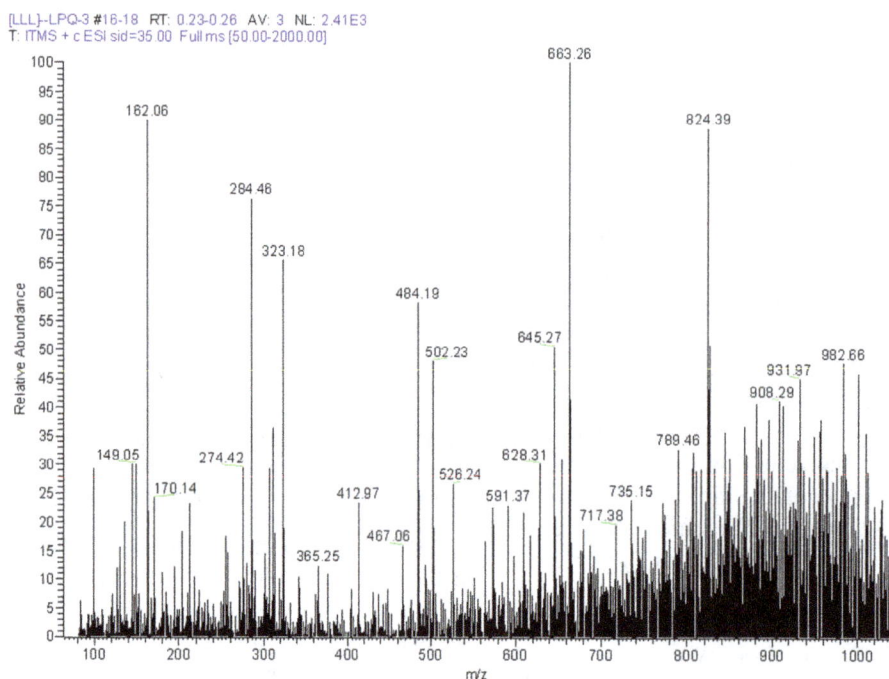

B

Figure 1. The positive ion ESI-MS spectra of OCHA (A) and OCHB (B).

Both of OCHA and OCHB were crude samples not pure compounds. Hence, we just conducted their ESI-MS and IR experiments. However, all the above analyses proved the oligochitosan nature of OCHA or OCHB.

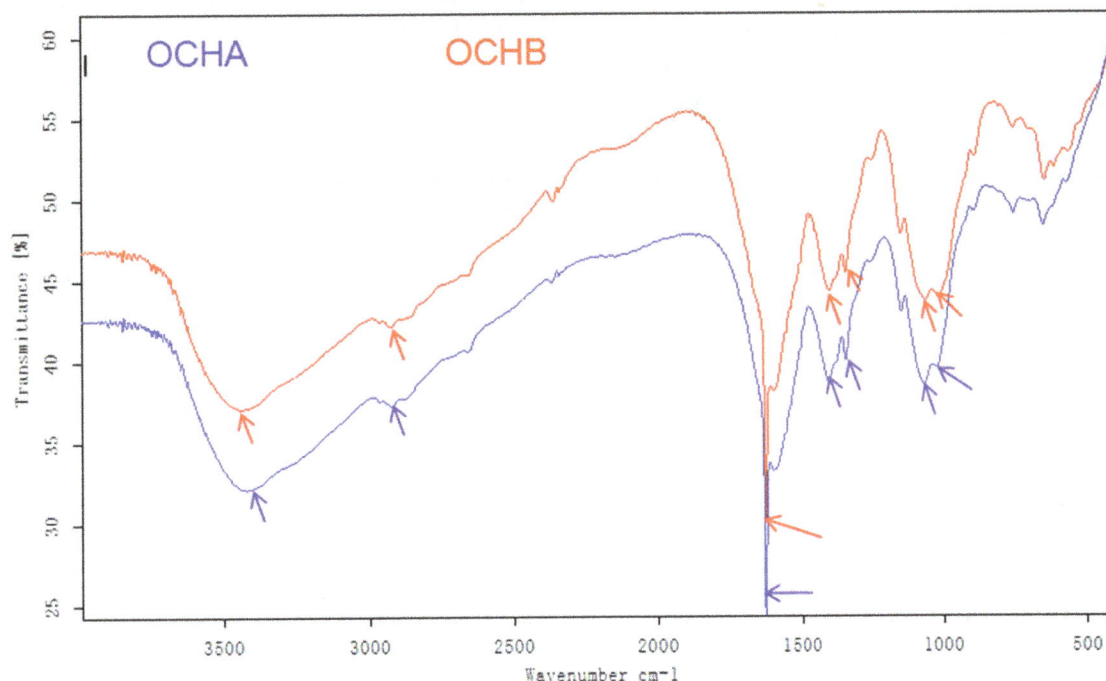

Figure 2. The Fourier transform infrared spectra of OCHA and OCHB.

3.2. Effects of OCHA and OCHB on Radial Colony Growth of Tested Pathogenic Fungi

The effects of OCHA and OCHB on the radial colony growth were summarized in Table 1, which displayed that both of OCHA and OCHB could inhibit the radial colony growth of all three tested pathogenic fungi. For P. zanthoxyliand F. sambucinum, OCHA showed more effective inhibition than that of OCHB when both of them were used at low concentrations. However, when both of OCHA and OCHB were applied at the concentration of 0.6 mg/mL or higher than o.6 mg/mL, OCHB exhibited stronger inhibitory effects on P. zanthoxyliand F. sambucinum. For P. boehmeriae, OCHA always displayed higher inhibitory effects than that of OCHB among all tested concentrations. The inhibitory effects of both of OCHA and OCHB on all tested pathogenic fungi showed concentration-dependent relationship, which became more obvious as the concentration of OCHA or OCHB was higher. The strongest inhibitory effects were induced by 1.0 mg/mL oligochitosan, when the corresponding highest RGI values induced by OCHA were respectively 40.18%, 78.92% and 77.85% for P. zanthoxyli, F. sambucinum and P. boehmeriae. For OCHB, the maximum RGI values for P. zanthoxyli, F. sambucinum and P. boehmeriae were separately 47.58%, 81.27% and 68.25%. By comprehensive analyzing Table 1, OCHA or OCHB showed the strongest inhibitory effects on the radial colony growth of F. sambucinum, P. boehmeriae subsequently, and the weakest inhibition on P. zanthoxyli.

Table 1. Effects of COHA and OCHB on radial colony growth of P. zanthoxyli, F. sambucinum and P. boehmeriae on the 7th day.

Oligo-chitosan	Concentration (mg/mL)	RGI (%)		
		P. zanthoxyli	F. sambucinum	P. boehmeriae
OCHA	0.2	26.52 ± 3.06f	41.23 ± 1.27g	36.00 ± 1.27h
	0.4	27.11 ± 2.01f	52.14 ± 1.79e	45.01 ± 0.56f
	0.6	29.26 ± 2.41e	63.58 ± 1.97d	61.04 ± 0.83d
	0.8	35.57 ± 2.31d	72.32 ± 2.71c	70.24 ± 1.01b
	1.0	40.18 ± 1.27c	78.92 ± 1.58b	77.85 ± 1.39a
OCHB	0.2	20.56 ± 0.37h	33.57 ± 1.57h	28.33 ± 0.56i
	0.4	25.00 ± 4.07g	49.25 ± 2.06f	37.96 ± 2.74g
	0.6	35.74 ± 1.98d	73.58 ± 2.27c	57.59 ± 2.32e
	0.8	42.35 ± 1.24b	78.27 ± 2.27b	61.34 ± 1.58d
	1.0	47.58 ± 1.45a	81.27 ± 3.80a	68.25 ± 1.21c

Note: Each value was expressed as mean ± standard deviation (n = 3). The significant difference analyses of RGI for each pathogen were respectively carried out at p = 0.05 level. Different letters indicated significant differences under different treatments for each pathogen.

3.3. Effects of OCHA and OCHB on Mycelial Biomass of Tested Pathogenic Fungi

To determine the effects of OCHA and OCHB on the mycelia biomass growth, the submerged liquid culture of all three tested pathogenic fungi were conducted in PDB medium supplemented with OCHA or OCHB, which was calculated by the indicator of BGI (%). Table 2 summarized all the BGI values caused by OCHA and OCHB respectively for P. zanthoxyli, F. sambucinum and P. boehmeriae.

Table 2. Effects of COHA and OCHB on mycelial biomass of P. zanthoxyli, F.sambucinum and P. boehmeriaeon the 7th day.

Oligo-chitosan	Concentration (mg/mL)	BGI (%)		
		P. zanthoxyli	F. sambucinum	P. boehmeriae
OCHA	0.2	$18.35 \pm 0.97h$	$33.33 \pm 1.03i$	$21.12 \pm 1.32i$
	0.4	$20.34 \pm 1.34g$	$57.35 \pm 2.38g$	$27.24 \pm 1.04g$
	0.6	$24.32 \pm 1.67f$	$67.78 \pm 1.65e$	$39.35 \pm 2.14e$
	0.8	$33.25 \pm 1.27d$	$76.25 \pm 2.56c$	$45.27 \pm 1.87c$
	1.0	$40.57 \pm 2.01b$	$81.25 \pm 3.57b$	$50.24 \pm 1.37a$
OCHB	0.2	$16.34 \pm 0.97i$	$39.78 \pm 1.97h$	$20.24 \pm 0.78i$
	0.4	$20.47 \pm 1.54g$	$64.32 \pm 2.76f$	$25.37 \pm 1.47h$
	0.6	$27.25 \pm 1.25e$	$72.21 \pm 2.13d$	$34.37 \pm 0.94f$
	0.8	$35.67 \pm 1.29c$	$82.35 \pm 2.47b$	$42.27 \pm 1.39d$
	1.0	$43.25 \pm 1.57a$	$87.25 \pm 3.04a$	$47.25 \pm 2.34b$

Note: Each value was expressed as mean ± standard deviation (n = 3). The significant difference analyses of RGI for each pathogen were respectively carried out at p = 0.05 level. Different letters indicated significant differences for every pathogen under different treatments.

Either OCHA or OCHB could inhibit the mycelia biomass growth of all tested fungi significantly. Furthermore, BGI showed a concentration-dependent relationship with OCHA or OCHB. The larger the concentration of OCHA or OCHB was, the higher the BGI was. The maximum BGI values induced by OCHA respectively for P. zanthoxyli, F. sambucinum and P. boehmeriae were 40.57%, 81.25% and 50.24% when OCHA was employed at 1.0 mg/mL. For OCHB, the highest BGI were separately 43.25%, 87.25% and 47.25% when 1.0 mg/mL OCHB was respectively treated P. zanthoxyli, F. sambucinum and P. boehmeriae. As demonstrated in Table 2, OCHA or OCHB showed the strongest inhibitory effects on the mycelia biomass growth of F. sambucinum, P. boehmeriae subsequently, and the weakest inhibition on P. zanthoxyli, which was coincident with their effects on radial colony growth.

3.4. Time Dynamics of RGI and BGI for Tested Pathogenic Fungi

To investigate the effects of OCHA and OCHB on the radial colony and mycelia biomass growth of tested pathogenic fungi on different cultured time, their RGI and BGI were respectively calculated on the cultured days of 2, 4, 6, 8, 10, 12 and 14. The time dynamics of RGI and BGI for the tested pathogenic fungi were graphed in Figure 3. As presented in Figure 3A, the radial colony growth of P. zanthoxyli was inhibited by OCHA or OCHB during examining period, and the RGI value was improved as the cultured time from the beginning to the 10 day and then decreased slightly. The variation tendency of BGI of P. zanthoxyli was similar to that of RGI, and the maximums of BGI were observed on the 8 day as shown in Figure 3B. The time dynamics of RGI and BGI of F. sambucinum were exhibited in Figure 3C and 3D. From Figure 3C, it was seen that when F. sambucinum was treated by OCHA or OCHB at the concentration of 0.5 mg/mL, the maximum of RGI was observed on the 4 day, and then kept relatively constant till to the 8 day, after which the RGI was decreased. However, when the concentration of OCHA or OCHB was increased to 1.0 mg/mL, RGI was significantly increased from the beginning and reached to the maximum on the day of 10. The variation trend of BGI of F. sambucinum was similar to its RGI as presented in Figure3D. The dynamics of RGI and BGI of P. boehmeriae were shown in 3E and 3F, which indicated RGI or BGI reached the maximum either on the 8 day or 10 day and OCHA exhibited stronger inhibitory effects than that of OCHB.

For P. zanthoxyli and F. sambucinum, the maximum RGI values were induced by 1.0 mg/mL OCHB on the 10 day, with their corresponding values as 51.25% and 95.69%, while the maximum BGI values were respectively got on the 8 day and 10 day still induced by 1.0 mg/mL OCHB with corresponding values as 44.76% and 92.34%. For P. boehmeriae, the maximum values of RGI and BGI were respectively obtained on the 10 day and 8 day induced by 1.0 mg/mL OCHA, which were separately 82.35% and 53.24%.

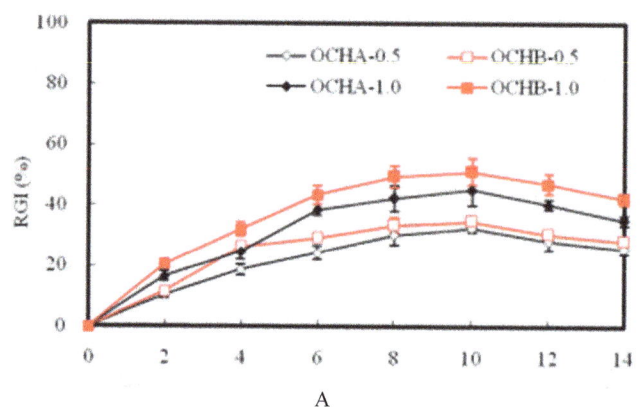

A

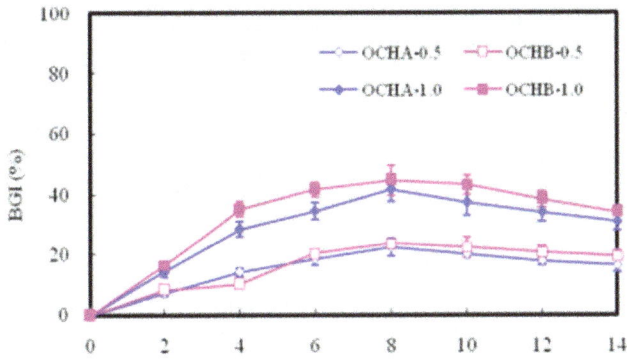

B

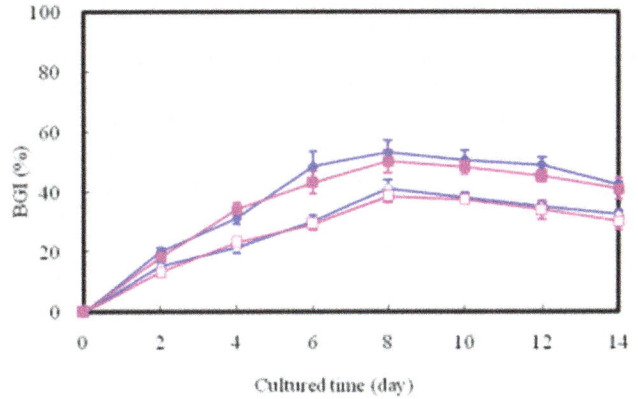

F

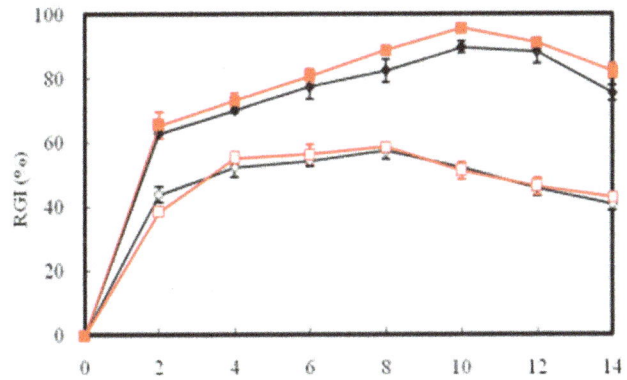

C

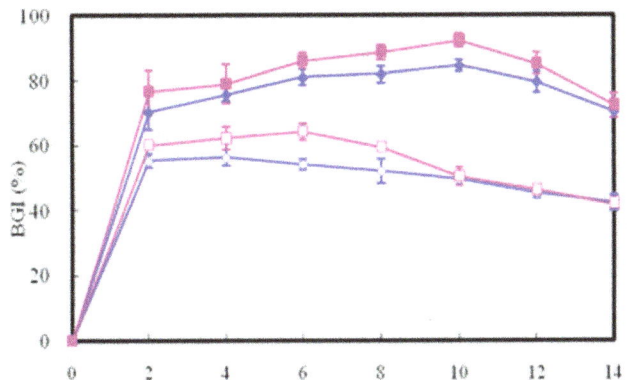

D

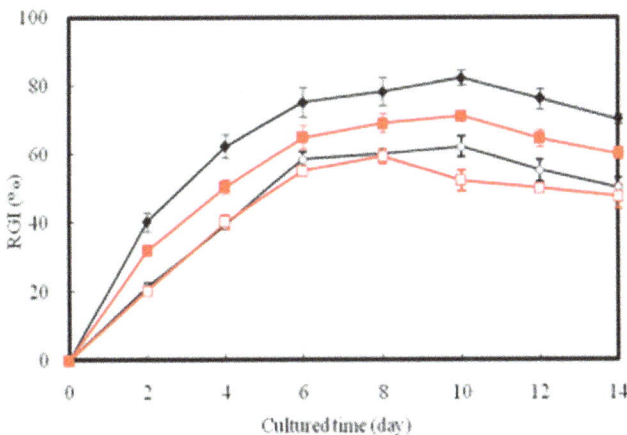

E

Figure 3. *Ime dynamics of RGI and BGI respectively for P. zanthoxyli (A and B), F. sambucinum (C and D) and P. boehmeriae (E and F).*

It has been extensively reported that oligochitosan exhibited excellent antimicrobial activities [25]. The present research first studied the inhibitory effects of oligochitosan on pathogenic fungi isolated from Z. bungeanum and desirable results were obtained. Both of OCHA and OCHB processed excellent inhibitory effects on the tested pathogenic fungi. However, the differences of inhibition capabilities between OCHA and OCHB were also observed, which might be attributed to their different preparation methods leading to their different components. Both of OCHA and OCHB had a molecular weight lower than 1000. The comprehensive analyses of their ESI-MS and IR spectra demonstrated the oligosaccharide property. Although both of OCHA and OCHB were not pure oligosaccharide, just crude samples, their obvious inhibitory effects on pathogenic fungi demonstrated their potential as substitute for pesticides to control plant diseases.

Furthermore, it was also observed that the inhibitory effects of OCHA or OCHB on tested pathogenic fungi were related to its concentrations and treated time. The optimal concentration was the highest concentration 1.0 mg/mL in designed ranges, which might indicate a higher concentration of oligochitosan will benefit plant disease control. The maximum inhibitory effects were observed either on the 8 day or 10 day, which demonstrated OCHA or OCHB has long-lasting effective duration. In the present research, OCHA exhibited stronger inhibitory effects on P. zanthoxyli and F. sambucinum than that of OCHB. But for P. boehmeriae, the inhibitory ability of OCHA was slightly weaker than that of OCHB. The present research might provide the illumination for the nuisanceless control of Z. bungeanum diseases. However, all the experiments of the present investigation were carried out in laboratory. The protective effects of OCHA and OCHB on Z. bungeanum in the field against plant diseases are worth further investigating.

4. Conclusion

In conclusion, the excellent inhibitory effects of OCHA and OCHB were observed in laboratory by examining the variations of RGI and BGI of the pathogenic fungi *P. zanthoxyli*, *F. sambucinum* and *P. boehmeriae* isolated from *Z. bungeanum*. The highest inhibitions on the growth of radial colony and mycelia biomass of *P. zanthoxyli* and *F. sambucinum were induced by* 1.0 mg/mL OCHB with the corresponding RGI values as 51.25% and 95.69%, and BGI values as 44.76% and 92.34%. For *P. boehmeriae*, the maximum values of RGI and BGI were induced by 1.0 mg/mL OCHA with the corresponding values as 82.35% and 53.24%. Desirable results obtained from the present research might establish the basement for the utilization of oligochitosan for the nuisanceless control of *Z. bungeanum* diseases.

Acknowledgements

This work was financially supported by the National Natural Science Fundation of China (NO.31300542), the Natural Science Foundation of Shaanxi Province (NO.2015JQ3084), the Fundamental Research Funds for the Central Universities of Northwest A&F University (NO.Z109021422; NO.2452015336) and the Scientific Research Start-up Funding supported by Northwest A&F University (NO.Z111021303).

References

[1]   Q. He, J. Huang, X. Yang, X. Yan, J. He, S. Li, and J. Jiang, "Effect of pesticide residues in grapes on alcoholic fermentation and elimination of chlorothalonil inhibition by chlorothalonil hydrolytic dehalogenase," Food Control vol. 64, pp. 70-76, 2016.

[2]   C. K. Bempah, A. A. Agyekum, F. Akuamoa, S. Frimpong and A. Buah-Kwofie, "Dietary exposure to chlorinated pesticide residues in fruits and vegetables from Ghanaian markets," J. Food Compos. Anal. vol. 46, pp. 103-113, 2016.

[3]   M. Bhanti, and A. Taneja, "Contamination of vegetables of different seasons with organophosphorous pesticides and related health risk assessment in northern India," Chemosphere vol. 69, pp. 63-68, 2007.

[4]   P. Messing, A. Farenhorst, D. Waite, and J. Sproull, "Influence of usage and chemical–physical properties on the atmospheric transport and deposition of pesticides to agricultural regions of Manitoba, Canada," Chemosphere vol. 90, pp. 1997-2003, 2013.

[5]   M. D. M. Oliveira, C. M. R. Varanda, and M. R. F. Félix, "Induced resistance during the interaction pathogen x plant and the use of resistance inducers," Phytochem. Lett. vol. 15, pp. 152-158, 2016.

[6]   A. Yacoub, J. Gerbore, N. Magnin, P. Chambon, M.-C. Dufour, M.-F. Corio-Costet, R. Guyoneaud, and P. Rey, "Ability of *Pythium oligandrum* strains to protect *Vitis vinifera* L., by inducing plant resistance against *Phaeomoniella chlamydospora*, a pathogen involved in Esca, a grapevine trunk disease," Biol. Control vol. 92, pp. 7-16, 2016.

[7]   P. Zou, X. Yang, J. Wang, Y. Li, H. Yu, Y. Zhang, and G. Liu, "Advances in characterisation and biological activities of

[8]   S. Bautista-Baños, A. N. Hernandez-Lauzardoa, M. G. Velazquez-del Vallea, M., Hernandez-Lópeza, E. Ait Barkab, E. Bosquez-Molinac, and C. L. Wilsond, "Chitosan as a potential natural compound to control pre and postharvest diseases of horticultural commodities," Crop Prot. vol. 25, pp. 108-118, 2006.

[9]   D. F. Kendra, and L. A. Hadwiger, "Characterization of the smallest chitosan oligomer that is maximally antifungal to *Fusarium solani* and elicits pisatin formation by *Pisumsativum*," Exp. Mycol. vol. 8, pp. 276-281, 1984.

[10] M. Sathiyabama, and R. Balasubramanian, "Chitosan induces resistance components in *Arachishypogaea* against leaf rust caused by *Puccinia arachidis* Speg," Crop Prot. vol. 17, pp. 307-331, 1984.

[11] M. V. B. Reddy, J. Arul, E. Ait-Barka, P. Angers, C. Richard, and F. Castaigne, "Effect of chitosan on growth and toxin production by *Alternaria alternata* f. sp. *lycopersici*," Biocontrol Sci. Technol. vol. 8, pp. 33-43, 1998.

[12] M. Plascencia-Jatomea, G. Viniegra, R. Olayo, M. M. Castillo-Ortega, and K. Shirai, "Effect of chitosan and temperature on spore germination of *Aspergillus niger*," Macromol. BioSci. vol. 3, pp. 582-586, 2003.

[13] W. T. Xu, K. L. Huang, F. Guo, W. Qu, J. J. Yang, Z. H. Liang, and Y. B. Luo, "Postharvest grapefruit seed extract and chitosan treatments of table grapes to control *Botrytis cinerea*," Postharvest Biol. Technol. vol. 46, pp. 86-94, 2007.

[14] R. Jia, Y. Duan, Q. Fang, X. Wang, and J. Huang, "Pyridine-grafted chitosan derivative as an antifungal agent," Food Chem. vol. 196, pp. 381-387, 2016.

[15] S. K. Paik, K. H. Koh, S. M. Beak, S. H. Paek, and J. A. Kim, "The essential oils from *Zanthoxylum schinifolium* pericarp induce apoptosis of HepG2 human hepatoma cells through increased production of reactive oxygen species," Biol. Pharml. Bull. vol. 28, pp. 802-807, 2005.

[16] J. H. Li, Y. H. Zhang, and L. H. Kong, "Research progress of *Zanthoxylum bungeanum*," China Condiment vol. 34, pp. 28-35, 2009.

[17] Z. M. Cao, C. M Tian., Y. M. Liang, and P. X. Wang, "Diseases investigation of *Zanthoxylum bungeanum* in Shaanxiand Gansu provinces," J. Northwest Forestry Univ. vol. 9, pp. 39-43, 1994.

[18] Y. Tang, Z. M. Cao, J. F. Wang, and P. Q. Li, "Morphology, biological characteristics and fungicide screening of the pathogen causing prickly ash leaf mold," Forest Pest. Dis. vol. 33, pp. 1-4, 2014.

[19] N. Xie, Z. Cao, C. Liang, Y. Miao, and N. Wang, "Identification of *Phytophthora* species parasiting on prickly ash," J. Northwest Forestry Univ.vol. 28, pp. 125-130, 2013.

[20] Z. M. Cao, Y. L. Ming, D. Chen, and H. Zhang, "Resistance of prickly ash to stem rot and pathogenicity differentiation of *Fusarium sambucinum*," J. Northwest Forestry Univ. vol. 25, pp. 115-118, 2010.

[21] M. F. He, and E. C. Li, "The occurrence regularity and control technology of main diseases and pests of *Zanthoxylum bungeanum*," Shaanxi J. Agri. Sci. vol. 2009, pp. 218-220, 2009.

chitosan and chitosan oligosaccharides," Food Chem. vol. 190, pp. 1174-1181, 2016.

[22] C. Norman, "EPA sets new policy on pesticide cancer risks," Sci. vol. 242, pp. 366-367, 1988.

[23] B. Prapagdee, K. Kotchadat, A. Kumsopa, and N. Visarathanonth, "The role of chitosan in protection of soybean from sudden death syndrome caused by *Fusarium solani*f. sp. *glycines*," Bioresource Technl. vol. 98, pp. 1353-1358, 2007.

[24] F. Yonni, M. T. Moreira, H. Fasoli, L. Grandi, and D. Cabral, "Simple and easy method for the determination of fungal growth and decolourative capacity in solid media," Int. Biodeter. Biodegr. vol. 54, pp. 283-287, 2004.

[25] Y. Peng, B. Han, W. Liu, and X. Xu, "Preparation and antimicrobial activity of hydroxypropyl chitosan," Carbohydr. Res. vol. 340, pp. 1846-1851, 2005.

25

Modification, calibration and validation of APSIM to suit maize (*Zeamays L.*) production system: A case of Nkango Irrigation Scheme in Malawi

John Mthandi[1, *], **Fredrick C. Kahimba**[1], **Andrew K. P. R. Tarimo**[1], **Baandah. A. Salim**[1], **Max W. Lowole**[2]

[1]Department of Agri. Eng. and Land Planning, Sokoine University of Agriculture, Morogoro, Tanzania
[2]Department of Crop Science, Bunda College of Agriculture, Lilongwe, Malawi

Email address:
johnmthandi@yahoo.com (Mthandi J.)

Abstract: Nitrogen (N) is the most important nutrient in maize production and its availability can affect the production potential of maize. Availability of nitrogen in soil largely varies with place and time. Models are some of practical methods used to evaluate and monitor availability and impact of nitrogen on maize production; APSIM is one of such models. APSIM has several modules that have different functionalities and one of such modules is SoilWat module. The study modified SoilWat module by incorporating Nitrogen Distribution model. Trial and error method was used in the calibration of the nitrogen distribution model that was incorporated in the APSIM model as subroutine. The initial values of nitrogen distribution were obtained from literature and these values formed the basis for development of the model. After development of model using parameters obtained from literature review, field experiment was conducted to collect data to be used in redefining the model. The simulated nitrogen distribution was compared with values obtained from the field experiment and their mean differences were initially high but the process was repeated until the mean difference was small. In field experiment, the study had two factor, each with four regimes. The Triscan Sensor (EnviroScan, Sentek Pty Ltd, Stepney, Australia) was used to measure total nitrogen concentration at lateral distances and vertical depths. Primary soil samples were collected and analysed at Bunda College Laboratory. The study inferred that Soil water percolates down to underlying layer only when proceeded layers are satisfied i.e. has reached its field capacity, above which excess water is left free to percolates down the soil profile. Before water arriving in last layer it had to satisfy the above-lying soil profiles. The study has shown that increase of nitrogen contents in underlying layers corresponds with decrease of the same in top layers due to advection movement. Consequently, the increases of soil water in a specific layer correspond to decrease of nitrogen content in that particular layer. The study has shown that APSIM under predicted during the latter stage of the maize growing season and over predicted in the early stage of the growing season, and it overestimates soil water contents in soil profile.

Keyword: APSIM, Soil Water Content, Nitrogen Concentration, Maize

1. Introduction

Nitrogen (N) is the most limiting factor in maize production [1]. Increasing N fertilization increases corn grain yield [2,3] and increasing soil moisture enhances maize yield response to N fertilization, especially when high N rates are applied [4,5]. The requirement of nitrogen fertilizer by maize varied temporally and spatially among and within seasons [6]. N fertilizer management practices that do not accommodate temporal and spatial variability may lead to lower yields and economic returns, poor N use efficiency, and detrimental environmental impacts due to excessive N inputs [7]. There are two primary approaches taken to understand and quantify temporal and spatial requirements and availability of N to plants in the soil. The methods are measurements and modeling. Although measurements can be one of method to quantify temporal and spatial N needs and availability by crops, it can be prohibitively time consuming, too costly due to equipment, chemicals, and frequency of data collection required, and too variable to be practical at the farm scale [8]. Modeling is one approach used to address the increasing need of understanding the implications of management decisions on N cycling and leaching. The models are built using set of equations that governs the transport of nutrients including N in soil. N movement in soil can therefore be described by the phenomenon of solutes transport in soil. The N movement in soil is predominantly by three basic mechanisms of solute transport namely Advection, diffusion, and dispersion [9, 10, 11]. Based on these three basic mechanisms several models have been developed to infer behaviour of N in the soil and some of these models are as follows: LEACHMN [12], RZWQM [13, 14], NLEAP [15], HYDRUS program [16], and APSIM [17].

The Agricultural Production Systems sIMulator (APSIM) is a multi-purpose and comprehensive model developed as a tool for exploring crop management strategies that can improve the economics of agricultural production systems and the consequences of the soil resources and environment [18]. APSIM is a modeling framework that allows the coupling of various one-dimensional models from separate research efforts into a single simulation. It is used to study interactions between plants, soil, water, and nutrients [17]. The APSIM is a centralized engine into which different modules that enable the simulation of systems covering a range of plant, animal, soil, climate and management interactions could be connected.

Figure 1 indicates different modules of APSIM, which have different functionalities and can be used depending on type of decisions to be used. Each module provides a small piece of simulation functionality with the 'engine' coordinating the flow of data/variables between the modules. The SoilN module describes the dynamics and processes of carbon and nitrogen in soil, and the SoilWater module

account for solute movement in the soil [18].

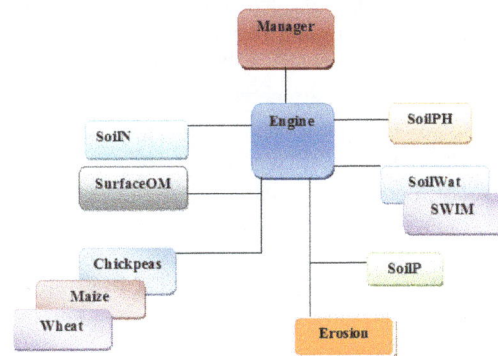

Fig. 1. *Different modules of APSIM (Source: www.apsim.ingo)*

1.1. Solutes Movement in the SoilWat Module

A SoilWat module calculates the redistribution of water and solutes throughout the soil profile using separate algorithms for saturated or unsaturated flow. The redistribution of solutes, such as nitrate- and urea-N, is carried out in this module. The algorithm assumes that all water and solutes entering or leaving a layer is completely mixed and flow has mixing efficiency factors of 1.0 meaning that solute movement can simply be calculated as the product of the water flow and the solute concentration in that water [18].

SoilWat module is an integral module of APSIM; it links other modules for APSIM to simulate properly (Figure 2). The SoilWater module is responsible for movement of water, solutes and heat from one place to the other within the system. The SoilWat module has five different Subroutines that perform various functionalities within the module as indicated in figure 3 below.

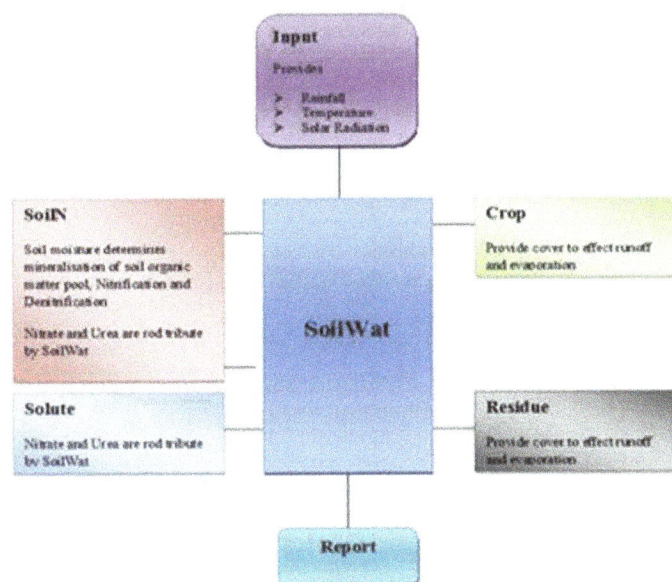

Fig. 2. *Flow diagram of communication between SoilWat and other APSIM modules (Source: www.apsim.ingo)*

The SoilWat module do not consider the spreading pattern of nitrogen and water in both lateral and vertical dimensions to establish the amount of nitrogen deposited in the rooting zone. [19] reported APSIM models do not consider lateral flow or horizontal heterogeneity. APSIM only account for the nitrogen content entering and leaving each layer but not distribution pattern of nitrogen in each layer. However, [20] reported that previous tests of APSIM have shown that modifications to APSIM's parameters that control the extent of mixing of percolating water may be required to improve estimates of the movement of N in soil. This paper undertakes modification and calibration of solute flux subroutine of SoilWater module in the ASPIM model so as to improve prediction capacity of model on the N movement in soil in Malawi. Figure 4 below shows the flow chart of SoilWat module of APSIM. It shows the input parameters, the processes and output parameters.

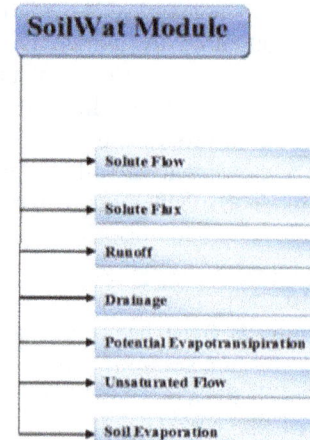

Fig. 3. Subroutine structure of SoilWat module in APSIM

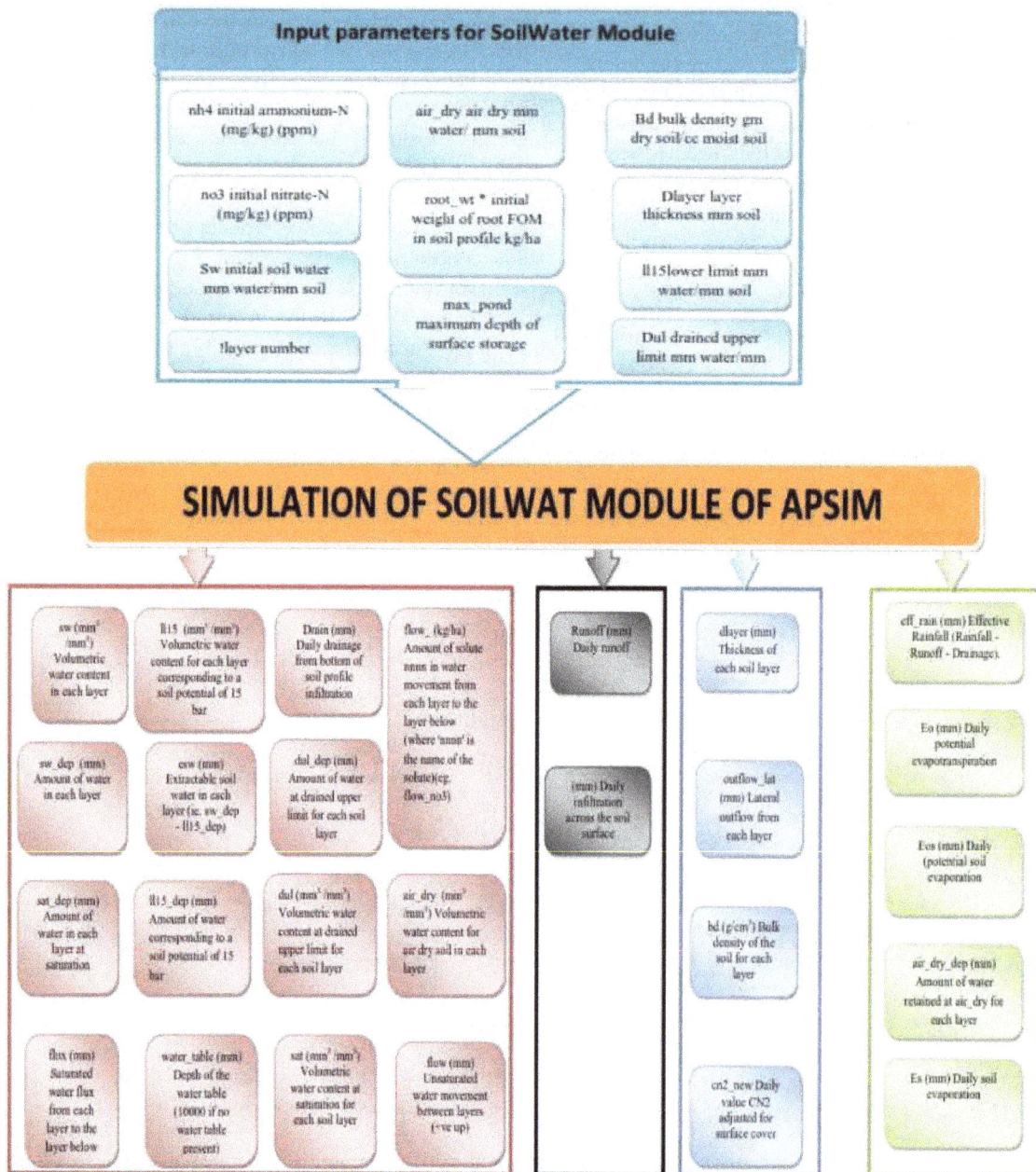

Fig. 4. Flow chart of SoilWat module of APSIM

2. Materials and Methods

2.1. Calibration of the Nitrogen Distribution Subroutine

Trial and error method was used in the calibration of the nitrogen distribution model that was incorporated in the APSIM model as subroutine. Trial and error is a method of solving problems and is characterised by repeated, varied attempts which are continued until success. [21] reported that trial and error method involves assigning an initial value usually from values found in literature or field experimental data relating to the parameter to be estimated by calibration.

The initial values of nitrogen distribution in the soil were obtained from literature and these values formed the basis for development of the model. After development of model using parameters obtained from literature review, field experiment was conducted to collect data to be used in redefining the model. The simulated nitrogen distribution was compared with values obtained from the field experiment and their mean differences were initially high. The process was repeated until the mean difference was small.

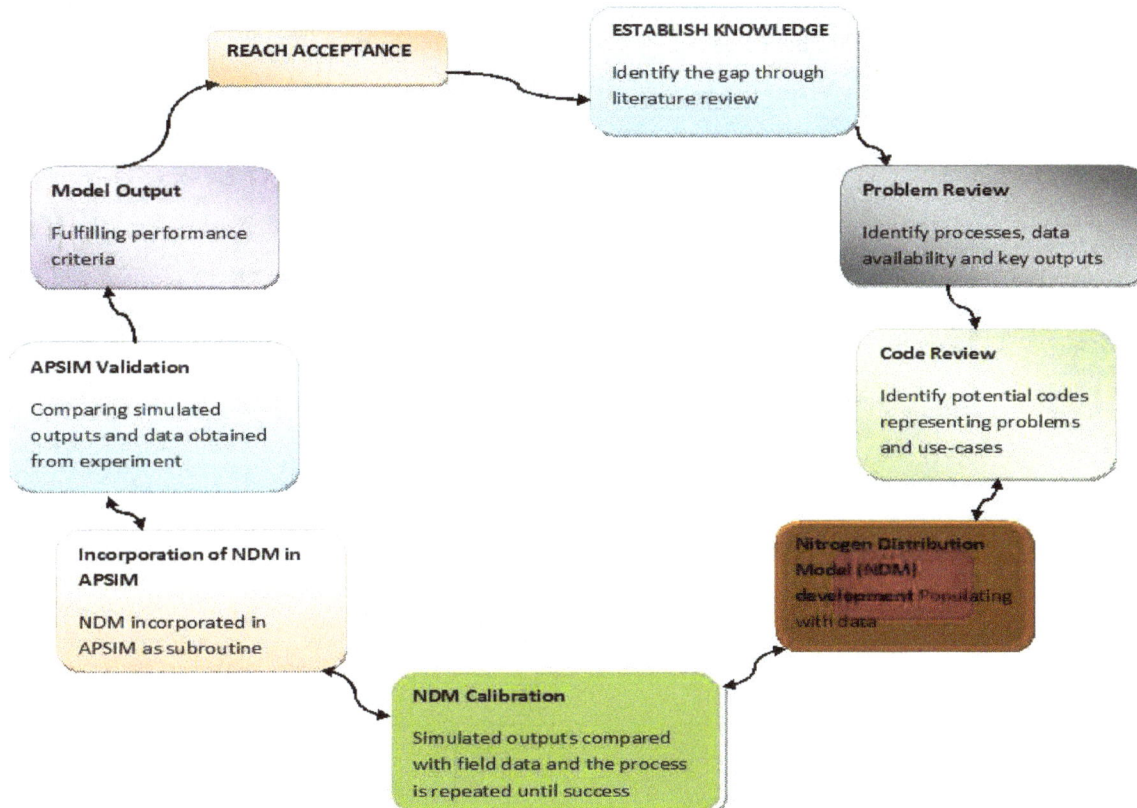

Fig. 5. *Model development process*

2.2. Field Experiments

The field experiment was carried out at Nkango Irrigation Scheme in Kasungu district, Malawi. Data were collected at growing seasons of 1st June to 8th September, 2012; and 10st September to 5th December, 2012. The scheme is situated at Latitude 12^035' South and Longitudes 33^031' East and is at 1186 m above mean sea level. The mean annual rainfall is about 800 mm. The site has coarse sandy loam as the dominant soil type. The study plots were 5 X 5 m in size and ridges were spaced at 75 cm apart. The distance between plot to plot were 2 metre for avoiding 'sharing' of responses, water and nitrogen (edge effects). Three maize seeds of hybrid maize (SC 407) were planted per hole at spacing of 25 cm. They were later on thinned to one seed per station 7 days after germination.

The trials consisted of factorial arrangement in a Randomised Complete Block Design (RCBD). The factors were water and nitrogen and both at four levels. Water had four application regimes *viz* farmers' practice regime; full (100%) water requirement regime (FWRR) of maize plant; 60% of FWRR and 40% of FWRR. A full maize water requirement was determined by the procudure described by Allen et al., (1998). Nitrogen also had four application regimes i.e. the typical nitrogen application rate in the area (TNPRA) of 92 kg N/ha was used as a basis to determine other dosage levels in the study[22]. The nitrogen dosage levels were TNPRA, 92 kg N/ha; 125% of TNPRA, 115 kg N/ha; 75% of TNPRA, 69 kg N/ha and 50% of TNPRA, 46 kg N/ha. The fertlizer was applied two times, basal and top dressings, 21 and 51 days after planting, respectively.

The Triscan Sensor (EnviroScan, Sentek Pty Ltd, Stepney, Australia), which has ability to monitor the direction and movement of nitrogen in the soil at instant time of inserting the monitoring probe in the soil, was used to measure total nitrogen concentration at lateral distances. The measurement of the sensor are in Volumetric Ion Concentration (VIC), but using standazation equation the concentration of total nitrogen on each point was known. The lateral distances at which measurements were taken were as follows: at point of application (represented by 0 cm), at 5 cm away from the plant (represented by -5 cm), at 5 cm towards the plant, 10 cm towards the plant (this point was maize planting station), and 15 cm (this point was 5 cm after planting station in the direction opposite from where N was applied). The lateral distances were taken based on spreading and elongation pattern of lateral roots of maize plants.

The lateral reading of nitrogen were respecively taken at five soil depths of 20, 40, 60, 80, and 100 cm. The soil depths were selected based on maize roots growth habits which extend down to 100 cm [23]. Confirmation of data was done through analyising soil samples collected from the respective points. The soil profile was dug to a depth of 120 cm using soil auger. Soil samples were collected from the lateral points for laborotary analysis.

Primary soil samples were collected and analysed at Bunda College Laboratory to know the values of wilting point or Lower Limit of soil water content at 15 Bar (LL15), Drained Upper Limit of soil water content (DUL) or Field Capacity, and Saturated soil water content (SAT). The hydraulic conductivities of soil were from the soil literature review.

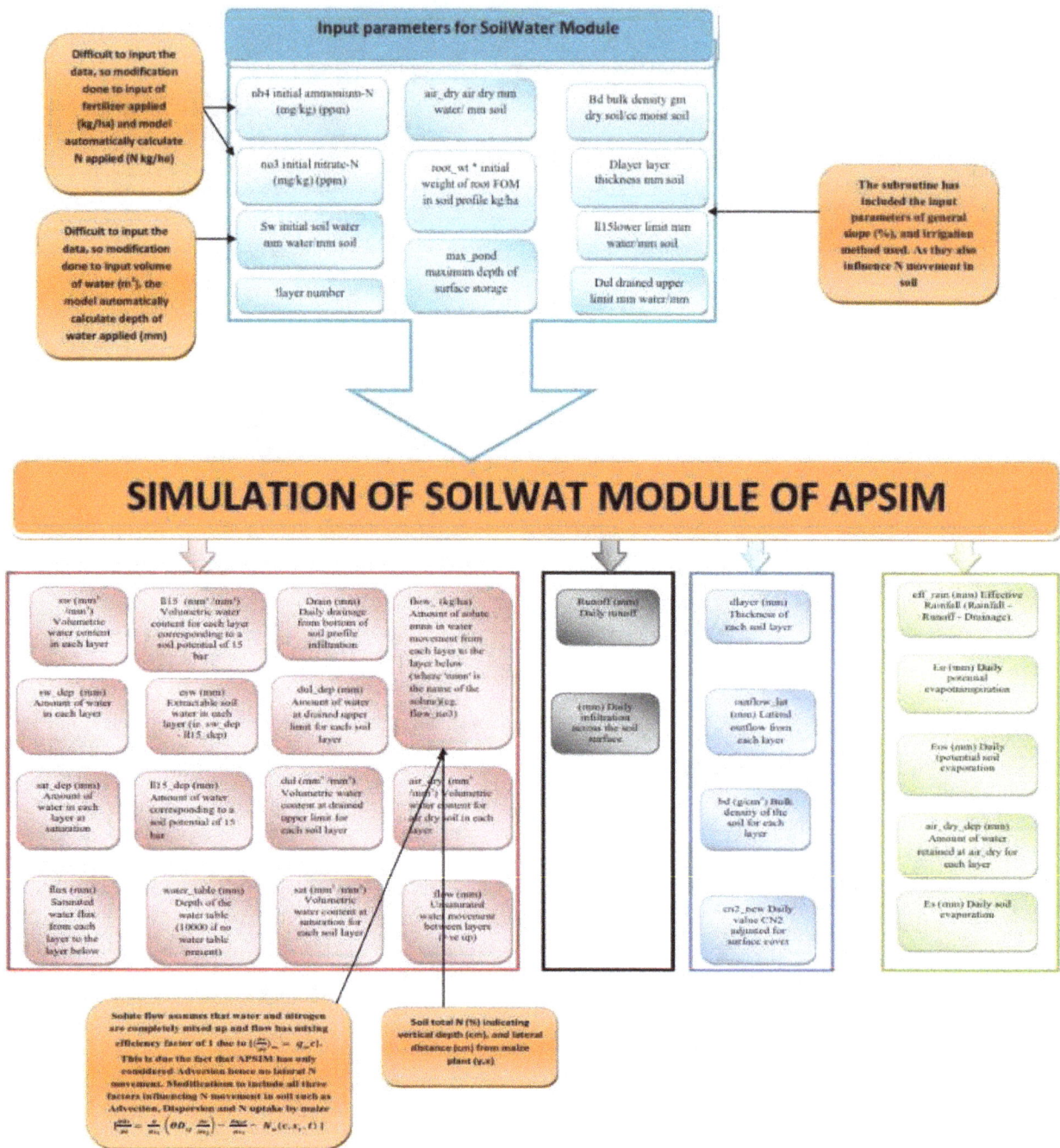

Fig. 6. Flow chart and parameters that have been incorporated in the SoilWat module

3. Results and Discussion

3.1. Total Nitrogen Contents in the Soil

Majority of nitrogen available to plants is in the form of inorganic NH_4^+ and NO_3-. Ammonium ions (NH_4^+) bind to the soil's negatively-charged Cation Exchange Complex (CEC), while nitrate ions (NO_3-) do not bind to the soil solids because they carry negative charges. Since none of the nitrate is adsorbed to soil particles it is abundant in the soil water and the movement of the nitrate to the root rarely limits its uptake.

Figure 7 shows the simulated temporal distribution of nitrogen within the soil profile of 100FWRR plot which received TNPRA. The no3(1) shows the concentration of nitrogen in the first layer (0-10 cm), no3(2) shows the nitrogen concentration in the second layer (10-20 cm), the nitrogen concentration in third layer (20-40 cm) is being represented by no3(3). The figure indicates that on 1/06/2012 the nitrogen concentration in first layer was above 65 N kg/ha while the concentration in underlying layers were all around 5 N kg/ha. This is because on 1/06/2012, NPK

(23:21:0+4s) fertilizer was basal dressed and hence high concentration of N at the point of application. The other underlying layers still had the residual nitrogen concentration which was 5 N kg/ha. On 10/06/2012, N concentration second layer (10-20 cm) and third layer (20-40 cm) had increased from 5 N kg/ha to above 20 N kg/ha, while that of first layer [no3 (1)] had reduced from above 65 N kg/ha to about 35 N kg/ha. This change of N concentrations can be explained by movement of water which carried N solute down to underlying (second and third) layers. This can further be validated by trend of soil water movement in figure (12) below. Another line of interest is no3 (6) which is N concentration in last layer (80-100 cm), figure 7 shows that N concentration remained relatively constant until after 16/07/2012 indicating that soil water arrived to this layer on this particular day. This can as well be confirmed by the trends of water movement in figure (12) below, which also indicates that soil water arrived in the last layer after six irrigation events. The increase of N concentration on 16/07/2012 is due to second application (top dressing) of urea fertilizer, which was applied 45 days after planting.

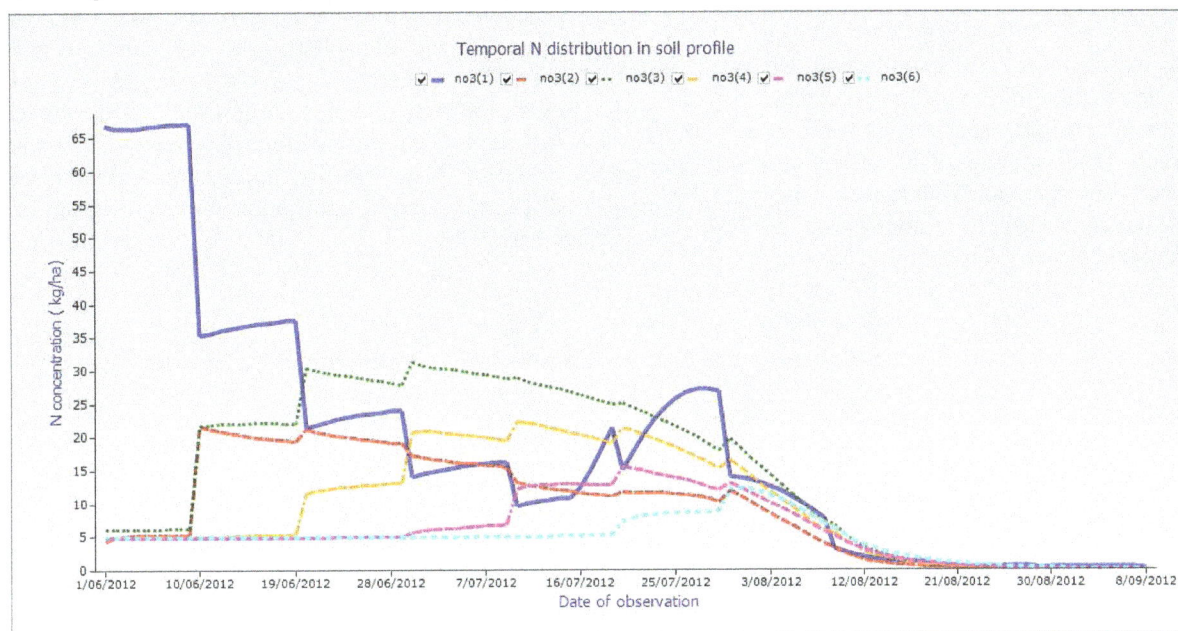

Fig. 7. Temporal distribution of nitrogen in 100FWRR plot which received TNPRA.

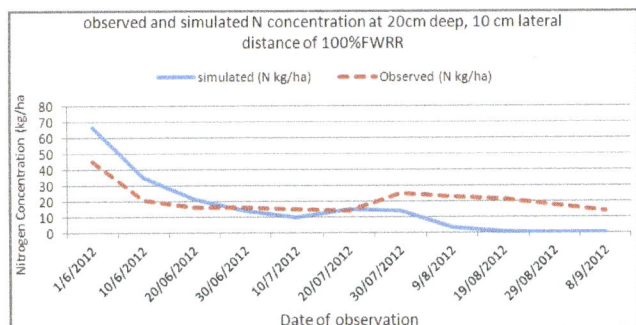

Fig. 8. Comparison between observed and simulated N concentration at top layer (0-20 cm).

Figure 8 shows the trend of changes of N concentration within the top layer of simulated and observed data in100FWRR plot which received TNPRA. From figure 7 it can be shown that while simulated N concentration was above 65 N kg/ha, the observed was below 45 N kg/ha. This indicates that the APSIM had overestimated N concentration in the first layer. The difference between observed and simulated N concentrations from 30/06/2012 to 20/07/2012 was relatively small meaning that observed concentrations were close to those predicted by APSIM. From 20/07/2012, simulated N concentration is low than observed N concentration indicating that APSIM is under predicting the

N concentration in first layer. The APSIM under predicted during the latter stage of the maize growing season and over predicted in the early stage of the growing season. This therefore suggests that APSIM does not perfectly predict concentration of nitrogen. [20] also reported that APSIM simulations over estimate the soil solution nitrate concentration.

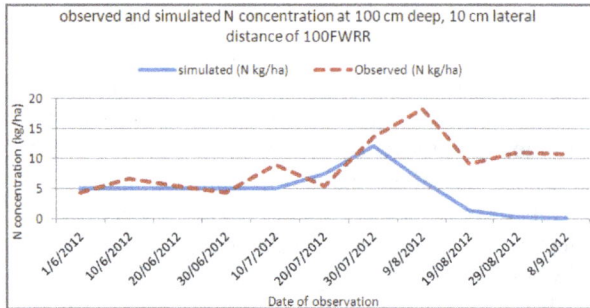

Fig. 9. *Comparison between observed and simulated N concentration at bottom layer (80-100 cm).*

Figure 9 shows changes in N concentrations within the bottom layer (80-100 cm) of plot that received TNPRA. From 1/06/2012 to 30/06/2012 the change of N concentrations between simulated and observed data is small. The trend indicates that in simulated data, there was no addition of new concentration of nitrogen while in observed data; there was a small addition of N concentrations. From 30/07/2012, the N concentration from observed data is higher than from simulated data indicating that APSIM is underestimating the N concentration in last layer.

3.2. Soil Water Contents in the Soil

In this section of the paper, soil water contents from simulated and observed data have been presented and explained. Unless specified the unit of soil water content in the soil is given as millimeter of water in millimeter of soil depth (mm/mm). Figure 10 presents the amount of water applied in the plot from which proceeding figures of soil water contents have been taken and discussed.

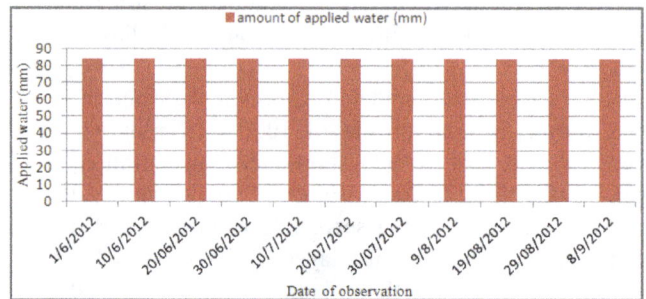

Fig. 10. *Amount of water applied in 100FWRR plot which received TNPRA.*

Figure 10 shows the time of water application and amount of water applied in a plot that received full water requirement regime presented as 100FWRR. This plot also received 92 N kg/ha of nitrogen presented as Typical Nitrogen Application Regime in the Area (TNPA). Amount of water applied in each irrigation event was 84 mm of water and was applied after every 10 days. The first application was done on 1st June 2012 and the last application event was done on 8th September, 2012.

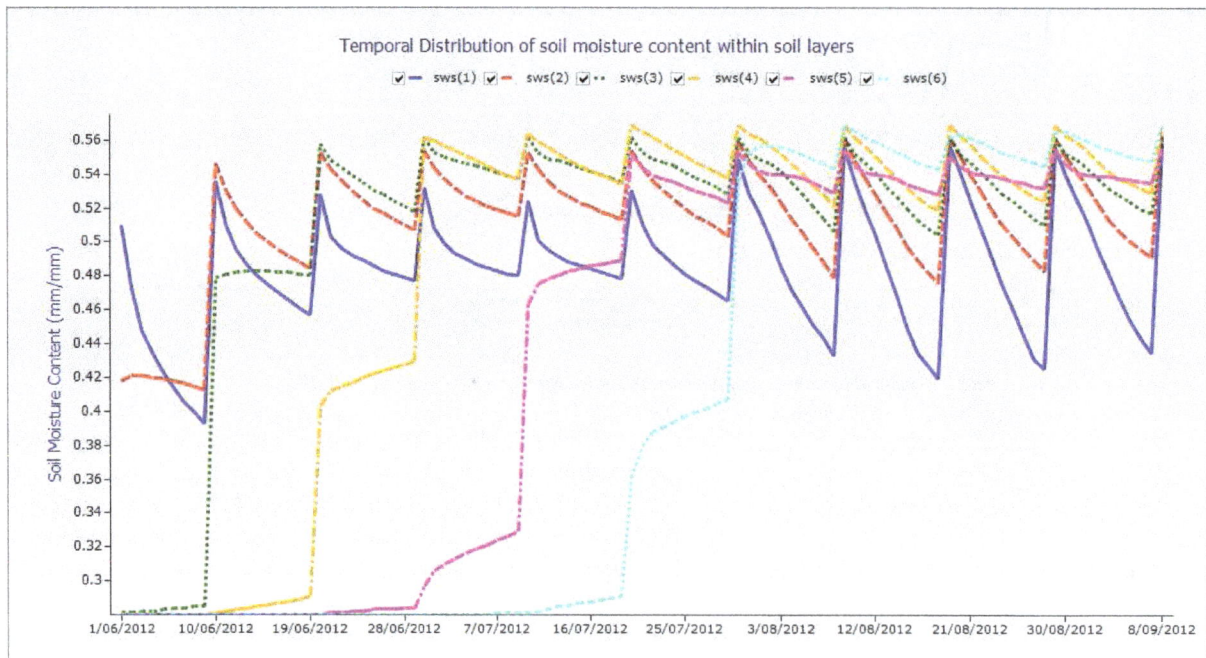

Fig. 11. *Simulated soil moisture distribution in the soil profile.*

Figure 11 shows soil water distribution through the soil profile as simulated by the by APSIM from 100FWRR plot which received TNPRA. From figure 12 it shows that soil layers received soil water at different times. The first soil layer (0-10 cm) registered the first soil water of above 0.5 mm/mm on 1st June which had declined to below 0.4

mm/mm before next irrigation on 10th June, 2012. This sharp declined can be explained as due to massive soil water loss through evaporation as during this time, crop cover had not fully covered the soil to reduce soil evaporation. From 19/06/2012, sws (1) line has its difference of soil water contents between when water was applied to the last day before next application reduced i.e. it is more flatter than compared to before 19/06/2012 and after 3/08/2012. The trend can be explained by crop cover which fully covered the soil and hence reduce soil evaporation. Another trend of interest is sws (6) line which is last section of soil profile (80-100cm), the line started to peak on 7/07/2012 which is almost 37days after first water application i.e. after 5 water applications. From 1/06/2012, the soil water content was stagnant at 0.28 mm/mm. Soil water percolates through soil profile only when proceeded soil is satisfied i.e. has reached its field capacity, above which excess water is left free to percolates down the soil profile. Before water arriving in last layer it had to satisfy the above-lying soil profiles. However, when water had reached in this layer, the soil water lose was small and this is the reason why the sws (6) line is flatter than compared to all other soil water lines. This observation is similar to that of many literatures which explain that fluctuations of soil water contents in bottom layers of soil profiles are small compared to top layer. Another observation from the graph is that even during irrigation time, soil water content never exceeded 0.56 mm/mm indicating that the saturation point of this soil is 0.56mm/mm. figures 12 and 13 below are comparing differences of simulated soil water contents to the observed soil water contents within the top layer (0-20cm) and bottom layer (80-100cm).

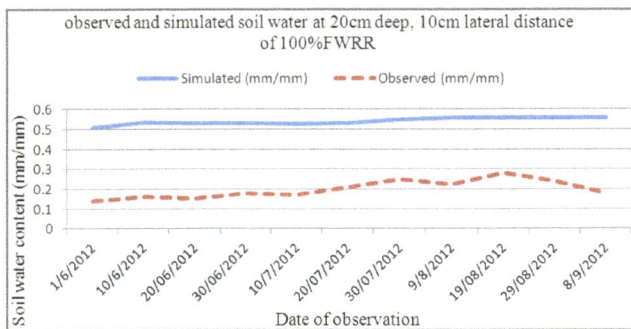

Fig. 12. Comparison of simulated and observed soil water distribution within top layer 0-20cm

Figure 12 shows the comparison between simulated and observed soil water distribution within the first layer (0-20 cm) of 100FWRR plot which received TNPRA. The figure 11 indicates that observed soil water contents were far less compared with simulated soil water contents. The soil water contents from observed graph line indicates that were within 0.14 to below 0.28 mm/mm of soil, while soil water contents from simulated graph line were within the range of 0.51 to 0.558 mm/mm of soil. Figure 12 shows that APSIM had overestimated soil water contents of top layer in 100FWRR plot of TNPRA.

Figure 13 shows soil water distribution within 80-100 cm layer over the growing season in 100FWRR plot which received TNPRA. From figure 13 it can be seen that observed graph line is below simulated graph line meaning that observed soil water contents were less to that of simulated soil water contents. However, both graph lines in figure 13 are showing similar trend in the way they are increasing and decreasing.

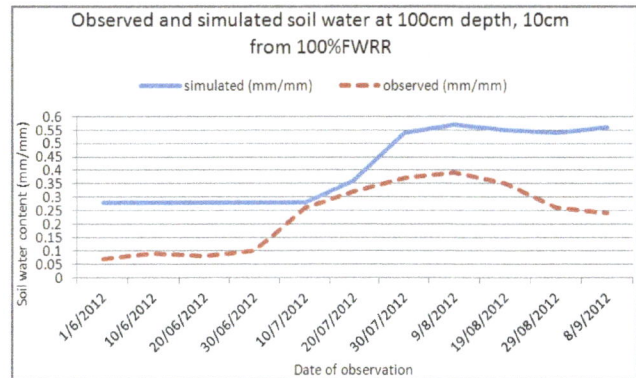

Fig. 13. Comparison of simulated and observed soil water distribution within bottom layer (80-100cm).

3.3. Relationship of Soil Water and Nitrogen Content in Soil

Figure 14 indicates the relationship between soil water and nitrogen contents in top layer of plot which received 100% of full water requirement regime with nitrogen application regime of 92 N kg/ha. From figure 14 it shows that on 01/06/2012, N concentration was over 65 N kg/ha and soil water content was about 0.43 mm/mm. Just 10 days later (10/06/2012), N concentration declined to about 34 N kg/ha while soil water content increased to over 0.55 mm/mm. The decrease in N concentration can be due to down movement to underlying layers. This can further be confirmed with increase of N concentration in third layer (20-40cm) in figure 16 below, where N concentration increased from around 5 N kg/ha on 01/06/2012 to over 20 N kg/ha on 10/06/2012 indicating that it had received additional nitrogen from overlying layers. Within APSIM's SoilWat module the saturated and unsaturated flows of soil water are used to calculate the redistribution of solutes throughout the soil using a 'mixing' algorithm (Probert et al., 1998). Efficiency factors adjust the effectiveness of mixing for either saturated or unsaturated flows which infer that nitrogen movement in calculated by the product of water flow and nitrogen concentration in that water. However, it has to be noted that concentration is function of solvent i.e. water and solute i.e. nitrogen meaning that with increase of water in the solution, the concentration of nitrogen will be decreasing. This might be one of the reason contributing to the trend of figure 14 where increase in soil water content is resulting into decrease in nitrogen concentration.

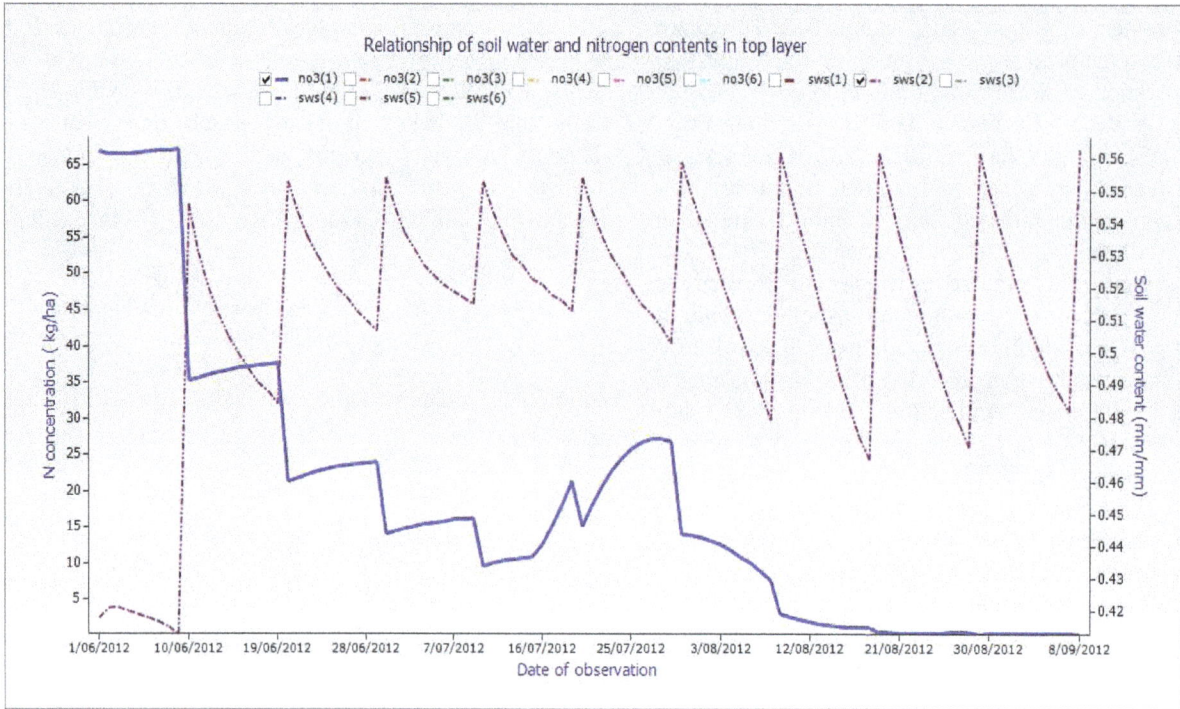

Fig. 14. Showing relationship of soil water and nitrogen content in top layer (0-10cm) of 100FWRR plot which received TNPRA

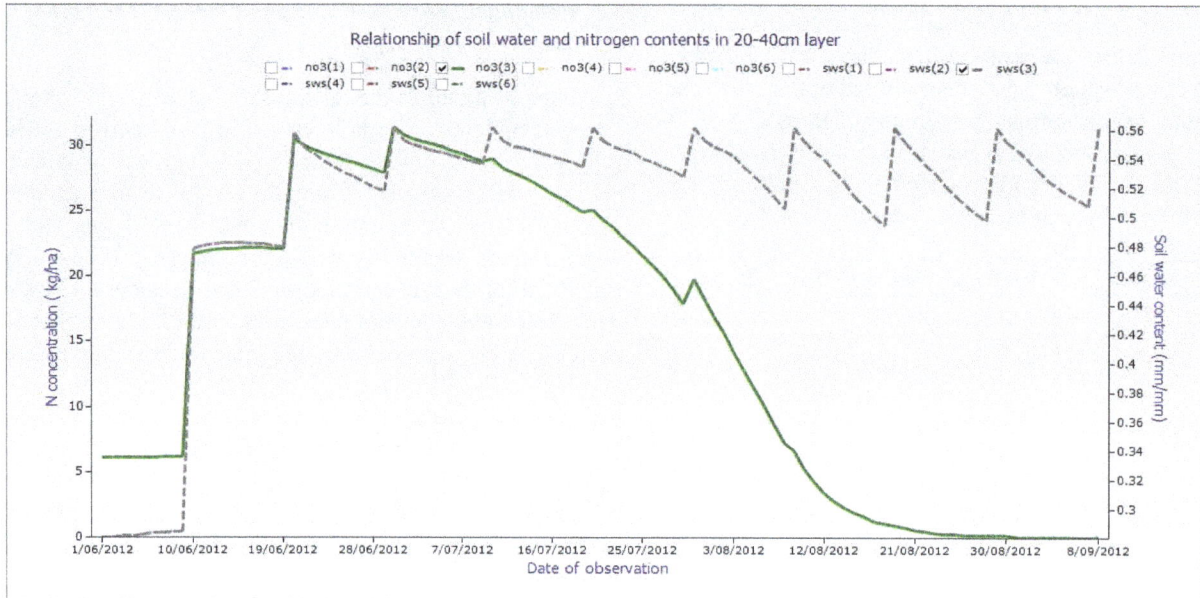

Fig. 15. Showing relationship of soil water and nitrogen content in third layer (20-80cm) of 100FWRR plot which received TNPRA.

Figure 15 indicates an increase in soil water and nitrogen contents from below 0.3 mm/mm and over 5 N kg/ha on 01/06/2012 to over 0.48 mm/mm and over 20 N kg/ha on 10/06/2012 respectively. The increase in soil water and nitrogen content in this layer corresponds with decrease in nitrogen content in top layer as shown in figure 14 above. The other interesting observation in figure 15 is the soil water and nitrogen contents trends from 10/06/2012 to 07/07/2012. The trend appears to show the similar pattern of increase and

decrease of soil water and nitrogen contents in the third layer. This trend can be explained as due to advection movement of nitrogen from overlying layers to this layer. Nitrogen is perfectly dissolved in water solution and as water cascades down the layers it moves with water nitrogen concentration. However, as more water is being applied (by irrigation), the decrease in N concentration is witnessed, hence nitrogen content starts to decline from 07/07/2012 and thereafter.

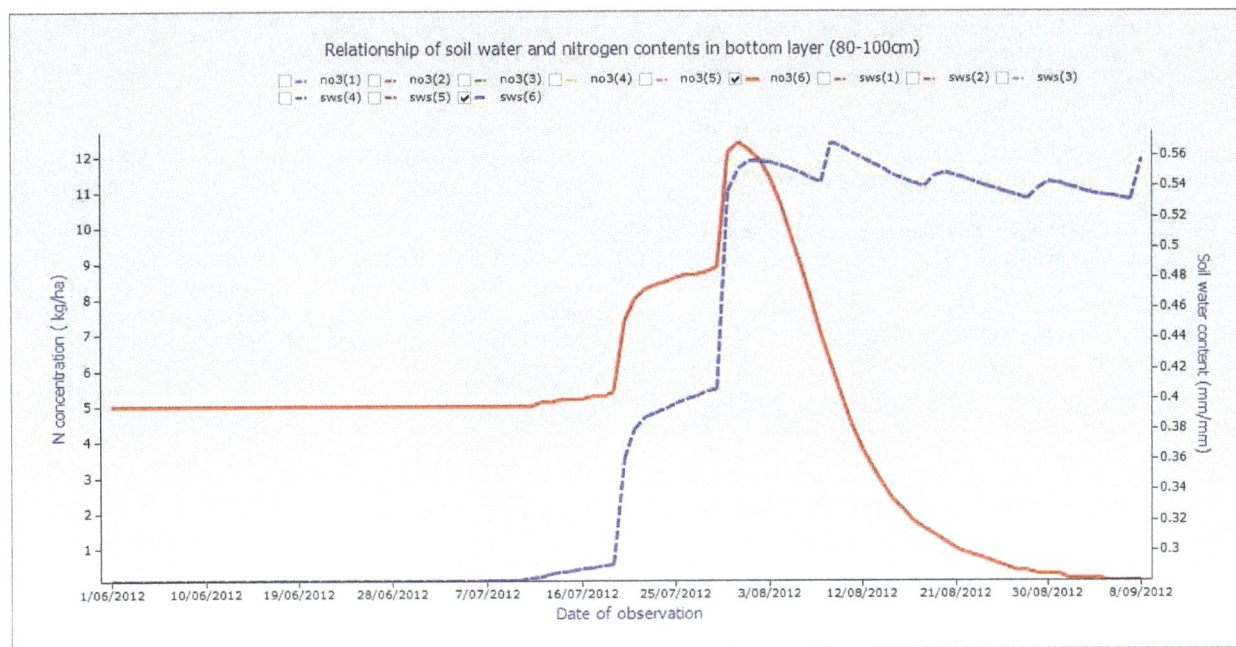

Fig. 16. *Showing relationship of soil water and nitrogen content in third layer (80- 100cm) of 100FWRR plot which received TNPRA.*

From 01/06to 16/07/2012 in figure 16, N concentration remained the same on about 5 N kg/ha, which increased to about 8 N kg/ha on 25/07/2012 and subsequently increased to its highest concentration of about 11 N kg/ha before it started to decline on 03/08/2012 to almost 0 on 08/09/2012. Soil water content increased in similar fashion from below 0.3 mm/mm to 0.38 mm/mm and subsequently to over 0.54 mm/mm in just 10 days from 16/07/2012, despite staying constant on 0 for 46 days from 01/06/2012. The sudden rise of soil water and nitrogen contents in this layer shows that overlying layers had their 'needs' satisfied as such additional application of water just moved to this bottom layer without any hindrance (being held by soil particles).

4. Conclusions

The following conclusions have been drawn from the study: Soil water percolates down to underlying layer only when proceeded layers are satisfied i.e. has reached its field capacity, above which excess water is left free to percolates down the soil profile. Before water arriving in last layer it had to satisfy the above-lying soil profiles. This therefore suggests that to avoid deep percolation and losing of nitrogen to layers below maize rootzone, water applied should correspond to the amount of water that can be retained in soil layers of maize rootzone. The study has shown that increase of nitrogen contents in underlying layers corresponds with decrease of the same in top layers due to advection movement. Consequently, the increases of soil water in a specific layer correspond to decrease of nitrogen content in that particular layer. This suggests that to maximise deposition of nitrogen within rooting zone of maize plant, amount of applied water should be reduced after nitrogen application. Movement of nitrogen through soil profile corresponds to amount of water applied. The study has

shown that APSIM under predicted during the latter stage of the maize growing season and over predicted in the early stage of the growing season, and it overestimates soil water contents in soil profile.

References

[1] Uribelarrea, M., Crafts-Brandner, S.J., Below, F.E., 2009. Physiological N response of field-grown maize hybrids (*Zea mays* L.) with divergent yield potential and grain protein concentration. Plant and Soil, 316, 151-160.

[2] Mullins, G.L., Alley, S.E., Reeves, D.W., 1998. Tropical maize response to nitrogen and starter fertilizer under strip and conventional tillage systems in southern Alabama. Soil Till. Res. 45, 1-15.

[3] Halvorson, A.D., Mosier, A.R., Reule, C.A., Bausch, W.C., 2006. Nitrogen and tillage effects on irrigated continuous corn yields. Agron. J. 98, 63-71.

[4] Eck, H.V., 1984. Irrigated corn yield response to nitrogen and water. Agron. J. 76, 421-428.

[5] Al-Kaisi, M.M., and X. Yin. 2003. Effects of nitrogen rate, irrigation rate, and plant population on corn yield and water use efficiency. *Agron. J.* 95: 1475 – 1482.

[6] Devienne-Barret, F., Justes, E., Machet, J.M., Mary, B., 2000. Integrated control of nitrate uptake by crop growth rate and soil nitrate availability under field conditions. Ann. Bot. 86, 995-1005.

[7] Russell, A.E., Laird, D.A., Mallarino, A.P., 2006. Nitrogen fertilization and cropping system impacts on soil quality in midwestern mollisols. Soil Sci. Soc. Am. J. 70, 249-255.

[8] Addiscott, T.M. 1995. Modelling the fate of crop nutrients in the environment: Problems of scale and complexity. *European Journal of Agronomy* 4: 413417.

[9]   Jury, W.A., Gardner, W.R. and Gardner, W.H. 1991. Soil physics. 5th edition. John Wiley and Sons. New York.

[10] Van Genuchten, M.Th., Leij, F.J., Yates, S.R., 1991. The RETC code for quantifying the hydraulic functions of unsaturated soils. USAEPA Rep. 600/2-91/065. R.S. Kerr Environmental Research Laboratory. Ada, OK 74820.

[11] Vitousek, P.M., Hattenschwiler, S., Olander, L. and Allison, S. 2002. Nitrogen and nature. Ambio. 31: 97-101

[12] Hutson, J.L. and Wagenet, R.J. 1993. A pragramatic field scale approach for modelling pesticides. *Journal of Environmental Quality* 22: 494–499.

[13] Hanson, J.D., Rojas, K.W. and Shaffer, M.J. 1999. Calibrating the root zone water quality model. *Agronomy Journal.* J. 91:171–177.

[14] Ahuja, L.R., Rojas, K.W., Hanson, J.D., Shaffer, M.J. and Ma L. (eds.). 2000. Root zone water quality model: Modelling management effects on water quality and crop production. Water Resources Publications LLC, Highlands Ranch, Co. 372 pp.

[15] Shaffer, M. J., Halvorson, A. D. and Pierce, F. J. 1991. Nitrate leaching and economic analysis package (NLEAP): Model description and application. Pages 285–322. *In* R. F. Folet et al. (eds.). Managing nitrogen for groundwater quality and farm profitability. SSSA, Madison WI.

[16] Simunek, J., Sejna, M. and van Genuchten, M.Th. 1999. The HYDRUS-2D software package for simulating two-dimensional movement of water, heat, and multiple solutes in variably saturated media. Version 2.0, IGWMC - TPS - 53, International Ground Water Modeling Center, Colorado School of Mines, Golden, Colorado, 251pp.

[17] McCown, R.L., Hammer, G.L., Hargreaves, J.N.G., Holzworth, D.P. and Freebairn, D.N. (1996). APSIM: A novel software system for model development, model testing, and simulation in agricultural research. *Agricultural Systems* 50: 255-271.

[18] Probert, M.E., J.P. Dimes, B.A. Keating, R.C. Dalal, and W.M. Strong. 1998.APSIM's water and nitrogen molecules and simulation of the dynamics of water and nitrogen in fallow systems. Agricultural systems, 56, 1-28,

[19] Sharp, J.M., Thomas S.M, Brown, H.E. 2011. A validation of APSIM nitrogen balance and leaching predictions. The New Zealand Institute for Plant & Food Research Limited, Private Bag 4704, Christchurch, 8140, New Zealand.

[20] Allen, R.G., L.S. Pereira, D. Raes, and M. Smith. 1998. Crop evapotranspiration. Guidelines for computing crop water requirements. Irrig. And Drainage Paper no. 56.

[21] Salim, B.A. (1999). Modelling and Measurement of Soil Moisture Content Based on a Remote Sensing Method for Application in Semi-Arid Tropics. PhD dissertation. Institut fur Landtechik der Rheinischen Friedrich-Wilhelms-Universitaet Bonn, pp. 148-149.

[22] Ministry of Agriculture and Food Security. (2011). Annual reports and notes. Lilongwe, Malawi.

[23] FOASTAT   (2000)   Maize. http://www.fao.org/nr/water/cropinfo_maize.html

Technological Gaps in Adoption of Improved Soybean Production Technology by Soybean Growers in Dahod District, Gujarat

R. G. Machhar, S. K. Patel*, **H. L. Kacha, U. M. Patel, G. D. Patel, R. Radha Rani**

Krishi Vigyan Kendra, Anand Agricultural University, Campus Dahod, Gujarat, India

Email address:

sppiari@gmail.com (S. K. Patel)

Abstract: The present study was conducted in Dahod district of Gujarat State, India. Fifteen villages of Dahod district and ten farmers from each village were selected randomly for the study. Thus, in all, 150 soybean growers constituted the sample for this investigation. The data for this study was collected by arranging personal interview and filling up of the survey performa. The results of the study revealed that the technological gap of all categories of respondents was found to be negatively and significantly related with the independent variables viz. level of knowledge, education, social participation, source of information utilized and cropping intensity. Further, it was found that technological gap among the marginal farmers' was negatively and significantly related with one independent variable i.e. farm implements. Age was positively and significantly related with the technological gap in recommended soybean production technology.

Keywords: Soybean, Adoption, Technological Gap, Dahod, Gujarat

1. Introduction

Soybean [Glycine max (L) Marril] belongs to family Leguminoseae, sub family Papilionaideae and genus Glycine. It is mainly grown in kharif season and occupies second place in the world, following groundnut in oilseed production. Soybean has now established as an economically important leguminous crop, known for its high valued protein, oil, food, feed and industrial application. It enriches the soil by fixing nitrogen in symbiosis with bacteria. In the international world trade markets, soybean is ranked number one among the major oil crops (Specht et al., 1999). In India, the consumption of oil has been increasing steadily as a result of rise in population and living standard of the people (Chung and Sing, 2008). Presently soybean is grown in many countries of the world like USA, China, India, Brazil and Argentina. In India, the total area under soybean cultivation is 106.95 lakh ha with the production and the productivity of 126.78 lakh tonnes and 1185 kg ha^{-1}, respectively (Anon., 2012). In India major States growing soybean are Madhya Pradesh, Maharashtra, Rajasthan, Andhra Pradesh, Karnataka, Chattisgarh and Gujarat. Major soybean growing districts of Gujarat state are Dahod, Sabarkantha, Bharuch, Vadodara, Panchmahals and Amreli. Farmers of Dahod district have changed their old cropping pattern to new cropping pattern i.e. diversified maize crop into soybean crop (Patel et al, 2014 and Kacha & Patel, 2015).The area of soybean crop in the Dahod Districtis increasing very fast since year 2005 to 2011 i.e. 0 to 35000 ha but the productivity of soybean crop is very low as compare to its potentiality (Anon., 2013 and Machhar et al., 2015). So, there is a need to find out the gap existing between technologies available and actually applied by farmers in their fields (Singh et al., 2006 and Bhatiya et al., 2006). Therefore, a study entitled "A study of technological gaps in adoption of improved soybean production technology by soybean growers in Dahod district" was undertaken.

2. Methodology

Fifteen villages of Dahod district and ten farmers from each village were selected randomly for the study. Thus, in all, 150 soybean growers constituted the sample for this investigation. The method used for collecting data was the well structured personal interaction schedule prepared in

English, keeping in view the objectives of the study by referring the review of related literature, popular articles and guidance provided by senior officials. Questions and statements on each and every aspect of the problem were framed in order to study with maximum possible accuracy, clarity and objectivity.

The data were collected through personal interaction with the soybean growers. Before administrating the schedule, the investigators have introduced themselves as soybean growers. The objectives of the study were explained to them with a view to facilitate them in giving correct responses. The majority of soybean growers were interviewed either at their residence or at farm. The questions from the interview schedule were asked one by one and their responses were recorded on the spot. Possible care was taken to maintain friendly atmosphere to get unbiased responses from the soybean growers. During data collection, investigators have also gathered the useful information through observation and informal discussion with the soybean growers for its implication at the time of interpretation of data.

The data of this study was collected by arranging personal interview and survey performa. To determine the extent of adoption of production technology of soybean cultivation, adoption quotient developed by Sharma and Intodia (1991) was used. The technological gap was computed by following formula developed by All India Coordinated Project in Extension Education (AICPEE) and Indian Agricultural Research Institute (IARI), New Delhi. The data was analyzed in light of objectives.

3. Results and Discussions

Table 1. Distribution of soybean growers according to age (n=150).

Sr. No.	Age group	Number	Percent
1.	Young age(Up to 30 year)	19	12.67
2.	Middle age(31 to 50 year)	110	73.33
3.	Old age(Above 50 year)	21	14.00
	Total	150	100.00

The data depicted in Table 1 shows that maximum number of soybean growers (73.33 per cent) were found in middle age group followed by old age 14.00 per cent. Remaining 12.67 per cent soybean growers were found in young age.

Table 2. Distribution of soybean growers according to their level of education(n=150).

Sr. No.	Level of education	Number	Percent
1.	Illiterate	18	12.00
2.	Primary education (Up to VII Std.)	43	28.67
3.	Secondary education (VIII to X Std.)	68	45.33
4.	Higher Secondary education (XI to XII Std.)	18	12.00
5.	College and above education	03	2.00
	Total	150	100.00

The data depicted in Table 2 indicate that 45.33 per cent soybean growers were educated up to secondary education followed by 28.67 per cent were educated up to primary

level. Percent of soybean growers that were illiterate was 12.00 percent and equal to the group with higher secondary education. Remaining only 2.00 per cent were having college and above education.

A look into Table 3 reveals that 64.00 per cent Soybean growers had no membership in any organization. While a little less than one-forth of them (24.67 per cent) had membership in more than one organization. Only 8.00 per cent soybean growers had membership in one organization. Remaining 3.33 per cent soybean growers were found holding position in various organizations.

Table 3. Distribution of soybean growers according to their social participation (n =150).

Sr. No.	Social participation	Number	Percent
1.	No membership	96	64.00
2.	Membership in one organization	12	08.00
3.	Membership in more than one organizations	37	24.67
4.	Holding position	05	03.33
	Total	150	100.00

Table 4. Distribution of soybean growers according to their size of land holding(n =150).

Sr. No.	Size of land holding	Number	Percent
1.	Marginal farmers (Up to 1.00ha)	56	37.33
2.	Small farmers (1.01 to 2.00 ha)	81	54.00
3.	Medium farmers (2.01 to 4.00 ha)	11	7.33
4.	Large farmers (Above 4.00 ha)	02	1.34
	Total	150	100.00

The data presented in Table 4 shows that more than half of the soybean growers (54.00 per cent) were found to have small size of farm followed by 37.33 and 7.33 per cent with marginal (up to 1.00 ha) and medium (2.01 to 4.00 ha) size of farm, respectively. Only a mere, 1.34 per cent soybean growers were found to have large size of farm.

Table 5. Distribution of soybean growers according to their cropping intensity(n =150).

Sr. No.	Cropping intensity	Number	Percent
1.	100	04	2.70
2.	125	01	0.07
3.	128	01	0.07
4.	150	06	4.00
5.	167	01	0.07
6.	193	01	0.07
7.	200	133	88.70
8.	217	01	0.07
9.	228	01	0.07
10.	300	01	0.07
	Total	150	100.00

A look into Table 5 shows that 88.70 per cent soybean growers obtained 200 per cent cropping intensity followed by 150 and 100 per cent cropping intensity got by 4.00 and 2.00 per cent soybean growers, respectively. Only 0.07 per cent soybean growers were having 125,128,167,193,217,228 and 300 per cent cropping intensity.

Table 6. *Distribution of soybean growers according to their occupation(n =150).*

Sr. No.	Occupation	Number	Percent
1	Only farming	08	5.33
2	Farming + Animal Husbandry	80	53.33
3	Farming + Animal Husbandry + Labour work	61	40.67
4	Farming +Animal Husbandry + Service	01	0.67
	Total	150	100.00

The data presented in Table 6 revealed that more than half (53.33 per cent) of the soybean growers were engaged in farming along with Animal Husbandry followed by 40.67 per cent of the respondents who were engaged in farming + animal husbandry + labour work. Remaining 5.33 and 0.67 per cent of the soybean growers were engaged in only farming and Farming + Animal Husbandry + Service, respectively.

Table 7. *Distribution of soybean growers according to their annual income (n =150).*

Sr.No.	Annual income	Number	Percent
1.	Less than 20,000	01	0.67
2.	20,000 to 30,000	07	4.67
3.	31,000 to 50,000	84	56.00
4.	50,000 to 70,000	41	27.33
5.	70,000 and above	17	11.33
	Total	150	100.00

The data presented in Table 7 shows that 56.00 per cent soybean growers were having annual income between 31,000 to 50,000 followed by 50,000 to 70,000 and 70,000 and above annual income with 27.33 and 11.33 per cent, respectively. Remaining 4.67 and 0.67 per cent soybean growers were found with 20,000 to 30,000 and less than 20,000 annual income, respectively.

Table 8. *Distribution of soybean growers according to their level of extension participation(n =150).*

Sr. No.	Extension participation	Number	Percent
1.	Low (<1.94 score)	02	01.33
2.	Medium (Between 1.94to16.79score)	126	84.00
3.	High (16.79>score)	22	14.67
	Total	150	100.00

Mean: 9.36, S.D. 7.42

The result of the study reported in Table 8 reveals that more than four-fifth (84.00 per cent) of the soybean growers had medium extension participation whereas less than one–fifth (14.67 per cent) had high extension participation. Only a mere, 1.33 per cent soybean growers had low extension participation.

Table 9. *Distribution of soybean growers according to their sources of information utilized(n =150).*

Sr. No.	Sources ofinformation	Number	Percent
1.	Low (< 9.04 score)	85	56.67
2.	Medium (Between 9.04to 17.99score)	42	28.00
3.	High (>17.99 score)	23	15.33
	Total	150	100.00

Mean: 13.52, S.D. 4.48

A perusal of data presented in Table 9 reveals that more than half (56.67 per cent) of the soybean growers had low sources of information utilized. Whereas one – forth (28.00 per cent) and 15.33 per cent of the soybean growers had high and low sources of information utilized, respectively.

Table 10. *Distribution of soybean growers according to their level of adoption(n =150).*

Sr. No.	Level of adoption	Number	Percent
1.	Low (< 6.69 score)	13	8.67
2.	Medium (Between 6.69 to 9.76 score)	119	79.33
3.	High (> 9.76 score)	18	12.00
	Total	150	100.00

Mean: 8.23, S.D. 1.54

The data presented in Table 10 reveals that more than three-fifth (79.33 per cent) of the soybean growers had medium level of adoption whereas 12.00 and 8.67 per cent of the soybean growers had high and low level of adoption, respectively.

Table 11. *Distribution of soybean growers according to their level of adoption and technological gap in package of practices (n =150).*

Sr. No.	Package of practices	Number	Adoption Per cent	Technological gap
1.	Soil	147	98.00	02.00
2.	Improved Variety	146	97.33	02.67
3.	Seed rate	140	93.33	06.67
4.	Seed Treatment	37	24.67	75.33
5.	Time of sowing	138	92.00	08.00
6.	System of sowing	133	88.67	11.33
7.	Sowing distance	36	24.00	76.00
8.	Manuring	91	60.67	39.33
9.	Chemical Fertilizer	110	73.33	26.67
10.	Weeding	108	72.00	28.00
11.	Plant protection	01	0.67	99.33
12.	Time of harvesting	147	98.00	02.00

The data presented in Table 11 reveals that maximum technological gap was observed in plant protection (99.33 per cent) followed by sowing distance (76.00 percent), seed treatment (75.33 per cent), use of manure (39.33 per cent) weeding (28.00 per cent), use of chemical fertilizer (26.67 per cent) and system of sowing (11.33 per cent).Less than 10.00 percent technological gap was found in time of sowing (8.00 per cent), seed rate(6.67 per cent), use of improved variety(2.67 per cent), time of harvesting (2.00 per cent and selection of soil(2.00 per cent) in descending order.

Table 12. *Constraints faced by soybean growers in adoption of improved cultivation practices (n =150).*

Sr.No.	Constraints	Number	Percent	Rank
1	Shattering loss in existing varieties	67	44.67	I
2	Unavailability of sufficient labour in time	59	39.33	II
3	Lack of technical guidance	35	23.33	III
4	High cost of insecticide/ pesticide and weedicide	32	21.33	IV
5	Fluctuating market price of soybean	30	20.00	V
6	High cost of fertilizer	20	13.33	VI

The data presented in Table 12 reveals that constraints faced by soybean growers in adoption of improved cultivation practices of soybean were shattering loss in existing varieties (44.67 per cent) followed by unavailability of sufficient labour in time (39.33 percent) and Lack of technical guidance(23.33 percent).Remaining constraints were high cost of insecticide/ pesticide and weedicide, fluctuating market price of soybean and high cost of fertilizer with 21.33, 20.00 and 13.33 per cent, respectively.

4. Conclusion

As per results, majority soybean growers were in age of 31 to 50 years, secondary education, no membership in any social organization, small farmers, having farming with animal husbandry as the main occupation, medium extension participation, low use of sources of information and medium level of adoption.

The technological gap of soybean growers was found to be the highest in plant protection measures followed by sowing distance, seed treatment and manuring.

Constraints faced by soybean growers in adoption of improved cultivation practices of soybean were shattering loss in existing varieties followed by unavailability of sufficient labour in time and Lack of technical guidance.

References

[1]   Anonymous (2012).Soybean Processors Associations of India (2012).

[2]   Anonymous (2013). Comprehensive District Agricultural Plan (2013).

[3]   Chung, G. and Singh. R.J. (2008). Broadening the genetic base of soybean. A multidisciplinary approach. *Plant Sci.*, 27: 295-341.

[4]   Patel, S.K., Machhar, R.G., Kacha, H.L., Trivedi, M.M. and Patel, U.M. (2014). Crop Diversification for Sustainable Development. Spring, 3 (4).

[5]   Sharma, F. L. and Intodia, S. L., (1991). Technological gap in adoption of improved animal husbandry practices. Maharashtra J. Extn, Edn., 10 (2): 128-132.

[6]   Kacha, H.L. and Patel, S.K. (2015). Impact of Frontline Demonstration on Okra (Abelmaschus esculentus (L.) Moench) Yield Improvements. Journal of AgriSearch, 2(1): 69-71.

[7]   Machhar, R.G., Sadhu, A.C., Patel, S.K. and Patel, V.J. (2015). Residual effect of organic manures, biofertilizers & fertilizers on soybean-wheat sequence under middle Gujarat. Green farming, Vol. 6 (5): 1042-1045.

[8]   Specht, J.E., D.J. Hume, and S.V. Kumudini. 1999. Soybean yield potential - a genetic and physiological perspective. Crop Sci. 39:1560-1570.

[9]   Piara Singh, Vijaya, D., Srinivas, K., and Wani, S.P. (2006). Potential Productivity, Yield Gap, and Water Balance of Soybean-Chickpea Sequential Systematic Selected Benchmark Sites in India. An Open Access Journal published by ICRISAT (SAT e Journal), 2(1): 1-50.

[10] Bhatia VS, Singh Piara, Wani SP, Kesava Rao AVR and Srinivas K. 2006. Yield Gap Analysis of Soybean, Groundnut, Pigeonpea and Chickpea in India Using Simulation Modeling. Global Theme on Agroecosystems Report no. 31. Patancheru 502 324, Andhra Pradesh, India: International Crops Research Institute for the Semi-Arid Tropics (ICRISAT). 156 pp.

Optimization of Minituber Size and Planting Distance for the Breeder Seed Production of Potato

Md. Altaf Hossain[1], Abdullah-Al-Mahmud[2], Md. Abdullah-Al-Mamun[3], Md. Shamimuzzaman[4], Md. Mizanur Rahman[5]

[1]Tuber Crops Research Centre (TCRC), Bangladesh Agricultural Research Institute (BARI), Joydebpur, Gazipur, Bangladesh
[2]International Potato Center (CIP), USAID Horticulture Project, Bangladesh
[3]Department of Agriculture Extension (DAE), Rangpur, Bangladesh
[4]Department of Crop Sciences, University of Illinois at Urbana-Champaign, Urbana, Illinois 61801, USA
[5]Department of Horticulture, Bangabandhu Sheikh Mujibur Rahman Agricultural University, Joydebpur, Gazipur, Bangladesh

Email address:
Altafmy@yahoo.com (Md. A. Hossain), A.Mahmud@cgiar.org (Abdullah-Al-Mahmud), Mamun.dae@gmail.com (Md. Abdullah-Al-Mamun)

Abstract: Six grades of potato minitubers (<5 mm, 5-10 mm, 10-15 mm, 15-20 mm, 20-25 mm and > 25 mm) and four planting distance (25 cm, 20 cm, 15 cm and 10 cm) with a potato variety Diamant were taken in an study during 2013-14 at the Tuber Crops Research Centre of Bangladesh Agricultural Research Institute, Gazipur, Bangladesh. The objective was to observe the effect of minituber grades and planting distance on growth, seed yield, increase ratio and seed potential of potato. The largest minitubers (>25 mm) planted at widest distance (25 cm) produced maximum number of tubers per plant (18.7). The highest number of tubers per m^2 (306.7) was obtained with largest minitubers (>25 mm) planted at the closest plant spacing 10 cm, while it was lowest (64.0) in smallest size minituber (<5 mm) with the widest distance 25 cm. A significant increase ratio was found ranged from 14 (largest minituber with the closest planting distance) to 297 (smallest minituber with widest planting distance). The maximum percentage (53%) of 'A grade' seed (28-55mm size) was obtained from the pea size (5-10 mm) minituber size planted at 15 cm distance. The highest seed potential (39.8) was in >25 mm size minituber planted at 10 cm distance. The lowest (4.4) was in <5 mm size minituber when planted at 25 cm distance. Seed sizes increasing from <5 mm to >25 mm had significant increase ratio ranged from 12 to 269. The highest economic return (9.4) would occur for the pea size (5-10 mm) minituber when planted at 15 cm spacing.

Keywords: Minituber, Breeder Seed and Potato

1. Introduction

Unavailability of certified seed tubers is a major constraint to potato production in Bangladesh. This compels most farmers to use planting materials from informal sources such as previous harvests from own field, local markets and neighbours. Bangladesh Agricultural Research Institute (BARI) has the national mandate to produce basic seed tubers (Breeder's seed) but can only supply less than 1% of the national requirements. BADC and other private seed potato producer can supply maximum 5% quality seed (Hossain *et al.*, 2008). It is recorded that in 2012-2013, BADC supplied 19,322 M tons of seed potato which is only 4.12% of total quality seed [1].

Minitubers are usually defined as the progeny tubers produced on *in vitro* derived plantlets [17]. The size of minitubers may range from 5-25 mm and a range in weight between 0.1-10 gm and sometimes higher. Larger mini-tubers also have become common ([8],[17]). Minituber production has significantly reduced the number of generations required to produce commercial seed potatoes. This has reduced exposure to pathogens during field multiplication, resulting in healthier tubers in seed crops. Experience in production of single-hill, first generation, seedling screening material in the variety development program has shown that minitubers as small as 1-2 g can produce viable productive plants [13]. The optimizing of plant density is one of the most important subjects of potato production, because, it affects to seed cost, plant development, yield, and quality of the crop [4]. In practice, plant density in potato crop is manipulated through the number and size of the seed tubers planted [3]. Widely spaced plants allow better separation at harvest to isolate tubers from individual hills. The low population provides individual plants an advantage in access to moisture, nutrients, and sunlight. To optimize production from pre-nucleus minitubers, plants populations should closely mirror populations typically used for seed production [16]. Therefore, many studies have been conducted to establish the optimal combination of seed size and planting distance for a certain environment ([18], [5], and [4]).

A lot of research work has been done to evaluate the performance of potato seed tubers but little information exists on the field performance of minituber for the breeder's seed production of potato. Therefore, the present study was conducted to evaluate the field performance of different size potato minituber and planting distance for the production of breeder's seeds of potato.

2. Materials and Methods

2.1. Site, Soil and Season of the Experiment

The experiment was conducted at the net house of Tuber Crops Research Centre, BARI, Gazipur during November, 2013 to March 2014. The location of the experimental site was to the 34 km north from Dhaka city (24.38^0 N latitude and 90.13^0 longitudes) at 8.4m above the sea level. The soil of the experimental field was grey terrace contained pH 6.4. This area is moderately drought prone, and face drought both winter and late winter season. The experimental site is situated in a sub- tropical climate zone and characterized by no rainfall during December to March.

2.2. Planting Materials and Date of Planting

Diseases free well sprouted seed potato minitubers of Diamant variety were used as planting material for the experiment. Mini-tubers were planted on 6th November, 2013.

2.3. Crop Management

The field was ploughed 3-4 times to a depth of 25 cm. Full doses of well rotten cow dung (10 t ha^{-1}), TSP (220 kg ha^{-1}), MP (270 kg ha^{-1}), Gypsum (120 kg ha^{-1}), Boric acid (6 kg ha^{-1}) and half doses of Urea (175 kg ha^{-1}) were applied at the time of final land preparation. The rest half dose of Urea (175

kg ha^{-1}) was applied as side dressings at 30 DAP followed by earthing-up and light irrigation. First earthing up was done at 30 DAP when the plant attained a height of about 15-20 cm from the base, second earthing-up was done after 20 days of first earthing up. Before first earthing up, Urea was applied. Irrigation was applied 3 times. First one was applied just after planting, second one was just after earthing up at 30 DAP, and last one was on 55 DAP. During land preparation, Furadan 5G was applied @10 kg ha-1 as basal during land preparation and Admire (0.2%) was sprayed in two installments at 45 and 60 DAP to control insects. The crops were also sprayed alternatively with Dithane-M 45 (0.2%) and Secure (0.1%) at 15 days interval to prevent the late blight infection of potato. The field was netted during the entire growing period to protect the plants from the insect infestation specially aphids which is the vector of different viruses. Seeds were planted at row distance of 60 cm row and planting distance of 25, 20, 15 and 10 cm. Haulm pulling was done at 75 DAP by hand. Hardening and setting up of skins of tubers were allowed for 10 days under the soil there after crop was harvested at 85 DAP. Tubers were collected carefully with the help of spade without any injury.

2.4. Design and Treatments of the Experiment

The experiment was laid out in two factors Randomized Complete Block Design (RCBD) with three replications. First of all the entire experimental field was divided into three blocks, representing three replications. Each block again divided into twenty four unit plots. The treatment was assigned randomly to unit plots of each block. The size of a unit plot was 3.0 m × 2.4 m. There were six grades of minitubers based on minituber diameter (S_1/Under size = <5 mm, S_2/pea size=5-10 mm, S_3/small size=10-15 mm, S_4/medium size = 15-20 mm, S_5/large size = 20-25 mm and S_6/extra-large= > 25 mm) based on minituber diameter and four planting distance (D_1=25 cm, D_2=20 cm, D_3=15 cm and D_4=15 cm) which formed twenty four treatment combinations. Treatment combinations were as follows- S_1D_1, S_1D_2, S_1D_3, S_1D_4, S_2D_1, S_2D_2, S_2D_3, S_2D_4, S_3D_1, S_3D_2, S_3D_3, S_3D_4, S_4D_1, S_4D_2, S_4D_3, S_4D_4, S_5D_1, S_5D_2, S_5D_3, S_5D_4, S_6D_1, S_6D_2, S_6D_3 and S_6D_4.

2.5. Climatological Data

Air temperature and humidity, precipitation, evaporation, soil temperature and ground water table were recorded throughout the crop period (Table 1).

Table 1. Climatological data of 2013-14 crop season.

Month	Air Temperature ($^\circ$C)			Humidity (%)	Rain Fall (mm)
	Max.	Min.	Av.		
2013-14					
November	26.60	22.43	24.52	80.47	8.44
December	19.90	15.45	17.68	89.05	0.00
January	15.20	11.58	13.39	90.80	0.00
February	23.85	19.08	21.47	89.89	8.43
March	31.16	26.09	28.62	76.70	29.84

2.6. Data Collection

Data on different growth and yield contributing characters were recorded from the sample plots of each plot during the course of experiment. The sampling was done randomly. The plants in the outer row were excluded during random selection. Five plants were randomly selected from each plot to record the data on the following parameters: Plant emergence, plant height, leaf area, number and weight of tubers per plant, yield (kg m^{-2}), and percentage of different grades of tuber by number, seed potentials and seed increase ratio.

2.7. Statistical Analysis

To find out the significance of experimental results, the collected data on different parameters were analyzed statistically by using MSTAT-C program. The mean for all the treatments were calculated and analysis of variance for each parameter was performed by F-test. The mean separation was done by DMRT at 5% level of probability.

3. Results and Discussion

3.1. Plant emergence (%)

Analysis of variance indicated that the interaction effect of minituber size and planting distance had significant influence on emergence rate at 30 DAP (Table 2). The highest emergence rate was found in large minituber with 25 cm planting distance (94.1%), which was statistically identical to other planting distance and extra-large. The lowest emergence rate (61.8%) was obtained from under-size minituber at 10 cm planting distance, which was statistically at par with 15 cm planting distance (62.7%). In an average emergence performance of the under-size minituber at any planting distance is very low compared to other sizes. These results support the findings of Rykbost and Charlton (2004); Karafyllidis et al. (1997) and El Amin et al. (1996).

Table 2. Interaction effect between minituber size and planting distance on percent plant emergence

Minituber size	Planting distance (cm)				Mean
	25	20	15	10	
Under size	69.9	68.9	62.7	61.8	65.8
Pea	79.4	75.8	77.8	78.2	77.8
Small	82.5	80.4	83.8	83.2	82.5
Medium	89.4	82.9	87.6	84.9	86.2
Large	94.1	89.2	90.5	91.4	91.3
Extra-large	91.1	89.4	88.1	88.8	89.4
Mean	84.4	81.1	81.8	81.4	
LSD (0.05)					
Planting distance (P)	8.72				
Minituber size (M)	7.99				
P x M	21.35				

3.2. Plant Height (cm)

The tallest plant (85.9 cm) was produced by extra-large size

minituber when planted at 25 cm distance; shortest (50.2 cm) plant was produced by under-size mintuber with 25 cm (Table 3). Jagroop et al. (1993) found taller plants with large size normal seed tubers planted at closer spacing. Probably the plant height was the highest in the larger size minituber due to the presence of more reserve food which caused rapid growth of plants earlier. Similar findings have also been reported by Zakaria (2003) but his research was on the effect different size of microtuber on plant height.

Table 3. Interaction effect between minituber size and planting distance on plant height (cm)

Minituber size	Planting distance (cm)				Mean
	25	20	15	10	
Under size	50.2	59.8	58.4	57.8	56.6
Pea	71.7	66.5	64.3	70.7	68.3
Small	65.7	68.7	62.0	67.7	66.0
Medium	70.4	69.5	65.5	70.7	69.0
Large	71.1	71.0	71.3	65.3	69.7
Extra-large	85.9	72.2	79.8	68.8	76.7
Mean	69.2	68.0	66.9	66.8	
LSD (0.05)					
Planting distance (P)	2.97				
Minituber size (M)	4.85				
P x M	9.69				

3.3. Leaf Area (cm²)

Generally, yield was positively correlated to leaf area and was increased linearly as leaf area increased. Planting distance had significant difference on the leaf area production of plants derived from minituber (Table 4). Extra-large minituber contributed the highest leaf area values across all population density levels in comparison to other seed minitubers. The trend was such that the bigger the seed piece, the greater the leaf area. There was increase in leaf area with increase in minituber size and planting distance. The leaf area was highest in largest minituber with planting distance 25 cm but lowest in smallest mini-tuber with closest planting distance 10 cm. These results are in conformity with the findings of Akhtar et al. (2010). Lower leaf area index and radiation interception in small seed size undoubtedly reduced production of assimilates.

Table 4. Interaction effect between minituber size and planting distance on leaf area

Minituber size	Planting distance (cm)				Mean
	25	20	15	10	
Under size	837.7	814.3	797.0	785.7	808.7
Pea	924.0	891.3	849.0	831.0	873.8
Small	966.0	948.3	933.7	912.0	940.0
Medium	1042	1031	1008	991.0	1018.0
Large	1233	1210	1173	1150	1191.5
Extra-large	1414	1385	1353	1280	1358.0
Mean	1069.5	1046.7	1019.0	991.6	
LSD (0.05)					
Planting distance (P)	120.6				
Minituber size (M)	130,2				
P x M	124.8				

3.4. Tuber Number per Plant and per m^2

Minituber size and planting distance interacted significantly and affected the number of tuber per plant (Table 5 & 6). The tuber number per plant increased with increase in minituber size in each planting spacing. The extra-large minituber planted at widest distance (25 cm) produced maximum number of tuber per plant (18.7) which was statistically similar to 20 cm (18.0)) and 10 cm (18.4) of same groups respectively. Number of tubers per plant was the lowest in smallest minituber with closest planting distance 10 cm (8.10). Larger minituber have higher amount of reserve food and interplant competition for space, light, water and nutrient is less in the wider spacing that can contribute the increase number of tuber per plant. The number of tubers per m2 was increased with increase in minituber size with closer planting distance (Table 16). The highest number of tuber per m2 was obtained with extra-large minitubers planted at the 10 cm distance (306.7), while it was lowest in under-size minituber with the widest distance 25 cm (64.0). The results were in conformity with the findings of Haverkort et al. (1991) who found increasing number of tubers per plant with increase in size of microtuber. The same information was reported by Rykbost and Charlton (2004) and Karafyllidis et al. (1997). These results are in conformity with the findings of Tuku (2000) who reported that higher yield was associated with proper nutrients and water availability to the plant and more tuber weight. Gopal et al. (2007) after conducting similar study also proposed that selection for tuber yield can be practiced at the minituber level in potato breeding processes. Zkaynak & Samanci (2006) worked on field performance of three weight classes of small minitubers ranging from 6.0 - 18.0 g was studied in two years at different planting dates. The heavy minitubers gave higher values than light minitubers for tuber yield, tuber weight, tuber number and stem number.

Table 5. Interaction effect between minituber size and planting distance on tuber number per plant

Minituber size	Planting distance (cm)				
	25	20	15	10	Mean
Under size	9.6	9.6	10.8	8.1	9.5
Pea	11.5	12.9	10.3	10.0	11.2
Small	12.5	10.7	10.4	9.6	10.8
Medium	11.5	10.4	11.5	13.9	11.8
Large	13.1	14.8	13.7	15.7	14.3
Extra-large	18.7	18.0	14.4	18.4	17.4
Mean	12.8	12.7	11.9	12.6	
LSD $_{(0.05)}$					
Planting distance (P)	0.85				
Minituber size (M)	1.40				
P x M	2.09				

Table 6. Interaction effect between minituber size and planting distance on tuber number per m^2

Minituber size	Planting distance (cm)				
	25	20	15	10	Mean
Under size	64.0	84.8	120.8	151.0	105.2
Pea	76.4	114.1	114.0	166.7	117.8
Small	83.5	94.1	90.3	173.4	110.3
Medium	76.4	91.8	127.3	231.0	131.6
Large	87.1	130.7	152.4	262.1	158.1
Extra-large	124.5	158.9	160.0	306.7	187.5
Mean	85.3	112.4	127.5	215.2	
LSD $_{(0.05)}$					
Planting distance (P)	14.42				
Minituber size (M)	18.28				
P x M	7.57				

3.5. Tuber Weight per Plant and per m^2

Minituber size and planting distance interacted significantly and there was a significant influence on tuber weight per plant (Table 7 & 8)). Tuber yield per plant increased significantly with increase in minituber size and planting distance. The maximum tuber weight per (985.0 g) plant was obtained from the extra-large minituber with planting distance 25 cm and minimum tuber weight per plant (119.0 g) was found from the under-size minituber with closest planting distance 10 cm which was statistically similar to other spacing of the same size. Different trend was observed in weight of tuber per m^2. The highest weight of tubers per m^2 (8.50 kg) was obtained from the largest minitubers with closest planting distance 10 cm which was at par with large size minituber with closest planting distance 10 cm (8.29 kg). The lowest weight of tuber per m2 (0. 99 Kg) was found in under-size with widest spacing 25 cm.

Table 7. Interaction effect between minituber size and planting distance on tuber weight per plant (g)

Minituber size	Planting distance (cm)				
	25	20	15	10	Mean
Under size	148.4	147.2	126.0	119.4	135.3
Pea	719.6	641.0	615.7	469.7	611.5
Small	670.2	584.4	552.7	476.9	571.1
Medium	688.2	600.2	495.9	390.5	543.7
Large	811.1	757.6	651.7	497.3	679.4
Extra-large	985.0	831.3	694.9	509.9	755.3
Mean	670.4	593.6	522.8	410.6	
LSD $_{(0.05)}$					
Planting distance (P)	3.09				
Minituber size (M)	3.88				
P x M	7.57				

Table 8. Interaction effect between minituber size and planting distance on tuber yield (kg m⁻²)

Minituber size	Planting distance(cm)				
	25	20	15	10	Mean
Under size	0.99	1.30	1.40	1.99	1.42
Pea	4.80	5.66	6.84	7.83	6.28
Small	4.47	5.16	6.14	7.95	5.93
Medium	4.59	5.30	5.51	6.51	5.48
Large	5.41	6.69	7.24	8.29	6.91
Extra-large	6.57	7.34	7.72	8.50	7.53
Mean	4.47	5.24	5.81	6.85	
LSD $_{(0.05)}$					
Planting distance (P)	0.22				
Minituber size (M)	0.27				
P x M	0.53				

3.6. Seed Potential

From the calculation, the seed potentials of minituber of the different sizes planted at different spacing ranged from 4.4 to 39.8 (Table 9). The highest seed potential (39.8) was found in extra-large size minituber planted at 10 cm distance. The lowest seed potential (4.4) was in under-size minituber

when planted at 25 cm distance. However, the yield (both number and weight per plant) of under-size and extra-large size minituber was negligible.

3.7. Increase Ratio

Results observed in this trial demonstrated that extremely high increases ratios as minituber size decreased and a large reduction in this ratio as minituber size increased (Table 9). Effects of increasing minituber size on yield are attributed to a combination of increases in both number and size of daughter tubers. Seed sizes increasing from under-size to extra-large size had significant increase ratio ranged from 12 to 269. Rykbost and Charlton (2004) reported 65 to 317 increase ratios from the minituber size ranged from 1.2 g to 13.6 g. They also reported that a typical seed increase expectation is 15 or 20 to 1 but in the irrigated production, the increase ratio is likely to be 20 to 1. The similar results were found by Masarirambi et al. (2012) and Islam et al. (2012).

Table 9. Interaction effect between minituber size and planting distance on seed potential and increase ratio

Minituber size	Seed potential				Increase ratio			
	Planting distance (cm)				Planting distance (cm)			
	25	20	15	10	25	20	15	10
Under size	0.99	1.30	1.40	1.99	130	119	93	72
Pea	4.80	5.66	6.84	7.83	269	223	210	155
Small	4.47	5.16	6.14	7.95	90	75	72	60
Medium	4.59	5.30	5.51	6.51	47	41	34	26
Large	5.41	6.69	7.24	8.29	26	25	21	16
Extra-large	6.57	7.34	7.72	8.50	22	19	16	12
Mean	4.47	5.24	5.81	6.85	97.33	83.67	74.33	56.83

3.8. Economic Analysis

Significant variation in partial budget analysis was observed in different treatment combinations (Table 10). In seed production of potato from mini tuber total variable cost (TVC) was highest in extra-large minituber when planted at 10 cm distance (Tk.161.88).The highest gross net return Tk. 224.09 was found in the same treatment but it BCR was 1.4 which was lower than other treatment combinations. The lowest net return Tk. 18.57 was found in under-size planted at 25 cm distance and its BCR was also lowest (1.5). Closer

planting required more labour involvement and higher seed rate/ha which resulting high TVC and lower the BCR in breeder seed production by using mini tuber as has been reported by Mamun (2012). The result suggests that the highest economic return (9.4) would occur for the 1-4 g small minituber when planted at 15 cm spacing if the price per kilogram of the minitubers is equal for all sizes. Pricing compensation for larger seed sizes would need to be large for much lower production potential. Production of basic seed from minituber is very costly.

Table 10. Partial budget analysis of potato for different treatment combinations of minituber size

Treatment		Total material cost (Tk.)	Total non-material cost (Tk.)	Total variable cost (Tk.)	Gross return (Kg/m²)		Net return (Tk/m²)	BCR
Minituber size	Planting distance				Seed	Non-seed		
Under size	60 × 25	5.33	6.75	12.08	13.01	5.56	18.57	1.5
	60 × 20	5.60	6.95	12.55	15.76	7.62	23.38	1.9
	60 × 15	5.89	7.15	13.04	15.57	8.81	24.38	1.9
	60 × 10	6.58	7.35	13.93	18.01	13.90	31.90	2.3
Pea	60 × 25	7.84	6.75	14.59	107.48	12.17	119.65	8.2
	60 × 20	8.92	6.95	15.87	118.21	17.20	135.41	8.5
	60 × 15	10.06	7.15	17.21	139.78	21.81	161.59	9.4
	60 × 10	12.84	7.35	20.19	155.22	26.56	181.78	9.0
Small	60 × 25	14.51	6.75	21.26	107.56	8.85	116.41	5.5
	60 × 20	17.75	6.95	24.70	119.92	11.63	131.55	5.3
	60 × 15	21.17	7.15	28.32	144.08	13.37	157.45	5.6
	60 × 10	29.51	7.35	36.86	181.45	19.02	200.47	5.4

| Treatment | | Total material cost (Tk.) | Total non-material cost (Tk.) | Total variable cost (Tk.) | Gross return (Kg/m²) | | Net return (Tk/m²) | BCR |
Minituber size	Planting distance				Seed	Non-seed		
Medium	60 × 25	24.51	6.75	31.26	112.79	8.30	121.09	3.9
	60 × 20	30.99	6.95	37.94	130.62	9.46	140.08	3.7
	60 × 15	37.83	7.15	44.98	136.75	9.52	146.27	3.3
	60 × 10	54.51	7.35	61.86	156.94	12.79	169.73	2.7
Large	60 × 25	44.52	6.75	51.27	127.21	11.70	138.91	2.7
	60 × 20	57.48	6.95	64.43	161.00	13.23	174.23	2.7
	60 × 15	71.16	7.15	78.31	170.31	15.63	185.94	2.4
	60 × 10	104.52	7.35	111.87	190.13	19.52	209.65	1.9
Extra-large	60 × 25	64.53	6.75	71.28	161.96	14.79	176.75	2.5
	60 × 20	83.97	6.95	90.92	179.18	15.66	194.84	2.1
	60 × 15	104.49	7.15	111.64	187.80	15.80	203.61	1.8
	60 × 10	154.53	7.35	161.88	207.88	16.21	224.09	1.4

Breeder's Seed price = 30 Tk/kg and non-seed price = 10 Tk./kg

4. Conclusion

Breeder seed production of potato was affected by minituber size and planting distance. Larger size of minituber produced more number of tubers with increased yield when it was planted in greater distance. But, the higher seed yield potential was found in larger sized with closer planting of minituber. So, it can be concluded that pea size minituber with 15 cm planting distance might have the highest economic return. However, based on the yield and net return, gross return, small size minituber at 10 cm planting distance may be used for cost effective production of breeders' seeds of potato.

Appendixes

Appendix 1. Meteorological conditions of the experimental site during January 2012 to December 2013

| Month | Air Temperature (°C) | | Humidity (%) | Rain Fall (mm) |
	Max.	Min.		
January	21.42	16.77	89.20	1.68
February	24.33	19.68	85.25	0.00
March	28.81	24.60	81.61	13.63
April	31.95	27.12	82.83	38.68
May	31.47	25.89	83.84	162.43
June	34.00	25.80	84.53	247.34
July	32.19	25.94	85.07	363.60
August	31.16	25.94	86.29	590.30
September	31.70	27.47	86.57	206.46
October	29.74	26.70	85.29	182.43
November	26.60	22.43	80.47	8.44
December	19.90	15.45	89.05	0.00
Ave./total	28.61	23.65	85.00	1814.94

Source: Weather station, BARI, Gazipur

Appendix 2. Meteorological conditions of the experimental site during January 2013 to December 2014

| Month | Air Temperature (°C) | | Humidity (%) | Rain Fall (mm) |
	Max.	Min.		
January	15.20	11.58	90.80	00.0
February	23.85	19.08	89.89	8.43
March	31.16	26.09	76.70	29.84
April	32.13	28.2	76.53	57.11
May	31.40	27.90	82.0	252.23
June	32.06	29.26	85.96	369.53
July	32.32	27.67	83.45	269.13
August	32.0	25.90	85.58	138.65
September	30.38	26.53	89.46	212.65
October	30.67	27.06	87.41	187.01
November	27.76	23.76	85.66	00.0
December	24.80	16.58	90.70	55.19
Ave./total	29.46	24.81	85.34	1579.77

Source: Weather station, BARI, Gazipur

References

[1] AIS. 2014. Seed supplied by BADC in 2012-13. *Krishi* Diary. Agriculture Information Service, Khamarbari, Farmgate, Dhaka, Bangladesh. P.7.

[2] Akhtar, Parveen., S. J. Abbas, M. Aziz, A. H. Shah and N. Ali. 2010. Effect of Growth Behavior of Potato Mini Tubers on Quality of Seed Potatoes as Influenced by Different Cultivars. Pak. J. Pl. Sci. 16 (1): 1-9.

[3] Allen, E. J. and D. C. E. Wurr. 1992. Plant density. In: P. M. Harris (Ed.), The Potato Crop. The scientific basis for improvement. Second edition. Chapman and Hall, London, UK, pp. 292-333.

[4] Bussan, A.J., P.D. Mitchell, M.E. Copas and M.J. Drilias, 2007. Evaluation of the effect of density on potato yield and tuber size distribution. Crop Sci. 47: 2462–2472.

[5] Creamer, N.G., C.R. Crozier and M.A. Cubeta, 1999. Influence of seed piece spacing and population on yield, internal quality and economic performance of Atlantic, Superior and Snowden potato varieties in Eastern North Carolina. *American J. Potato Res.*, 76: 257–261.

[6] El-Amin, S. M., B. Adam, E. Varis and E. Pehu. 1996. Production of seedling tubers. Experimental Agric. 32 (4): 419-426.

[7] Gopal, J., R. Kumar and G. S. Kang. 2007. The effectiveness of using a minituber crop for selection of agronomic characters in potato breeding programmes *Potato Journal*, 34 (1 & 2): 145-151.

[8] Hassanpanah D., A. A. Hosienzadeh and N. Allahyari. 2009. Evaluation of planting date effects on yield and yield components of Savalan and Agria cultivars in Ardabil region. *Journal of Food, Agriculture & Environment,* 7 (3&4): 525-528.

[9] Haverkort, A. J., M. Van de Waart & J. Marinus.1991. Field Performance of Potato Microtubers as Propagation Materials. Potato Research. 34: 353-364.

[10] Hossain, M. A., A. U. Hauque, M. S. Alam, M. Hossain, M. M. Khatun, M. M. Hasan and S. N. Begum. 2008. Disease free Minituber Production of Potato Using Tissue Culture Methods (In Bangla), TCRC, BARI, Joydebpur, Gazipur-1701, Bangladesh. P.1.

[11] Islam, M. S., S. Moonmoon, M. Z. Islam, H. Waliullah and M. S. Hossain. 2012. Studies on Seed Size and Spacing for Optimum Yield of Potato in Northern Region of Bangladesh. Bangladesh J. Prog. Sci. & Tech. 10(1): 113-116.

[12] Karafyllidis, D.I., D.N. Georgakis, N.I. Stavropoulos, E.X. Nianiou, and I.A. Vezyroglou. 1997. Effect of planting density and size of potato seed-minitubers on their yielding capacity. Acta Hort. 462:943–949.

[13] Kenneth, A. R. and B. A. Charlton. 2005. Effects of prenuclear minituber seed size on production of Wallowa Russet Seed. Annual Report. Klamath Experiment Station. USA. 31-38 pp.

[14] Mamun, A. A. 2012. Effect of Planting time and spacing of top shoot cutting for breeder's seed production of potato. MS thesis. BSMRAU. Gazipur-1706. 1-66 pp.

[15] Masarirambi, M. T., F. C. Mandisodza, A. B. Mashingaidze and E. Bhebhe, 2012. Influence of plant population and seed tuber size on growth and yield components of potato (*Solanum tuberosum*). Int. J. Agric. Biol. 14: 545–549.

[16] Rykbost, K. A. and B. A. Charlton.2004. Effects of Prenuclear Minituber Seed Size on Production of Wallowa Russet Seed. Annual Report. Klamath Experiment Station (KES), Klamath Falls, Oregon, USA. pp 38-43.

[17] Struik, P. C. 2007. The canon of potato science: Minitubers. Potato Res. 50(3-4):305-308.

[18] Sultana N and Siddique A. 1991. Effect of cut seed piece and plant spacing on the yield and profitability of potato. Bangladesh Horticulture. 19(1): 37-43.

[19] Tuku, B. T. 2000. The utilization of true potato seed (TPS) as an alternative method of potato production. Indonesian J. Agric. Sci. 1 (2): 29-38.

[20] Zakaria, M. 2003. Induction and Performance of Potato Microtuber. Ph D Dissertation. Department of Horticulture. BSMRAU, Gazipur-1701. Bangladesh. 144-159 pp.

[21] Zkaynak E. and B. Samanci. 2006. Field performance of potato minituber weights at different planting dates. Archives of Agronomy & Soil Scienc. 52 (3): 333-338.

28

Bio-organic Fertilizer on Pechay Homegarden in Cotabato

Mosib B. Tagotong[1], Onofre S. Corpuz[2]

[1]College of Agriculture, Cotabato Foundation College of Science and Technology, Doroluman Arakan, Cotabato, Philippines
[2]Research and Development Office, Cotabato Foundation College of Science and Technology, Doroluman Arakan, Cotabato, Philippines

Email address:
nfr_uplb@yahoo.com

Abstract: This study focus on the determination of bio-organic fertilizer and levels of application that could gave favorable response on the growth and yield of pechay planted in a home garden at Arakan, Cotabato. Bio-organic farming is a key to sustainable agriculture leading to sustainable development. The bio-organic fertilizers used in this study include fermented plant and fruit juice (FPJ and FFJ). Result of the study revealed that FFJ responded significantly better as compared to FPJ on plant height, number of plant leaves and yield in grams per plant. In the case of the levels of application, higher application levels (6tbsp/lit. H_2O) excelled on all parameters tested (plant height, number of plant leaves and yield). This implies that higher concentration of FPJ diluted on water will promotes better growth and development of pechay plant emphasizing development of the physical and biological properties of the soil.

Keywords: Bio-organic Fertilizer, Pechay, Fermented Plant Juice, Fermented Fruit Juice

1. Introduction

Pechay (*Brassica pekinensis* L.) is one of the common leafy vegetable crops grown in the Philippines belonging to family cruciferae. Vegetables are important and substantial source of food that significantly contributes to the quality of our diet because it provides variety of nutrients. Various parts of enumerable kinds and vegetables make meals of staple food appetizing because of their flavor and even pleasing to the eyes. The major nutritional contributions of vegetables to the human diet are vitamins A and C as source of iron and calcium needed by the human body.

Present farmers of limited area preferably those that are in the upland always decided to earn their income for living through vegetable gardening. Take the case of the marginal upland farmers in some selected barangay's of Arakan, Cotabato in Southern Philippines like Gambodis, Katipunan, San Miguel, and Napalico, most of them grown high valued vegetable crops particularly leafy vegetables like Pechay. The vegetable product were even brought to neighboring provinces such as; Bukidnon, Cagayan, Butuan, and even to Visayas Provinces as revealed by the vegetables farmers in the Bukidnon-Davao boundary (BUDA) area. Pechay is a shallow – rooted plant which grows in a loosely high fertile, well–drained soil, friable and rich in organic matter. At present there are notable source of organic matter which

contribute much to the attainment of higher level of organic fertility of the soil like farm manures which are highly recommended for crop production because of their nutrients contents which ensures good field to crops especially vegetable crops. Abbey *et al*. (2001) reported that animal and plant wastes at various stages of decomposition constitute soil organic manure. Soil organic manure come from dead plant roots, crops residues, green manure, dead soil microorganisms and farmyard manure (Abbey *et al*., 2001).Dunn (1994) identified compost; plant materials (straw and dry leaves), garden waste and green manure as forms of organic manure that are commonly used by farmers to improve soil fertility. He further pointed out that green manure is derived from leguminous crops, which are grown as cover crops and ploughed into the soil.

A side from farm manures there are organic fertilizers such as Fermented Plant Juice (FPJ) and Ferment Fruit Juice (FFJ) which known to promotes growth and development of vegetable crops.

However, as to which kinds and levels of organic fertilizers is beneficial to pechay plants, vegetable growers still looking, thus, the researcher attempted to investigate the effect of these farm manures and fermented plants/fruit juice and its levels on pechay plant, one of the high valued vegetable crops in the country.

The result of this study may provide an appropriate cost

effective alternative production technology for the vegetable growers in the area and the region in general.

2. Materials and Method

2.1. Materials

The materials used in this study were Fermented Plant Juice (FPJ) and Fermented Fruit Juice (FFJ), Pechay seeds (*Brassica pekinensis*), seed box, garden soil as germination media, shoots of squash for FPJ and ripen banana (cardava) and mascovado for the FFJ.

2.2. Methods

The experiment used split – plot design with four treatments replicated three times.

Main plot (kinds of organic fertilizer)

Fermented Plant Juice

Fermented Fruit Juice

For Fermented Plant Juice (FPJ) and Fermented Fruit Juice (FFJ)

L_0 – control

L_1 – 2 tbsp/Lit.of water

L_2 – 4 tbsp/Lit.of water

L_3 – 6 tbsp/Lit.of water

Land preparation.the total area of 85sqm. was thoroughly prepared by flowing and harrowing three times with an interval of one week to properly decomposed organic matters and to have ease control of weeds and improved the soil texture of the area for favorable growth of the pechay plant.

Preparation of seed box and sowing of pechay seeds was done two weeks before transplanting. After one week of sowing, hand picking of germinant was done prior to transplanting.

The plot dimension was one point five (1x 1.5) meters with three furrows per plot. The planting distance was 30x30cm. to allow five plants per furrows to have 15 plants per plot.

2.3. Preparation of Fermented Plant Juiceand Fermented Fruit Juice

One kilo of squash shoots was chapped into small pieces until become fine. Mixed with one kilo of crude sugar or mascovado. Place in a clay jar and pert the rock on top for the contents to settle at the bottom.

On the next day removed the rock and cover the jar with a clean sheet of paper and tie with string. Put the jar in a cool and shaded place. The fermentation process were completed within seven days.

For the FFJ chop one kilo of ripe banana fruit (cardava) the chopped banana fruit was place inside the clay jar and mixed with one kilo crude sugar or mascovado. The jar was covered with clean sheet of paper (manila paper) and tie with string. The jar was stored in a cool and shaded area. Fermentation was lasted for seven days.

Application was done ten days after transplanting using the different levels or treatments/dosage of FPJ and FFJ which

was prescribed in the experiment using clean knapsack sprayer.

2.4. Care and Management

Watering was done every other day until 2 weeks after transplanting. After two weeks watering was done every three days after until 25 days after transplanting.

Weeding was done regularly to control needs and to maintain the cleanliness of the experiment.

Harvesting was done after 35 days of planting and was done by cutting the plants from the base.

3. Results and Discussion

3.1. Plant Height

Fermented fruit juice significantly produces taller pechay (22.5cm) compared to FPJ of 18.27cm. This result supported the findings of Juane C.G 2004 which states that applying fermented plant juice (FPJ) to vegetable crops will promote good plant growth and vitality. He further mentioned that mixing 1 tbsp per liter of natural water will have a very convening result on the plant growth and its vitality including the physical and biological properties of the soil for it increases the water holding capacity of the farm area.

Tamhean R.B et.al 1980 mentioned that organic fertilizer likes fermented fruit juice is a good source of plant nutrients to improve the physical properties of soils. They further states that application of organic fertilizer will improve the essential properties of the farm or the soil that are responsible for the vigor growth and development of the plants.

Parnes (1990), indicated that, both plant and animal sources of organic manure contain macro and micronutrients.

Among the levels of fermented plant/fruit juice applied, it was found out the 6 tbsp per liter of water significantly responded on plant height at 25.69cm (Table 1). According to Tamhean R.B et.al 1980, as you increased the level of application of fermented fruit juice diluted with natural water there was a corresponding increased of soil properties that are essential for growth and development.

Table 1. Plant height as affected by kinds and levels of organic fertilizer.

Levels of Organic Fertilizer	Mean Plant Height (cm)
FPJ	
Control	13.921
2 tbsp/lit H_2O	18.38b
4 tbsp/lit H_2O	20.17c
6 tbsp/lit H_2O	20.60c
Mean	18.27a
FFJ	
Control	17.63a
2 tbsp/lit H_2O	23.01b
4 tbsp/lit H_2O	23.68b
6 tbsp/lit H_2O	25.69c
Mean	22.50b

Means with same letter subscript are insignificantly different at 1%

3.2. Number of Leaves

Analysis of Variance reveals that fermented fruit juice promotes and develop more number of pechay leaves (12.79) as compared to FPJ of 7.32 (Table 2). This difference is highly significant implying that the later is better than FPJ

Of the three levels of application, it was found out that highest application of 6 tbsp per liter of water seems better in producing and developing pechay leaves (14.28). However, the three levels did not differ significantly in terms of number of leaves. IRRI 1990 mentioned that when levels of FFJ application increased it serves as an effective dilution to water which subsequently improved the growth and yield performance of leafy vegetables.

Table 2. *Data on average number of leaves of Pechay applied with kinds and levels of organic fertilizers.*

Levels of Organic Fertilizer	Mean No. of leaves
FPJ	
Control	5.84a
2 tbsp/lit H$_2$O	7.20b
4 tbsp/lit H$_2$O	7.67b
6 tbsp/lit H$_2$O	8.57c
Mean	7.32a
FFJ	
Control	10.73a
2 tbsp/lit H$_2$O	12.28ab
4 tbsp/lit H$_2$O	13.88b
6 tbsp/lit H$_2$O	14.28b
Mean	12.79b

Means with same letter subscript are insignificantly different at 1%

3.3. Yield

It reveals in the Analysis of Variance that fermented fruit juice still performing significantly better in terms of yield (161.55g/plant) as compared to FPJ of only an average of 77.93g/plant (Table 3).

Table 3. *Yield of Pechay per plant applied with kinds and levels of organic fertilizers.*

Levels of Organic Fertilizer	Mean weight
FPJ	
Control	56.85a
2 tbsp/lit H$_2$O	66.27b
4 tbsp/lit H$_2$O	69.48b
6 tbsp/lit H$_2$O	77.93c
Mean	67.63a
FFJ	
Control	121.67a
2 tbsp/lit H$_2$O	163.92b
4 tbsp/lit H$_2$O	177.19c
6 tbsp/lit H$_2$O	183.84d
Mean	161.55b

The three levels of application significantly differ its other. The highest application of 6 tbsp per liter of water gave the best result of 183.84g/plant followed by 177.19g/plant (4 tbsp/lit H$_2$O) and the least was found with the control treatment of no application (121.67g/plant). According to IRRI 1990, increased fermented fruit juice application serves as an effective dilution to water which subsequently improved the growth and yield performance of vegetable crops especially those leafy vegetables.

4. Conclusion

In terms of Fermented application it was concluded that as far as height is concerned, FFJ performed better compared with FPJ. This conclusion was similar to the findings of Juane, C.G (2004) when he said that application of FPJ to vegetable crops will give good plant growth and vitality.

However in terms of the number of leaves and weight of pechay it was concluded that FFJ showed better performance compared to FPJ. This conclusion is similar to the findings of Tagotong, M.M (2009) when he said that organic fertilizers like Fermented Fruit Juice (FFJ) is good source of plant nutrients to improve the physical properties of soil that were responsible for the vigor growth and development of plants. He further mentioned that when levels of FFJ application increased it was an effective dilution of water to improve the growth and yield performance of vegetable crops those leafy vegetables.

Recommendations

1. Use the Fermented Fruit Juice in pechay production.
2. Result of this study maybe use as baseline data on a succeeding similar study.
3. Go for bio-organic vegetable production for health, nutrition and low cost inputs reason.

References

[1] Abbey, T. K. Essiah, J.W., Alhassan, A., Fometu, E., Ameyibor, K., and Wiredu, M. B. (2001).Integrate Science for Senior Secondary School. Unimax MacMillan Ltd. Ring Road South industrial Area Accra-North Ghana. Pp 177-178.

[2] Baco, M. et al. 2005. Soil Chemistry and fertility in tropical Asian New Delhi, Prentice Hall of India private.

[3] BednilL. 1985. The plant is found capable of absorbing nutrients

[4] Bradley, N.C. 2005.The nature and properties of soil.MC Millan Publishing Company Inc. pp. 350-352.

[5] Buskman and lyon. 1987. The nature and properties of soil. New York. MC Millan publishing company Inc. p. 541.

[6] Dunn, A. R. (1994). Professor and Extension Nematologist, Entomology and Entomology department, cooperative Extension Service, institute of food and Agricultural Science, university of Florida, Gainesville, FL 32611.

[7] Fall, RC Jr. 1985.Production of Fields Crops. New York. 5th Edition. New York. MC Grow Hill Inc. pp 167.

[8] Juane, C.G. 2004.Pechay in the Philippines. Agri review 14. Pp. 307.

[9]   Mahul, et al. 1987. Guide for Vegetable Production. Horticulture Division, Introduction Center, Elliptical Road Quezon City. Philippines.

[10] Nile, C. and Brady. 1990. Macro Nutrients Elements in the nature of soil properties, N.Y. pp. 372-526.

[11] Parnes, R. (1990). Fertile soil: A grower's guide to organic and inorganic fertilizers. Eg. Access.603 fourth St. Davids, A95616.

[12] Sangatangan, P.D. 1990.Animal Manure. Soil Management and Fertility properties and Fertilizers. New York. Mc Millan Book Company pp. 644.

[13] Tagotong, M.M. 2014. Pechay applied with kinds and levels of organic fertilizer. Unpublished Masteral Thesis, Cotabato Foundation College of Science and Technology, February 2014.

[14] Tisdale, S. L. and Nelson W.L. (1978), Soil fertility and fertilizers, Macmillan Publishing Co. Inc., New York, USA.

29

Utilization of Urea Super Granule in Raised Bed Versus Prilled Urea in Conventional Flat Method for Transplanted Aman Rice (Oryza Sativa)

Md. Halim Mahmud Bhuyan[1, *], Most. Razina Ferdousi[2], Md. Toufiq Iqbal[1], Ahamed Khairul Hasan[3]

[1]Department of Agronomy and Agricultural Extension, University of Rajshahi, Rajshahi, Bangladesh

[2]Islamia Academy High School & Agriculture College, Bagha, Rajshahi, Bangladesh

[3]Department of Agronomy, Bangladesh Agricultural University, Mymensingh, Bangladesh

Email address:

drhalim.bhuyan@gmail.com (Md. H. M. Bhuyan)

[*]Corresponding author

Abstract: Bed planting with urea super granule (USG) application of rice production systems is very new and research on it is still at an introductory phase. Impact of granular urea application on growth and yield of transplanted aman rice as well as evaluation of water and fertilizer use efficiency of rice-fallow-rice cropping system was investigated under raised bed cultivation method. Result showed that the USG in bed planting method increased grain yield of transplanted aman rice up to 12.32% over prilled urea (PU) broadcasting in conventional method. The USG application in bed planting method increased the number of panicle m^{-2}, number of grains panicle^{-1} and 1000-grains weight of rice than the PU in conventional method. Sterility percentage and weed infestation were lower at USG application in bed planting method than the PU in conventional method. Forty percent irrigation water and time for application could be saved through the USG in bed planting than the PU broadcasting in conventional method. Water use efficiency for grain and biomass production was higher by the USG application in bed planting than the PU broadcasting in conventional method. Likewise, agronomic efficiency of the USG in bed planting was higher than the PU broadcasting in conventional method. This study concluded that the USG in bed planting method is a new approach to get better fertilizer and water use efficiency as well as higher yield compared to the existing agronomic practice in the world.

Keywords: Agronomic Efficiency, Bed Planting, Prilled Urea, Rice, Urea Super Granule

1. Introduction

Paddy rice crop requires massive amounts of mineral nutrients especially nitrogen (N) for its growth, development and yield (Goswami and Banerjee, 1978; Sahrawat, 2000). However, the efficiency of added N is very low, generally around 30-40% and in many cases even lowers (De Datta, 1978; Choudhury and Khanif, 2001). The low utilization efficiency is attributed to losses like volatilization, denitrification, leaching and surface run-off (Ponnamperuma, 1972; De Datta, 1981).

These losses can be reduced by management practices like use of modified forms of urea. This modified form of urea is called urea super granule (USG). The USG can be produced locally in the factory by using roll press machine. It is provided in deep point placement and has some advantages. These include (i) reduced N loss by runoff, volatilization and denitrification, (ii) delayed N uptake and (iii) reduced ammonium fixation (Westsellar, 1985).

Research showed that the superiority of USG over conventional prilled urea (PU) in rice production regarding N use efficiency (Cao et al., 1984; De Datta, 1987). Generally farmers are accustomed to using N fertilizer in the form of PU which is very easy to apply though rice plant can receive only 25 to 30% of applied fertilizer (BRRI, 2007). To reduce nitrogen loss, it is strongly considered that application of the USG is an important alternative that can increase the efficiency of N to about 20 to 25% and yield by 15 to 20% (BRRI, 2008).

Our previous study showed that water use efficiency for grain and biomass production was higher in bed planting than the conventional method (Bhuyan et al., 2012a, Bhuyan et al., 2012b). However, no study was undertaken for the USG application on bed planting as compared to the PU on conventional flat method. Therefore, this study was conducted to determine the role of USG on bed planting as compared to the conventional cultivation method. The hypothesis of this study is that the USG technique on raised bed will produce higher aman rice yield than the conventional PU technique.

2. Materials and Methods

2.1. Experimental Site

The experiment was conducted at the farmer's field located in the village of Daulatdiar of sadar Upa Zilla in Chuadanga district high Ganges river flood plain in Bangladesh. The experimental field is located at 23°39′N latitude and 88°49′E longitudes at a mean elevation of 11.58 m above the sea level. The soil of the experiment plot was silt loam with a pH of 7.30. The average air temperature, relative humidity, rainfall and sunshine hours were 21.16°C, 86.06%, 136.50 mm and 149.50 hours, respectively on the experimental field site.

2.2. Raised Bed Preparation and Its Advantages

Raised bed was prepared manually for the experiment. It can also be prepared through raised bed planting machine. Paddy rice crops were transplanted in two rows on top of the raised bed and irrigation water was applied within the furrows between the beds.

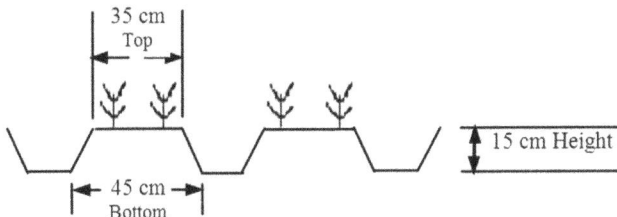

Fig. 1. Two rows rice on raised bed.

Water moves horizontally from the furrows into the beds. The height of the beds was maintained 15 cm, top was 35 cm and bottom was 45 cm with 6 beds per plot (Fig. 1). The furrow width was 25 cm. The major concern of raised bed technique is to enhance the productivity and save the irrigation water. Potential agronomic advantages of raised beds also include improved soil structure due to reduced compaction through controlled trafficking, reduced water logging and timelier of machinery operations due to better surface drainage.

2.3. Conventional Flat Land Preparation and Its Drawbacks

Land was prepared by puddling the soil in conventional method for transplanting paddy rice. The final land was prepared through ploughing and cross ploughing by two wheel power tiller with two laddering before two days of transplanting. The main difference between conventional flat and raised bed was furrow and transplanted aman rice on top of the bed in two rows. Conventional flat method also maintains continuing inundation until maturity. This affects soil physical, chemical and biological properties that influence the growing condition of paddy rice. In contrast, conventional flat method has some problems such as destruction of soil structure that leading to higher bulk density, higher soil penetration resistance and enhanced surface cracking.

2.4. Experimental Design and Procedure

Two planting methods for comparison of utilization of N fertilizer under conventional flooded and raised bed as well as furrows conditions were established across a soil. The combinations of treatments were deep placement of USG in raised bed and PU broadcasting in conventional flat method. Plots were 4 m wide × 2 m long. Two 30 day old seedlings hill^{-1} of Swarna, a popular aman (July to November growing period) rice variety, were transplanted on 9th August 2011 at a row to row and hill to hill spacing of 20 cm on both beds and the flat. Only two rows of rice were planted on each bed and plant density was much higher in the conventional treatment. Fertilizer was applied at the following rates: N=66, P=5, K=18, S=6 and Zn=0.5 kg ha^{-1} applied as urea, triple super-phosphate (TSP), murate of potash (MP), gypsum and Zinc Sulphade (ZnSO$_4$), respectively. Whole of TSP, MP, gypsum and ZnSO$_4$ were applied at the time of final land preparation as basal dose in the plots with conventional treatment. In the plots with bed planting treatments, the basal doses were applied before transplanting on the top of the beds. The conventional PU was top dressed in three equal splits at 15, 30 and 50 days after transplanting (DAT) in conventional plot. In bed planting within a week after transplanting of aman rice, the USG briquettes are (1.8 gm weight) inserted into the puddled soil by hand, being placed to a depth of 7-10 cm (deep placement) in the middle of alternating squares of four hills of rice (BRRI, 2009).

2.5. Crop Harvesting and Measurements

Rice was harvested at 11th November, 2011. Twenty randomly selected hills from each plot were used for agronomic measurements. A 1 m^2 area from centre portion of each plot was harvested for determination of grain yield. Rice was threshed by using pedal thresher. Immediately after harvest grain moisture content and weight were recorded.

Grain yield was adjusted to 14% moisture content.

2.6. Water Management

Conventional plots were continuously flooded to a depth of approximately 8 cm until drainage at 80% grain maturity about 2 weeks before harvest. In the raised bed-furrow system, enough irrigation water was kept in the furrows to just submerge the beds for seedling establishment and weed control during the first two weeks after transplanting. Thereafter, irrigation was scheduled to maintain a water head of about half the height of the beds never allowing the beds to be submerged throughout the growing season.

2.7. Irrigated Water Measurement

Irrigation water was measured by using a delivery pipe and water pan. A plastic delivery pipe was connected from the water pump to the experimental field. A water pan with 300 litre volume was filed by irrigation through the delivery pipe and time required was recorded. Then plots with different methods of planting were irrigation through the delivery pipe and times required were recorded. The amount of irrigation water applied in different plots was calculated as follows:

Amount of water applied per plot=

$$\frac{\text{Volume of water pan (L)} \times \text{Time required to irrigation the plot (sec)}}{\text{Time required filling the water pan (sec)}}$$

2.8. Weeding

Manual weeding was done twice in the transplant aman rice field during growth period. The plots were weeded at 15 and 30 DAT. Weed samples from each plot were collected at the time of weeding for comparing weed population and dry biomass yield of different treatments.

2.9. Pest Control

The rice was infested by stem borer at tillering stage and by rice bug at grain filling stage. Furadan 5G at the rate of 10 kg ha^{-1} was applied at 40 DAT and Malathion 57 EC 5G at the rate of 1 litre ha^{-1} was applied at grain filling stage to control stem borer and rice bug, respectively.

2.10. Water Use Efficiency Calculation

Water use efficiency for grain and biomass production was calculated by the following equations:

Water use efficiency for grain production (kg ha^{-1}cm^{-1}) = grain yield (kg ha^{-1})/ total water required (cm)

Water use efficiency for biomass production (kg ha^{-1}cm^{-1}) = [grain yield (kg ha^{-1}) + straw yield (kg ha^{-1})]/ total water required (cm)

2.11. Agronomic Efficiency of Fertilizer

Agronomic efficiency (AE) of fertilizer was calculated by the following equation:

$$AE = GY_{NA} - GY_{N0}/N_R$$

Where GY_{NA} = Grain yield (kg /ha) with addition of nutrient

GY_{N0} = Grain yield (kg/ha) without addition of nutrient

N_R = Rate of added nutrient (kg/ha)

2.12. Statistical Analysis of Data

Data were analysed following standard statistical procedure and means of treatments were compared based on the least significant difference test (LSD) at the 0.05 probability level.

3. Results

3.1. Grain Yield and Yield Components

The yield increased by 12.32% when urea super granule (USG) was used in bed planting over prilled urea (PU) broadcasting in conventional method. A similar finding was also found for the panicles, grains per panicle and 1000 gm grain wt. The results were significantly different when compared to conventional method. Whereby the USG in bed planting had 9 panicle number m^{-2}, 20 grain number per panicle and 0.24 gm in 1000 grain wt more than PU in conventional method (Table 1).

Table 1. Grain yield and yield components with respect to urea super granule (USG) in raised bed and conventional prilled urea (PU) technique.

Method of Fertilizer application	Yield and yield components			
	Grain yield (t ha^{-1})	panicles m^{-2} (no)	Grains panicle^{-1} (no)	1000 grain wt (gm)
USG in bed planting	4.92a	285a	160a	23.12
PU broadcasting in conventional planting	4.38b	276b	140b	22.88
LSD at 5%	0.013	3.215	2.618	0.840
Level of significance	**	**	**	n.s.

Where n.s. and ** represents probability of > 0.05 and ≤ 0.01, respectively. Values were means of three replicates.

3.2. Other Plant Attributes

Planting method affected plant height, panicle length, non-bearing tillers m^{-2}, sterility percentage, straw yield and harvest index of rice. Plant height, panicle length and harvest index were higher by USG in bed planting than PU in conventional method. On the contrary, non-bearing tillers m^{-2}, and sterility percentage were higher PU broadcasting in conventional method than the USG in bed planting. Likewise, lower number of non-bearing tillers m^{-2} was recorded by the USG in bed planting treatments than the PU in conventional method. The USG in bed planting significantly reduced the sterility percentage compared to the PU in conventional planting. In bed planting sterility was lower. The lower sterility might be accountable for higher grains in bed planting. The bed planting resulted in higher harvest index than conventional method (Table 2).

Table 2. Plant biomass with respect to USG in raised bed and PU application in conventional planting.

Method of fertilizer application	Plant height (cm)	Panicle length (cm)	Non-bearing tiller (no-m^{-2})	Sterility (%)	Straw yield (tha^{-1})	Harvest index
USG in bed planting	90.24a	24.46	55b	9.78b	5.32a	0.48a
PU broadcasting in conventional planting	86.38b	24.30	59a	12.41a	4.92b	0.47b
LSD at 5%	1.114	0.281	2.225	0.379	0.125	0.109
Level of significance	**	n.s.	**	**	**	**

Where n.s. and ** represent probability of > 0.05 and ≤ 0.01, respectively. Values were means of three replicates.

3.3. Tiller Production

Transplanting of aman rice under different planting method affected the number of tillers m^{-2} of rice. The increasing trend of tiller m^{-2} was continued to 40 days after transplanting. At 40 days after transplanting both planting method attained the highest number of tiller m^{-2} and then started declining up to 100 days after transplanting (Table 3).

Table 3. Effect of tiller production by USG in raised bed and PU in conventional planting.

Method of fertilizer application	Tiller (no. m^{-2}) at days after transplanting								
	20	30	40	50	60	70	80	90	100
USG in bed planting	91	210	383	370	359	355	346	341	340a
PU in conventional plot	88	205	371	363	354	349	342	339	335b
LSD at 5%	2.069	4.984	9.662	5.396	4.139	14.19	2.18	9.438	2.34.
Level of significance	n.s.	n.s.	n.s.	n.s.	n.s.	n.s.	*	n.s.	**

Where n.s.,* and ** represent probability of > 0.05, ≤ 0.01 and ≤ 0.001, respectively. Values were means of three replicates.

3.4. Leaf Area Index

Planting method affected the leaf area index (LAI) of transplant aman rice recorded at different days after transplanting (DAT) (Table 4). Plant-to-plant distance in rows also influenced the leaf area index measured at different stages of crop growth. The highest leaf area index was achieved at 60 DAT by USG in bed planting method. After 60 DAT the leaf area index started declining and continued to 100 DAT by USG in bed planting. It was also revealed that at early stage of crop growth the leaf area index by USG in bed planting treatments was lower than conventional planting. The highest LAI was achieved by PU in conventional planting at 80 DAT. After 80 DAT the LAI started to decline and continued to 100 DAT by PU in conventional method. However, LAI differ significantly ($P \leq 0.01$) between two methods from 20 to 80 DAT (Table 4).

Table 4. Effect of leaf area index by USG in raised bed and PU in conventional planting.

Method of fertilizer application	LAI at different DAT				
	20	40	60	80	100
Urea super granule (USG) in bed planting	0.13b	2.38b	5.22a	5.01a	3.80
Prilled urea broadcasting in conventional planting	0.38a	2.47a	4.92b	4.97b	3.78
LSD at 5%	0.013	0.013	0.093	0.013	0.185
Level of significance	**	**	**	*	n.s.

Where n.s.,* and ** represents probability of > 0.05, ≤ 0.01 and ≤ 0.001, respectively. Values were means of three replicates.

3.5. Dry Matter Production

Planting method affected the dry matter production of transplanted aman rice recoded at different days after transplanting (DAT) (Table 5). In the first date of measurement (20 DAT) it was observed that the PU in conventional method produced higher dry matter yield than USG in bed planting. Likewise, at the final date (100 DAT) highest dry matter production was also recorded by USG in bed planting method than PU in conventional planting. However, dry matter production differs significantly ($P \leq 0.01$) at 20 to 100 DAT in both planting method except 80 and 90 DAT.

Table 5. Effect of dry matter production by USG in raised bed and PU in conventional planting.

Method of fertilizer application	Dry matter production (g m^{-2}) at different days after transplanting (DAT)								
	20	30	40	50	60	70	80	90	100
USG in bed planting	31b	73b	268a	433a	662a	860a	1066	1196	1293a
Prilled urea (PU) broadcasting in conventional planting	64a	127a	251b	350b	621b	851b	1062	1190	1260b
LSD at 5%	2.07	2.93	2.07	19.08	21.53	2.07	13.09	5.93	10.35
Level of significance	**	**	**	**	*	**	n.s.	n.s.	**

Where n.s.,* and ** represents probability of > 0.05, ≤ 0.01 and ≤ 0.001, respectively. Values were means of three replicates.

Crop growth rate

Results from the crop growth rate are shown in Table 6. At the initial stage (20 to 30 DAT) the crop growth rate (CGR) by USG in bed planting is lower than PU in conventional planting. The greatest CGR was observed at 50 to 60 DAT for both USG in bed planting and PU conventional planting method. In contrast, the lowest CGR was observed at 20 to 30 DAT for both by USG in bed planting and PU conventional planting method. However, crop growth rate significantly ($P \leq 0.01$) differed between both planting methods at all DAT except 70 to 80 and 80 to 90 DAT.

Table 6. Effect of crop growth rate by USG in raised bed and PU in conventional planting.

Method of fertilizer application	Crop growth rate (g m^{-2} day^{-1}) at different days after transplanting (DAT)							
	20-30	30-40	40-50	50-60	60-70	70-80	80-90	90-100
Urea super granule in bed planting	4.20b	19.5a	16.5a	22.9b	19.8b	20.6	13.0	9.30a
Prilled urea broadcasting in conventional planting	6.3a	12.4b	9.9b	27.1a	23a	21.1	12.8	7.0b
LSD at 5%	0.33	0.59	0.47	0.13	1.86	0.56	0.49	0.96
Level of significance	**	**	**	**	*	n.s.	n.s.	*

Where n.s.,* and ** represents probability of > 0.05, ≤ 0.01 and ≤ 0.001, respectively. Values were means of three replicates.

3.6. Weed Population

Weed population and dry biomass were greatly influenced by different planting methods of transplanted aman rice (Table 7). The USG in bed planting method reduced weed population resulting in lower dry biomass than PU in conventional planting. The PU in conventional method had significantly ($P \leq 0.01$) higher weed vegetation than raised bed planting.

Table 7. Effect of weed growth by USG in raised bed and PU in conventional planting.

Method of fertilizer application	Weed vegetation	
	Weed vegetation population (no- m^{-2})	Dry biomass (kg- ha^{-1})
Urea super granule in bed planting	110b	103.31a
Prilled urea broadcasting in conventional planting	380a	337b
LSD at 5%	4.984	1.873
Level of significance	**	**

Where ** represent probability of ≤ 0.001. Values were means of three replicates.

3.7. Irrigation Water

Amount of water required for different irrigations differed remarkably between the conventional and bed planting methods (Table 8). The PU in conventional planting received a higher amount of water at every irrigation and the total amount was 142.66 cm. The total amount of irrigation water received by USG in bed planting was 103.44 cm. Result showed that total water savings by USG in bed over PU in conventional method was 40%.

Table 8. Irrigation water savings by USG in raised bed and PU in conventional planting method.

Method of fertilizer application	Water required at different times of irrigation (cm)					Water saved over conventional method (%)
	Land preparation	Transplanting	Reproductive stage	Rainfall	Total	
USG in bed planting	-	5.62b	45.32b	52.50	103.44b	40
PU broadcasting in conventional planting	13.06	6.20a	70.90a	52.50	142.66a	
LSD at 5%	0.059	0.262	0.296	0.654	2.094	
Level of significance	-	*	**	n.s.	**	

Where n.s.,* and ** represents probability of > 0.05, ≤ 0.01 and ≤ 0.001, respectively. Values were means of three replicates. "-"indicates data not available.

3.8. Water Use Efficiency

Water use efficiency for grain and biomass production by USG in bed planting was 35.40 kg ha^{-1}cm^{-1} and 98.99 kg ha^{-1}cm^{-1}, respectively (Table 9). In contrast, water use efficiency for grain production and biomass production in conventional planting was 30.63 kg ha^{-1}cm^{-1} and 65.11 kg ha^{-1}cm^{-1}, respectively. However, water use efficiency for grain production and biomass production by USG bed planting over PU in conventional was 49% and 40.88%, respectively.

Table 9. *Water use efficiency by USG in raised bed and PU in conventional planting method.*

Method of fertilizer application	Water use efficiency savings by foliar spray in bed planting of rice over conventional method	
	Water use efficiency for grain production (kg ha^{-1}cm^{-1})	Water use efficiency for biomass production (kg ha^{-1}cm^{-1})
USG in bed planting	35.40a	98.99a
PU in conventional planting	30.63b	65.11b
LSD at 5%	1.083	2.121
Level of significance	**	**

Where ** represent probability of ≤ 0.001. Values were means of three replicates.

3.9. Agronomic Efficiency of N Fertilizer

Agronomic efficiency of N fertilizer by the USG in raised bed was 55.04% (Table 10). On the other hand, agronomic efficiency for PU in conventional planting was 43.67%. Agronomic efficiency of N fertilizer by USG in raised bed was significantly ($P \leq 0.01$) higher than the PU in conventional planting method.

Table 10. *Agronomic efficiency of fertilizer by USG in raised bed and PU in conventional planting method.*

Method of Fertilizer application	Agronomic efficiency of fertilizer (%)
USG in bed planting	55.04a
PU in conventional method	43.67b
LSD at 5%	4.225
Level of significance	**

Where ** represent probability of ≤ 0.001. Values were means of three replicates.

4. Discussions

4.1. The USG Techniques in Bed Planting Increases Plant Growth Parameters than the PU in Conventional Flat Methods

The increase in plant height by the USG in raised bed over the PU in conventional method was 4.46% (Table 2). Other studies also found that the USG techniques in bed planting produced taller plants than the PU in conventional cultivation techniques (Singh and Singh, 1986; Reddy and Mitra, 1985; Roy, 1988). This could be due to effective utilization of nitrogen fertilizer by rice plant during their vegetative growing period. Likewise, Puckridge et al, (1991) found that the paddy rice plant N uptake increased immediate after super granule placement into the soil during their vegetative growth period. They suggested that this increase in N uptake by rice plant during their vegetative growth period could be due to variation of stress on rice plants.

The USG treated plot in bed planting significantly recorded higher number of tiller and panicles m^{-2} compared to the PU in conventional planting method (Table 2). Similarly, Joseph et al, (1991) found that tiller number m^{-2}

was 265 and 309 for PU and USG, respectively. Similarly, Thakur (1991) found that panicles m^{-2} was 170 and 289 for PU and USG, respectively. However, they conducted their experiment in conventional flat method. This indicates that the USG is always superior to the PU in any condition for rice production. This could be due to the fact that the increase in N levels was responsible for increased number of leaves, resulting in higher photosynthesis, metabolic activity and cell division, which consequently increased growth and hence yield attributes (Jaiswal and Singh, 2001). Likewise, Jaisal and Singh (2001) also suggested that deep placement of the USG played a vital role for adequate nutrient supply with minimizing losses of nitrogen, NH$_3$ volatilization, nitrification and denitrification which ultimately increased the plant growth characteristics in raised bed plot.

4.2. Grain and Straw Yields Attributed to Higher in Raised Bed for USG than Conventional PU Technique

Higher grain and straw yields for the USG treated plots in bed planting were attributed to higher number of tiller and panicle (Tables 2 and 3). The superiority of the USG in bed planting over the PU in conventional planting regarding grain and straw yields of rice was found in many other investigations (De Datta, 1987; Bhuiyan et al, 1985; Choudhury and Bhuiyan, 1994). Similarly, Jena et al, (2003) found that deep placement of the USG significantly improved grain yield, straw yield and nitrogen use efficiency of rice and reduced the volatilization loss of ammonia relative to the application of the PU. Regardless of that, Lal et al, (1988) found that transplanted rice grain yield was 3.5 and 4.8 t ha^{-1} for PU and USG, respectively. This finding indicated that the USG is also suitable in conventional plot other than raised bed. They speculated that a high yield with the USG application seems to be associated with high N uptake due to higher N availability. Other study suggested that high yield of rice for placement of the USG in reduced soil layer could be due to decrease in N loss by volatilization, denitrification and aquatic weed competition including algal immobilization (Thomas and Prasad, 1982).

Our speculation is that existing conventional practice by farmers in Bangladesh often use urea inefficiently either because of this optimum is unknown, or the recommended doses is misleading or because of limited access to fertilizer. Likewise, Tran et al, (1989) reported that fertilizer and water use inefficiencies have substantially contributed to low yield in rice. Similarly, Pillai and Vamadevan (1978) speculated that the apparent reason for increased grain yield may be due to slower rate of urea hydrolysis of large sized urea super granule. Further, as the material was placed in the reduce zone of soil, losses of nitrogen due to nitrification-denitrification and leaching were minimized. Thus, the efficiency of applied nitrogen was increased which in turn contributed to increased grain yield (Pillai and Vamadevan 1978).

4.3. Nitrogen Use Efficiency Was Greater in the USG in Raised Bed than the PU in Conventional Method

The USG in raised bed had significantly ($P \leq 0.01$) lower nitrogen fertilizer utilization efficiency than the conventional plot (Table 10). Wes Emmott et al, (2013) reported that at basal dose N required 50 kg ha^{-1}. At active tillering stage N required by the rice plant was 89 kg ha^{-1}. The N required at the panicle initiation stage by the rice plant was 104 kg ha^{-1} at heading stage N required by rice plant was 30 kg ha^{-1}. The highest N required by the rice plant was observed at panicle initiation stage. However, the USG was applied 7 days after transplanting at rate of 81 kg ha^{-1} in our experiment. So N required by the rice plant was 37 kg ha^{-1}. In the method of deep placement (7-10 cm) of USG in bed planting the fertilizer was applied at active tillering stage. Because of large size of USG allow slower hydrolysis rate of urea fulfilled the highest demand of N at the panicle initiation stage. This may save the amount of N fertilizer required by the rice plant. Joseph et al, (1991) also found that nitrogen use efficiency was higher in USG than the conventional PU technique. Likewise, BRRI (2008) reported that nitrogen fertilizer use efficiency can be increased up to 20-25% and 30% of urea fertilizer can be saved by the USG application than the PU conventional cultivation method. Similarly, Thakur (1991) found that nitrogen use efficiency for PU and USG were 26.1 and 32.8 kg grain kg^{-1}N^{-1}, respectively. This might be due to the placement of USG in the reduced zone and its bigger size may have increased its efficiency by minimizing loss of N through ammonia volatilization and denitrification (Nommick, 1976). There also may be possibility that the USG were placed below soil surface that minimize N loss through runoff and volatilization. Similarly, Cao et al, (1984) speculated that with deep placement of urea, the losses due to N_2 denitrification and NH_3 volatilization can be reduced, which accounts for the superiority of deep placement of USG to PU application. Likewise, Thakur (1991) opined that the size of USG and less losses or better utilization through deep placement of the USG is responsible for higher nitrogen use efficiency than the PU in conventional flat method. Regardless of that, the apparent reason for increased nitrogen use efficiency for the USG may be due to slower rate of urea hydrolysis from

the large size USG and this release is in synchrony with the rice plant requirements. Further, as the USG was placed in the reduced zone of soil, losses of nitrogen due to nitrification-denitrification and leaching were minimized. Thus, the efficiency of applied nitrogen was increased which in turn contributed to increased grain yield (Pillai and Vamadevan, 1978).

4.4. The USG in Raised Bed Attributed Better Agronomic Efficiency of N Fertilizer and Suppress the Activities of Weed Growth than the PU in Conventional Method

Agronomic efficiency of N fertilizer was 12% higher using USG in bed planting over PU in conventional planting (Table 10). Higher agronomic efficiency of N fertilizer with USG over PU was also found in some other investigations (Choudhury and Bhuiyan, 1994; Choudhury et al, 1994). Our speculation is that the higher agronomic efficiency of N fertilizer using USG over PU could be due to the high recovery of applied N in the field. This might possibly be due to different losses like denitrification, ammonia volatilization, runoff and immobilization were too low in the USG than the PU. Further, hence the placement of USG on raised bed can provide agronomic benefits and minimize N application rate and loss.

Weed population were significantly ($P \geq 0.01$) higher in the conventional PU application than the USG in raised bed planting method (Table 7). This may be better agronomic management practices and deep placement of USG in raised bed than conventional PU application. However, Mohanty et al, (1999) speculated that the USG application in rice field retarded weed growth due to limited nitrification-denitrification process and biological nitrogen fixation is promoted under reduced floodwater N concentration.

4.5. The USG in Raised Bed Use Less Water and Has Higher Water Use Efficiency than the Conventional Flat Method

The USG in bed planting saved 40% irrigation water than the conventional flat method. This water saving mainly occurred in transplanting and reproductive stage (Table 8). Likewise, water use efficiency was significant by higher ($P \leq 0.001$) in raised bed for USG than the conventional flat method (Table 9). Other study like Kahlown (2006) found that raised bed system generally use 10 to 30% less water than the amount applied to farmers to their flooded basins. Thompson et al, (2003) also showed that irrigation water savings of about 14% using bed compared with flat method. Likewise, Beecher et al, (2006) found that raised bed (17.2 ML ha^{-1}) use significantly less water than the conventional flat (18.7 ML ha^{-1}) method. They speculated that saving in irrigation water use is related to the amount of time a crop is intermittently irrigated, thus the longer ponding occurs, the lesser it will be the water will be saving. These findings concluded that continuous submergence is not a must for rice production. Saturated soil condition is optimum for rice production.

5. Conclusions

The findings of this study demonstrated that the USG in bed planting is superior to PU broadcasting in conventional planting method for rice production. Placement of USG in raised bed produced higher number of panicles per unit area, panicle weight, number of grains per panicle and 1000-grain weight, which ultimately gave the highest grain yield than the PU in conventional plot. This study also concluded that the USG application within raised bed proved beneficial over conventional PU technique, especially with respect to grain yield, yield attributes, agronomic efficiency and water use efficiency. Deep placement of the USG in raised bed effectively increased N-use efficiency as compared to conventionally applied prilled urea. The placement of the USG below plough layer (5cm) is considered the best method to decrease N losses and thereby to increase fertilizer use efficiency. This study also suggests that the deep placement of the USG in raised bed is feasible for water and nitrogen use efficiency and reduction in soil compaction.

It was found from our previous studies that raised bed system for rice production achieve higher yield than conventional flat method (Bhuyan et al, 2012a; Bhuyan et al, 2012b). The deep placement of urea super granule for raised bed planting, new rice based farming systems, offers many potential advantages. Therefore, deep placement of the USG at 7-10 cm depth (reduced soil layer) on raised bed is one of the most efficient N management techniques developed for rice production. There is a good prospect of utilization of this technology to benefit the rice farmers. Further experiment will be conducted to distinguish role of urea super granule application on raised bed as compared to conventional method for transplanted boro/winter rice (completely depends on irrigation) production.

Acknowledgements

The authors are thankful to the farmer who leases his land for this experiment. The author also thanks to the pump operator for his kind help to irrigate the land. The authors are deeply indebted to the authorities of Bangladesh Rice research institute (BRRI), Local weather office, Chuadanga and seed processing unit, Bangladesh Agricultural Development Corporation (BADC), Chuadnaga for their assistance and constructive suggestion.

References

[1] Bangladesh Rice Research Institute (BRRI). 2008. ``Adhunik Dhaner Chash''- Bengali Version, Bangladesh Rice Research Institute, Gazipur-1701.

[2] Bangladesh Rice Research Institute (BRRI). 2007. Impact of LCC and USG on rice production in same selected areas. Annual Research Review Report, Agricultural Economics Division, Bangladesh Rice Research Institute, Gazipur, 14-17 January.

[3] BARC.2005. Bangladesh Agricultural Research Council, Fertilizer Recommendation Guide, Ministry of Agriculture, Governments of the People's Republic of Bangladesh, Dhaka, Bangladesh, p. 174-175.

[4] Beecher, H. G., Dunn, B. W., Thompson, J. A., Humphreys, E., Mathews, S. K., and Timsina, J. 2006. Effect of raised beds, irrigation and nitrogen management on growth, water use and yield of rice in south-eastern Australia. Australian Journal of Experimental Agriculture 46: 1363-1372.

[5] Bhuiyan, N. I., Mazid, M. A., Saleque, M. A.1985. Fertilizer nitrogen deep placement status of research by BRRI. IN Proceedings Fertilizer Nitrogen Deep Placement for Rice. Dhaka, Bangladesh: Bangladesh Agricultural Research Council. PP. 27-36.

[6] Bhuyan, M. H. M., Ferdousi, M. R., Iqbal, M. T. 2012a. Yield and growth response to transplanted aman rice under raised bed over conventional cultivation method. ISRN Agronomy (doi:10.5402/2012/646859).

[7] Bhuyan, M. H. M., Ferdousi, M. R., Iqbal, M. T. 2012b. Foliar spray of nitrogen fertilizer on raised bed increases yield of transplanted aman rice over conventional method. ISRN Agronomy (doi:10.5402/2012/184953).

[8] BRRI. 2008. Adhunik Dhaner Chash, 14th Edition, Bangladesh Rice Research Institute, Gazipur-1701, Bangladesh, pp.13-15.

[9] BRRI. 2009. Bangladesh Rice Research Institute, Fact sheet; soil & fertilizer management. pp. 3-5.

[10] Cao, Z. H., De Datta, S. K., Fillery, I. R P. 1984. Effect of placement methods on floodwater properties and recovery of applied N (^{15}N- labeled urea) in wetland rice. Soil Science Society of America Journal 48(1): 196-203.

[11] Choudhury, A. T. M. A., Bhuiyan, N. I. 1994. Effect of rates and methods of nitrogen application on grain yield and nitrogen uptake of wetland rice. Pakistan Journal of Scientific and Industrial Research 37 (3): 104-107.

[12] Choudhury, A. T. M. A., Byuiyan, N. I., Hashem, M. A., Matin, M. A.1994. Nitrogen fertilizer management in wetland rice culture. Thai Journal of Agricultural Science 27(3): 259-267.

[13] Choudhury, A. T. M. A., Khanif, Y. M. 2011. Evaluation of effects of nitrogen and magnesium fertilization on rice yield and fertilizer nitrogen efficiency using ^{15}N tracer technique. Journal of Plant Nutrition 24(6): 855-871.

[14] De Datta, S. K.1978. Fertilizer management for efficient use in wetland rice soils. In Soils and Rice. Los Banos, Phlippines: International Rice Research Institute. p. 671-679.

[15] De Datta, S. K.1981. Principles and Practices of Rice Production. New York, USA: John Wiley Sons, Inc.

[16] De Datta, S. K.1987. Advances in Soil Fertility Research and Nitrogen Fertilizer Management for Lowland Rice. In Efficiency of Nitrogen Fertilizers for Rice. Los Banos, Philippines: International Rice Research Institute. P. 27-41.

[17] Goswami, N. N., Banerjee, N. K. 1978. Phosphorus, potassium and other microelements. in Soils and Rice. Los Banos, Philippines: International Rice Research Institute. PP. 561-580.

[18] Hobbs, P. R.2001. Tillage and Crop Establishment in South Asian Rice-Wheat Systems, Journal of Crop Production 4(1): 1-22.

[19] Jaisal, V. P., Singh, G. R.2001. Performance of urea super granule and prilled urea under different planting methods in irrigated rice (*Oryza sativa*). Indian Journal of Agricultural Sciences 71(3): 187-189.

[20] Jat, M. L., Singh S, Rai., H K R S, Chhokar., Sharma S. K., Gupta, R K.2005. Furrow Irrigated Raised Bed (FIRB) Planting Technique for Diversification of Rice-Wheat System in Indo-Gangetic Plains. Japan Association for International Collaboration of Agriculture and Forestry 28 (1): 25-42.

[21] Jena, D., Misra, C., Bandyopadhyay, K K.2003. Effect of prilled urea and urea super granules on dynamics of ammonia volatilisation and nitrogen use efficiency of rice. Journal of the Indian Society of Soil Science 51(3): 257- 261.

[22] Joseph, K., Tomy, P. J., Nair, N. R. 1991. Effect of sources and levels of nitrogen on the growth, yield and nitrogen use efficiency of wetland rice. Indian Journal of Agronomy 36(1): 40-43.

[23] Lal, P., Gautam, R. C., Bishat, P. S., Pandey, P. C.1988. Agronomic and economic evaluation of urea super granule and sulphur coated urea in transplanted rice. Indian Journal of Agronomy 33(2): 186-190.

[24] Kahlown, M A. 2006. Low cost, innovative water conservation practices in irrigated agriculture. The 2nd International Conference on Water Resources and Arid Environment. pp. 9-10.

[25] Kuma, r D., Devakumar, C., Kumar, R., Das, A., Panneerselvam, P., Shivay, Y. S. 2010. Effect of neem-oil coated prilled urea with varying thickness of neem-oil coating and nitrogen rates on productivity and nitrogen-use efficiency of lowland irrigated rice under indo-gangetic plains. Journal of Plant Nutrition 33: 1939-1959.

[26] Mohanty, S. K., Singh, U., Balasubramanian, V., Jha, K P. 1999. Nitrogen deep-placement technologies for productivity, profitability, and environmental quality of rainfed lowland rice systems. Nutrient Cycling in Agroecosystems 53: 43–57.

[27] Nommik, H.1976. Further observations on ammonia loss from urea applied to forest soil with special reference to the effect of pellet size. Plant and Soil 45: 279-282.

[28] Pillai, K. G. Vamadevan, VK.1978. Studies on the integrated nutrient supply system for rice. Fertilizer News 23(3): 11-14.

[29] Ponnamperuma, F. N.1972. The Chemistry of submerged Soils. Advances in Agronomy 24: 29-96.

[30] Puckridge, D. W. Wiengweera., A Thongbai, P Sangtong, P Sattarasart, A Kongchum, M Runtan S.1991. Nitrogen uptake by deepwater rice during the preflood and flooded phases, in relation to dry matter and grain production. Field Crop Research 27: 315-336.

[31] Ram, J., Y, Singh., D. S Kler., K. Kumar., I, Humphreys., J. Timsina.2005. Performance crops and alternative cropping systems on permanent raised beds in the Indo-Gangetic plains of north-western India. Presented in ACIAR Wordshop on Permanent Bed Planting Systems. 1-3 Mar. 2005. Griffith, NSW, Australia.

[32] Reddy, MD Mitra, B N.1985. Response of rice to different forms of urea and phosphorus fertilization under intermediate deepwater conditions. Plant and soils 84(3): 431-435.

[33] Roy, B.1988. Coated and modified urea materials for increasing nitrogen use efficiency of lowland rice in heavy clay soils. Fertilizer Research 15(2): 101-109.

[34] Sahrawat, K L. 2000. Marco and micronutrients removed by upland and lowland rice cultivars in West Africa. Communications in Soils Science and Plant Analysis 31(5 & 6): 717-723.

[35] Singh, B. K., Singh, R P.1986. Effect of modified urea materials on rain fed low land transplanted rice and their residual effect on succeeding wheat crop. Indian Journal of Agronomy 31(2): 198-200.

[36] Singh, S., Ladhab JK Gupta., RK, Bhushana L., Raob, A. N. 2008. Weed management in aerobic rice systems under varying establishment methods. Crop Protection 27: 660-671.

[37] Thakur, R. B.1991. Relative efficiency of prilled urea and modified urea fertilizers on rainfed lowland rice. Indian Journal of Agronomy 36(1): 87-90.

[38] Thomas, J., Prasad R.1982. On the nature of mechanism responsible for the higher efficiency of the urea super granules for rice. Plant and Soil 69: 127-130.

[39] Thompson J., Griffin D., North, S.2003. Improving the water use efficiency of rice. Co-operative research centre for sustainable rice production, Final report No. 1204, Yanco, NSW.

[40] Tran, D. V., Marathée, J. P.1994. Major Issues in Asian Rice-Wheat Production Systems. Sustainability of Rice-Wheat Production Systems in Asia. RAPA Publication: 1994/11, 61-67.

[41] Wetsellar, R.1985. Deep point-placed urea in a flooded soil: a mechanistic view. Special Publications, International fertilizer Development Centre (IFDC), U. S. A. SP-6. pp. 7-14.

[42] Wes Emmott., Tyler Kernighan, Christine Tobin, Hugo Gonzalez, Blair Freeman.2013. Growth stages of rice-plant Agriculture. Source: www.plant.uoguelph.ca /courses/pbio-3110/documents/Rice_08.pdf

[43] Zia, M S Aslam., M Gill., M A.1992. Nitrogen management and fertilizer use efficiency for lowland rice in Pakistan. Soil Science and Plant Nutrition 38(2): 323-330.

Morphological and Molecular Analysis Using RAPD in Biofield Treated Sponge and Bitter Gourd

Mahendra Kumar Trivedi[1], Alice Branton[1], Dahryn Trivedi[1], Gopal Nayak[1], Mayank Gangwar[2], Snehasis Jana[2, *]

[1]Trivedi Global Inc., Henderson, USA
[2]Trivedi Science Research Laboratory Pvt. Ltd., Bhopal, Madhya Pradesh, India

Email address:

publication@trivedisrl.com (S. Jana)

Abstract: Plants are known to have sense and can respond to touch, electric and magnetic field. The present study was designed on the sponge gourd (*Luffa cylindrica*) and bitter gourd (*Momordica charantia*) seeds with respect to biofield energy treatment. The seeds of each crop were divided into two groups, one was kept control, while the other group was subjected to Mr. Trivedi' biofield energy treatment. The variabilities in growth contributing parameters were studied and compared with their control. To study the genetic variability after biofield energy treatment, both the seeds were analyzed for DNA fingerprinting using RAPD method. After germination, the plants of sponge gourd were reported to have uniform colored leaves and strong stem. The leaves and fruits of sponge gourd showed no infection, with anti-gravity properties during early stage of fruiting as compared with the control group. Similarly, treated bitter gourd showed uniform color of leaves, strong stem, with disease free fruits in biofield treated seeds as compared with the control. The true polymorphism (%) observed between control and treated samples of sponge and bottle gourd seed sample was an average value of 7.8% and 66% respectively. In conclusion, Mr. Trivedi's biofield energy treatment has the ability to alter the plant growth rate that may by interacting with plant genome, which resulted in high yield of crops.

Keywords: *Luffa cylindrica*, *Momordica charantia*, Biofield energy, Growth Attributes, DNA Fingerprinting, Polymorphism

1. Introduction

The gourd is generally used to describe the crop plants in the family *Cucurbitaceae*. The term gourd refers to around 825 species derived from tropical and subtropical regions, out of which approximately 26 species are cultivated as vegetables [1]. One of the important and common fruit of cucurbitaceous crop is sponge gourd (*Luffa cylindrica*), originating in India and southern Asia. Besides its vegetative importance, members of this group were grown for ornament purposes. Sponge gourd fruits and their seeds are used in the traditional medicine as stomachic, antipyretic, anti-helmintic, and other related medicinal importance [2].

Another important and commonly available plant is bitter gourd (*Momordica charantia*) of family *Cucurbitaceae*, valued for its nutritional and several medicinal properties. It is also known as bitter melon, karela, or balsam pear, and is

the most popular plant for diabetes management [3]. This crop also originates from India, and diversity has been reported in China but widely cultivated and used in India, Malaysia, Africa, and South America [4, 5]. Its fruit has been used from centuries in ancient traditional medicinal purpose due to its antimicrobial, antidiabetic, antiviral and anti-oxidant activities [6, 7]. Based on the historical reports [4] and RAPD molecular analysis, maximum diversity was reported in eastern India [8].

Identification of genetic diversity based on phenotypic character is very limiting, as an environmental factors and plant developmental stage will affect the morphological characters of plant. However, DNA polymorphism based upon molecular markers are independent of environmental conditions, which shows a high level of polymorphism. Molecular markers shows variation in the genome which may be expressed or not, while morphological markers reflects variation in expressed regions [9]. Randomly amplified polymorphic DNAs (RAPD) analysis shows maximum genetic relatedness among plant genome due to their simplicity, speed and low-cost [10].

To improve the crop yield, growth characters, and protect the crop from infections, different methods have been adopted such as genetic engineering, plant growth hormones, tissue culture, altered environmental parameters, use of fertilizers, pesticides, and many more [11]. Apart from these traditional approaches, recent reports suggest increased germination, yield, growth and its related parameters with the use of electric and magnetic field on seeds before germination [12-14]. Some alternative techniques such as exposure of electromagnetic field on seeds have shown good results with improved yield, as biofield energy is a form of low intensity electromagnetic field [15]. Authors have intended to study the impact of biofield energy treatment on the seeds of sponge and bitter gourd. Energy treatment as an alternative integrative medicine approach has been recently introduced to promote human wellness by National Center for Complementary and Alternative Medicine (NCCAM) [15]. Biofield is the name given to the electromagnetic field that permeates and surrounds living organisms. It is the scientifically preferred term for the biologically produced electromagnetic and subtle energy field that provides regulatory and communication functions within the organism. The objects always receive the energy and responding to the useful way that is called biofield energy treatment. Mr. Trivedi's unique biofield treatment is known as The Trivedi Effect®. Mr. Trivedi is having the unique biofield energy, which has been reported in several research areas [16-19].

Due to the importance of sponge and bitter gourd as vegetable and medicinal importance, and previous results of biofield treatment, the present study was designed to evaluate the effect of biofield energy treatment on sponge and bitter gourd. Genetic variability parameters of both the crops were studied using RAPD (DNA fingerprinting).

2. Materials and Methods

Sponge gourd (*Luffa aegyptiaca*) and bitter gourd (*Momordica charantia*) were selected for present study due to their high vegetative importance. Nirmal 28 variety of sponge gourd, and Nirmal 167 (Savitri) of bitter gourd were procured from Nirmal Seeds, Jalgoan, Maharashtra, India. Each variety of the seeds was divided into two parts, one part was considered as control, while other part was coded as treated and subjected to Mr. Trivedi's biofield energy treatment. Seeds from each group were cultivated in Shahapur agricultural land in Maharashtra for analysis. However, the control plants were given standard cultivation parameters such as proper irrigation, fertilizers, pesticides and fungicides; while the treated plants were given only irrigation, without any supportive measure. DNA fingerprinting of both the plants were performed using random amplified polymorphic DNA (RAPD) techniques

using Ultrapure Genomic DNA Prep Kit; Cat KT 83 (Bangalore Genei, India) to study the genetic relationship before and after treatment.

2.1. Biofield Treatment Strategy

The treated groups of seeds were subjected to Mr. Trivedi's biofield energy treatment under standard laboratory conditions. Mr. Trivedi provided the unique biofield treatment through his energy transmission process to the treated group of both the seeds without any touch. The treated samples were assessed for the growth germination of seedlings, leaves, length of plant, and rate of infections. Variability in different growth contributing parameters and genetic relatedness using RAPD of control and treated crops were compared [16].

2.2. Analysis of Growth and Related Parameters of Crops

Control and treated seeds of sponge gourd and bitter gourd were cultivated under similar conditions. Vegetative growth of the crops with respect to plant height, canopy, the shape of leaves, flowering conditions, infection rate, *etc.* were analyzed and compared with respect to the control group [20].

2.3. DNA Fingerprinting in Sponge and Bitter Gourd

2.3.1. Isolation of Plant Genomic DNA Using CTAB Method

Leaves disc of each plants was harvested after germination when the plants reached the appropriate stage. Genomic DNA from both plant leaves was isolated according to the standard cetyl-trimethyl-ammonium bromide (CTAB) method [21]. Approximately 200 mg of plant tissues were grinded to a fine paste in approximately 500 μL of CTAB buffer. The mixture (CTAB/plant extract) was transferred to a microcentrifuge tube, and incubated for about 15 min at 55°C in a recirculating water bath. After incubation, the mixture was centrifuged at 12000 g for 5 min and the supernatant was transferred to a clean microcentrifuge tube. After mixing with chloroform and iso-amyl alcohol followed by centrifugation the aqueous layers were isolated which contain the DNA. Then, ammonium acetate followed by chilled absolute ethanol were added, to precipitate the DNA content and stored at -20°C. The RNase treatment was provided to remove any RNA material followed by washing with DNA free sterile solution. The quantity of genomic DNA was measured at 260 nm using spectrophotometer [22].

2.3.2. Random Amplified Polymorphic DNA (RAPD) Analysis

DNA concentration was considered about 25 ng/μL using distilled deionized water for polymerase chain reaction (PCR) experiment. The RAPD analysis was performed on the each treated seeds using RAPD primers, which were label as RPL 2A, RPL 7A, RPL 12A, RPL 14A, RPL 18A, and RPL 23A for sponge gourd and RPL 4A, RPL 5A, RPL 6A, RPL 13A, and RPL 19A for bitter gourd. The PCR mixture including 2.5 μL each of buffer, 4.0 mM each of dNTP, 2.5

μM each of primer, 5.0 μL (approximately 20 ng) of each genomic DNA, 2U each of *Thermus aquaticus* (*Taq*) polymerase, 1.5 μL of $MgCl_2$ and 9.5 μL of water in a total of 25 μL with the following PCR amplification protocol. For sponge gourd, initial denaturation at 94°C for 5 min, followed by 40 cycles of annealing at 94°C for 1 min, annealing at 36°C for 1 min, and extension at 72°C for 2 min, while final extension was carried out at 72°C for 10 min. For bitter gourd, initial denaturation at 94°C for 7 min, followed by 8 cycles of annealing at 94°C for 45 sec, annealing at 35°C for 1 min, and extension at 72°C for 1.5 min. Further, 35 cycles was carried out at 94°C for 45 sec, 40°C for 60 sec, and 72°C for 60 sec. While, final extension was carried out at 72°C for 7 min. Amplified PCR products (12 μL of each) from control and treated samples were loaded on to 1.5% agarose gel and resolved by electrophoresis at 75 volts. Each fragment was estimated using 100 bp ladder (Genei™; Cat # RMBD19S). The gel was subsequently stained with ethidium bromide and viewed under UV-light [23]. Photographs were documented subsequently. The following formula was used for calculation of the percentage of polymorphism.

$$\text{Percent polymorphism} = A/B \times 100$$

Where, A = number of polymorphic bands in treated plant; and B = number of polymorphic bands in the control plant.

3. Results and Discussion

3.1. Effect of Biofield Treatment on Different Growth Contributing Parameters of Sponge Gourd

The untreated seeds of sponge gourd after germination showed that leaves were thicker than the treated group and the color was not uniform. This might be the symbol of initiation of infections resulted in poor yield of the crop. A large number of leaves in control plants had a rough and bubble-like surface due to the infections. The stem was tender and could be easily bent in control crops, which resulted in loss of crop yield. The plants showed high incidence of disease with dots on the leaves, and they were attacked by insects.

The biofield treated seeds of sponge gourd plants showed that the leaves were thinner than the control group and the color was uniform. The leaves had very smooth surfaces with no disease or insect attack. The fruits showed anti-gravity properties during early stage of fruiting. Normally, the fruit gravitates downwards due to weakness in the stem of the fruit, but the biofield treated plant's stem was so strong that the fruit was held up and pointed towards the sun.

Biofield energy treated fruits of sponge gourd was reported to be healthy with respect to quality and shape of fruit. Sponge fruit inheritance and its quality traits has been previously reported [24, 25]. Reports suggest that quantitative traits of fruit shape can be controlled by its gene expression and regulation during the maturation period of plant. This qualitative trait could control be controlled by different genetic mechanisms at different developmental stages. However, correlations has been well reported at different developmental stages with various genetic effects [26]. So, it can be assumed that biofield energy treatment might alter the genetic mechanism of sponge gourd, which is responsible for the better growth of sponge plant and fruits.

Figure 1. The Trivedi Effect® on sponge gourd (a) leaf of control plants were reported as pale yellow depicts the symptoms of infection, (b) control sponge fruit showed infection and unhealthy fruits, (c) biofield treated leaves of sponge gourd were free from any kind of disease or pest attack, (d) biofield treated seeds showed healthy plant growth and fruits free from any infections result in high yield. C: Control; T: Treated.

3.2. Effect of Biofield Treatment on Different Growth Contributing Parameters of Bitter Gourd

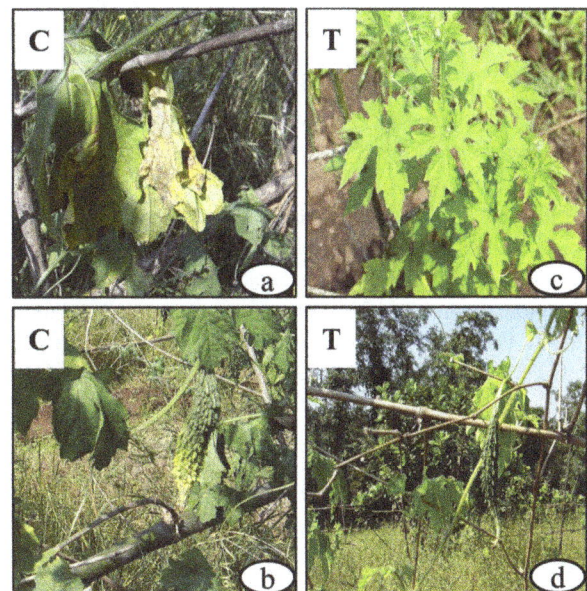

Figure 2. Effect of biofield energy treatment on bitter gourd (a) control leaves of bitter gourd reported with irregular yellowish patches, infections and circular patches or spots appear, (b) control seeds of bitter gourd plant showed fruits with curved shape fruit do not develop fully and remain small, (c) biofield treated group showed healthy leaves and all are free from pest attack, (d) biofield treated group showed that bitter gourd fruits are long, fresh, and free from infection. C: Control; T: Treated.

The growth in case of control bitter gourd was good, but the stem was very tender. Many leaves were diseased with yellow spots, and there were signs of insect attack on both the leaves and fruits of bitter gourd in the control group. The color of the leaves were yellowish-green and it was not uniform throughout the control group in bitter gourd.

The growth in biofield treated bitter gourd plants was also comparable, the leaves were of uniform color and the stem of bitter gourd plant was quite strong. The plants were absolutely disease-free and even the fruits were not attacked by any kind of insects. Therefore, it can be concluded that biofield treated bitter gourd might yield more healthy fruits than compared with the control. According to the recent report of Mahajan *et al*, the seeds of bitter gourd respond differently, when exposed to varying level of electric and magnetic field. They showed that seeds become polarized in the presence of altered field, while after removal of electromagnetic field seeds retain some level of polarization, known as remnant polarization. During germination, these seeds when come in contact with the water dipoles, an interaction between the water dipoles and seed dipoles taken place. These dipole interaction, results in the level of water uptake by the seeds, which improves the germination time and rate [13]. It can be assumed that Mr. Trivedi's biofield energy treatment might affect the polarization of bitter gourds seeds in treated group, and results in better dipole interaction of seeds and water. Hence, biofield treated seeds were reported with high germination rate, better growth, and less infections as compared with the control.

According to a latest report, the plants respond to the different environmental conditions such as geomagnetism, gravity response, electric signal, touch effect, wavelength of light, and many more. These all factors contributes and effect the final growth of plants at different stages. A report suggest that magnetic field exposure will increase the growth rate and have effect on plant roots and shoots [27]. Another report suggests that geomagnetic field has effect on activated state of cryptochromes, which resulted in modification of function and alters the growth yield [28]. Magnetic field exposure has been reported with increased level of photosynthesis, stomatal conductance and chlorophyll content in corn plants compared to the control under similar irrigated and mild stress condition [29]. It can be hypothesized that on exposure of biofield energy treatment on sponge and bitter gourd seeds, energy provided might change their paramagnetic behaviors, and orient themselves in the direction of positive energy, which leads to improved plant growth. It can be suggested that improved overall growth of plants and fruits of sponge and biter gourd could be due to improved level of photosynthesis and chlorophyll content.

Biofield energy is a type of energy medicine, under the category of complementary and alternative medicine (CAM), which basically involves low-level energy field interactions, and includes energy therapy, bio-electromagnetic therapy, *etc*. The improved growth contributing parameters of sponge and bitter gourd after biofield treatment may be the new alternative approach for better growth of pants and fruits.

3.3. RAPD Analysis of Biofield Treated Sponge and Bitter Gourd

Genetic analysis using RAPD molecular markers has been widely accepted technique in agriculture for improvement in vegetable crops [30]. By designing different RAPD markers related to samples, important information for genetic diversity can be evaluated for different plant species samples. Besides genetic diversity, population genetics study, pedigree analysis and taxonomic discrimination can also be correlated [23]. However, it has been proposed to be a powerful tool to evaluate differences between inter- and intra-population organisms including plants [31]. Biofield energy treatment might be a new approach in agriculture science, which can alter the genetic variability of the plant, improved yield of crops, growth, and high immunity along with change in chlorophyll content. Biofield treatment was reported with high genetic variability among species using RAPD fingerprinting [16]. However, the effect was also reported in case of biofield treated ginseng, blueberry [32], and lettuce, tomato [33] with an improved overall agronomical characteristics.

Biofield energy treated sponge and bitter gourd were analyzed and compared with their respective control for their epidemiological relatedness and genetic characteristics. Genetic similarity or mutations between the biofield treated and the control group was analyzed using RAPD. Both the samples required a short nucleotide random primers, which were unrelated to known DNA sequences of the target genome. DNA polymorphism can be efficiently detected using PCR primers and identify inter-strain variations among plant species after the biofield energy treatment. The degree of relatedness and genetic mapping can be correlated between similar or different treated samples.

Table 1. DNA polymorphism of sponge gourd analyzed after biofield treatment using random amplified polymorphic DNA (RAPD) analysis.

S. No.	Primer	Band Scored	Common bands in control and treated	Unique band Control	Treated
1.	RPL 2A	17	17	1	-
2.	RPL 7A	15	15	1	-
3.	RPL 12A	11	11	-	-
4.	RPL 14A	16	16	-	-
5.	RPL 18A	13	13	-	-
6.	RPL 23A	12	12	-	-

Random amplified polymorphic-DNA fragment patterns of control and treated sponge gourd samples were generated using six RAPD primers, and 100 base pair DNA ladder. The results of DNA polymorphism in control and treated samples are presented in Figure 3. The DNA profiles of treated group were compared with their respective control. The polymorphic bands observed using different primers in control and treated samples were marked by arrows. The results of RAPD patterns in biofield treated sponge gourd sample showed some unique, common and dissimilar bands as compared with the control. DNA polymorphism analyzed by RAPD analysis, showed different banding pattern in terms of total number of bands, and common, and unique bands, which are summarized in Table 1. The percentage of polymorphism between samples were varied in all the primers, and were ranged from 6.6 to 9% between control and treated samples. However, level of polymorphism was only detected using the primer RPL 2A and RPL 7A was found to be 6.6 and 9%, respectively, while rest of the used primers did not show any level of polymorphism.

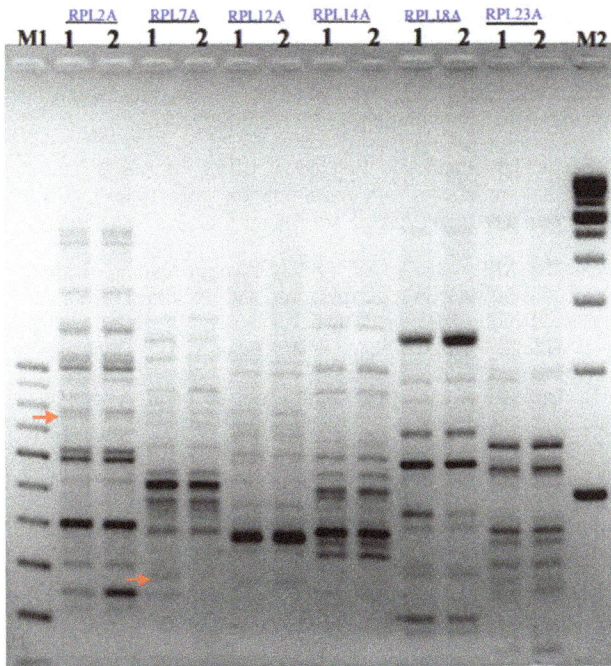

Figure 3. *Random amplified polymorphic-DNA fragment patterns of biofield treated sponge gourd generated using 6 RAPD primers, RPL 2A, RPL 7A, RPL 12A, RPL 14A, RPL 18A and RPL 23A. M: 100 bp DNA Ladder; Lane 1: Control; Lane 2: Treated.*

Table 2. *DNA polymorphism of bitter gourd analyzed after biofield treatment using random amplified polymorphic DNA (RAPD) analysis.*

S. No.	Primer	Band Scored	Common bands in control and treated	Unique band	
				Control	Treated
1.	RPL 4A	12	2	7	1
2.	RPL 5A	12	1	2	1
3.	RPL 6A	10	2	4	5
4.	RPL 13A	8	-	4	1
5.	RPL 19A	9	1	4	6

Similarly, a very high level for polymorphism was detected in biofield treated bitter gourd samples using five primers. Different banding pattern was observed using RAPD DNA polymorphism in terms of total number of bands, and common, and unique bands, which are summarized in Table 2. The polymorphic bands observed using different primers in control and treated samples of bitter gourd were marked by arrows in Figure 4. The level of polymorphism percentage in bitter gourd samples were varied in all the primers, and were ranged from 8 to 100% between control and treated samples. However, level of polymorphism was detected as 8%, 42%,

100%, 83%, and 100% using the primer RPL 4A, RPL 5A, RPL 6A, RPL 13A, and RPL 19A respectively. Highest level of polymorphism was detected using two primers namely RPL 6A, and RPL 19A *i.e.* 100%, while minimum level of polymorphism (8%) was detected using primer RPL 4A.

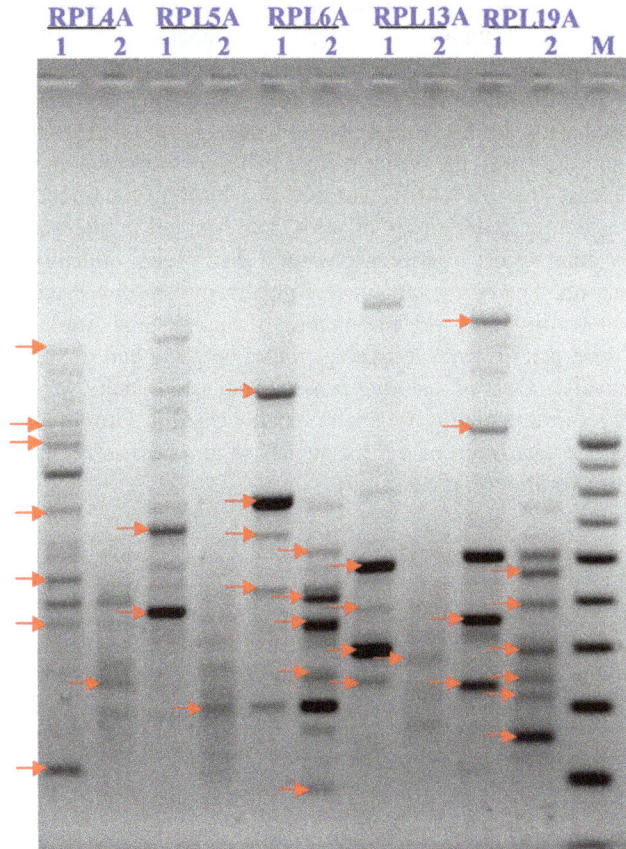

Figure 4. *Random amplified polymorphic-DNA fragment patterns of biofield treated bitter gourd generated using 5 RAPD primers, RPL 4A, RPL 5A, RPL 6A, RPL 13A and RPL 19A. M: 100 bp DNA Ladder; Lane 1: Control; Lane 2: Treated.*

RAPD analysis using different primers explains the relevant degree of genetic diversity among the tested samples. Overall, RAPD showed that polymorphism was detected between control and treated samples. The percentage of true polymorphism observed between control and treated samples of sponge and bottle gourd seed sample was an average value of 7.8% and 66%, respectively.

However, RAPD is a tool which will detect the potential of polymorphism throughout the entire tested genome. After biofield treatment, higher number of polymorphic bands in bitter gourd sample than sponge gourd sample indicated that the genotypes selected in bitter gourd possess a higher degree of polymorphism compared with sponge gourd. Molecular analyses and genetic diversity of bitter gourd have been well defined [34]. After biofield treatment, level of polymorphism was reported in both the crops, which suggested that Mr. Trivedi's biofield energy treatment might have the capability to alter the genetic character of plants, which might be useful in terms of productivity.

4. Conclusions

Biofield energy treatment on the sponge and bitter gourd was reported with improved growth characteristics such as healthy leaves, fruits, and control from pest attack as compared to their respective control. Biofield treated sponge plants were strong, thin leaves, and uniform color, which suggested higher immunity of plant as compared with the control. Further, the fruits of sponge gourd showed anti-gravity property at an early stage, along with strong stem as compared with the control plants. However, biofield treated bitter gourd showed the uniform color of leaves, strong stem, with disease free fruits. It is assumed that biofield treatment might affect the polarization of seeds, and resulted in altered dipole interaction between water and seed during germination. The percentage of true polymorphism observed between control and treated samples of sponge and bottle gourd seed sample was an average value of 7.8% and 66%, respectively. Overall, study results suggested that Mr. Trivedi's biofield energy treatment has the capability to alter the genetic character of plants, which might be useful in terms of overall crop productivity.

Acknowledgements

Authors thanks to Bangalore Genei Private Limited, for conducting DNA fingerprinting using RAPD analysis. Authors are grateful to Trivedi science, Trivedi testimonials and Trivedi master wellness for their support throughout the work.

References

[1] Henriques F, Guine R, Barroca MJ (2012) Chemical properties of pumpkin dried by different methods Croat. J Food Technol Biotechnol Nutr 7, 98-105.

[2] Oboh IO, Aluyor EO (2009) *Luffa cylindrica* - an emerging cash crop. Afr J Agric Res 4: 684-688.

[3] Leung L, Birtwhistle R, Kotecha J, Hannah S, Cuthbertson S (2009) Anti-diabetic and hypoglycaemic effects of *Momordica charantia* (bitter melon): A mini review. Br J Nutr 102: 1703-1708.

[4] Miniraj N, Prasanna KP, Peter KV (1993) Bitter gourd *Momordica* spp. In: Kalloo G, Bergh BO (edns) Genetic improvement of vegetable plants. Pergamon Press, Oxford.

[5] Singh AK (1990) Cytogenetics and evolution in the *Cucurbitaceae*. In: Bates DM, Robinson RW Jeffrey C (eds) Biology and utilization of *Cucurbitaceae*. Cornell Univ Press, Ithaca, New York.

[6] Welihinda J, Karunanayake EH, Sheriff MHH, Jayasinghe KSA (1986) Effect of *Momordica charantia* on the glucose tolerance in maturity onset diabetes. J Ethnopharmacol 17: 277-282.

[7] Raman A, Lau C (1996) Anti-diabetic properties and phytochemistry of *Momordica charantia* L. (*Cucurbitaceae*). Phytomedicine 2: 349-362.

[8] Dey SS, Singh AK, Chandel D, Behera TK (2006) Genetic diversity of bitter gourd (*Momordica charantia* L.) genotypes revealed by RAPD markers and agronomic traits. Sci Hort 109: 21-28.

[9] Thormann CE, Ferreira ME, Camargo LEA, Tivang JG, Osborn TC (1994) Comparison of RFLP and RAPD markers to estimating genetic relationships within and among *Cruciferous* species. Theor Appl Genet 88: 973-980.

[10] Rafalski JA, Tingey SV (1993) Genetic diagnosis in plant breeding: RAPDs, microsatellites and machines. Trends Genet 9: 275-280.

[11] Tuteja N, Gill SS, Tuteja R (2012) Improving crop productivity in sustainable agriculture. Wiley VCH.

[12] Maffei ME (2014) Magnetic field effects on plant growth, development, and evolution. Front Plant Sci 5: 445.

[13] Mahajan TS, Pandey OP (2015) Effect of electric and magnetic treatments on germination of bitter gourd (*Momordica charantia*) seed. Int J Agric Biol 17: 351-356.

[14] Alexander MP, Doijode SD (1995) Electromagnetic field, a novel tool to increase germination and seedling vigour of conserved onion (*Allium cepa* L.) and rice (*Oryza sativa* L.) seeds with low viability. Plant Genet Resour Newslett 104: 1-5.

[15] NIH, National Center for Complementary and Alternative Medicine. CAM Basics. Publication 347. [October 2, 2008]. Available at: http://nccam.nih.gov/health/whatiscam/

[16] Lenssen AW (2013) Biofield and fungicide seed treatment influences on soybean productivity, seed quality and weed community. Agricultural Journal 8: 138-143.

[17] Nayak G, Altekar N (2015) Effect of biofield treatment on plant growth and adaptation. J Environ Health Sci 1: 1-9.

[18] Trivedi MK, Patil S, Shettigar H, Bairwa K, Jana S (2015) Phenotypic and biotypic characterization of *Klebsiella oxytoca*: An impact of biofield treatment. J Microb Biochem Technol 7: 203-206.

[19] Trivedi MK, Patil S, Nayak G, Jana S, Latiyal O (2015) Influence of biofield treatment on physical, structural and spectral properties of boron nitride. J Material Sci Eng 4: 181.

[20] Shinde VD, Trivedi MK, Patil S (2015) Impact of biofield treatment on yield, quality and control of nematode in carrots. J Horticulture 2: 150.

[21] Green MR, Sambrook J (2012) Molecular cloning: A laboratory manual. (3rdedn), Cold Spring Harbor, Cold Spring Harbor Laboratory Press, NY.

[22] Borges A, Rosa MS, Recchia GH, QueirozSilva JRD, Bressan EDA, et al. (2009) CTAB methods for DNA extraction of sweet potato for microsatellite analysis. Sci Agric (Piracicaba Braz) 66: 529-534.

[23] Welsh J, McClelland M (1990) Fingerprinting genomes using PCR with arbitrary primers. Nucleic Acids Res 18: 7213-7218.

[24] Zalapa JE, Staub JE, McCreight JD (2006) Generation means analysis of plant architectural traits and fruit yield in melon. Plant Breeding 125: 482-487.

[25] Sun Z, Lower RL, Staub JE (2006) Analysis of generation means and components of variance for parthenocarpy in cucumber (*Cucumis sativus* L.) Plant Breeding 125: 277-280.

[26] Ye ZH, Lu ZZ, Zhu J (2003) Genetic analysis for developmental behavior of some seed quality traits in Upland cotton (*Gossypum hirsutum* L.) Euphytica 129: 183-191.

[27] Florez M, Carbonell MV, Martinez E (2007) Exposure of maize seeds to stationary magnetic fields: Effects on germination and early growth. Environ Exp Bot 59: 68-75.

[28] Xu C, Wei S, Lu Y, Zhang Y, Chen C, et al. (2013) Removal of the local geomagnetic field affects reproductive growth in Arabidopsis. Bioelectromagnetics 34: 437-442.

[29] Anand A, Nagarajan S, Verma AP, Joshi DK, Pathak PC, et al. (2012) Pre-treatment of seeds with static magnetic field ameliorates soil water stress in seedlings of maize (*Zea mays* L.). Indian J Biochem Biophys 49: 63-70.

[30] Raj M, Prasanna NKP, Peter KB (1993) Bitter gourd *Momordica* ssp. In: Berg BO, Kalo G (eds) Genetic improvement of vegetable crops. Pergmon Press, Oxford.

[31] Archak S, Karihaloo JL, Jain A (2002) RAPD markers reveal narrowing genetic base of Indian tomato cultivars. Curr Sci 82: 1139-1143.

[32] Sances F, Flora E, Patil S, Spence A, Shinde V (2013) Impact of biofield treatment on ginseng and organic blueberry yield. Agrivita J Agric Sci 35: 22-29.

[33] Shinde V, Sances F, Patil S, Spence A (2012) Impact of biofield treatment on growth and yield of lettuce and tomato. Aust J Basic Appl Sci 6: 100-105.

[34] Dalamu, Behera TK, Gaikwad A, Saxena S, Bharadwaj C, et al. (2012) Morphological and molecular analyses define the genetic diversity of Asian bitter gourd (*Momordica charantia* L.). AJCS 6: 261-267.

Traditional Rice Farming Ritual Practices of the Magindanawn in Southern Philippines

Saavedra M. Mantikayan, Esmael L. Abas

College of Agriculture, Cotabato Foundation College of Science and Technology, Doroluman, Arakan, Cotabato Philippines

Email address:

saavedramantikayan@gmail.com

Abstract: The study is purely a documentation of the traditional rice farming rituals and practices of the Magindanawn in Southern Philippines. Specifically aims: to identify the farming rituals of the Magindanawn farmers; to descriptively analyzed the rationale behind the practice of rituals in farming; and to determine the factors that made rituals persistence amidst the prevalence of the modern Agriculture. The researchers found that farming rituals are based on Magindanawn beliefs as acts or ways of communicating the soul of *uyag-uyag* (life sustenance) which is the elder brother of the spirit (soul) of human being as narrated by Tawalang Kalting. The rationale behind farming rituals retention among Magindanawn were based on their highest one belief that "ALLAH" is the most extremely super power. In such the Traditional Magindanawn found out the following belief: a.) Belief in the competence of ALLAH; b.) Palay (Rice), as one of the bounties given by ALLAH, followed the order of nature. c.) According to the Islamic point of view, Angel Michael is one of the Angels of ALLAH who was instructed as in-charge of bounties all over the World; d.) Human beings are enjoined to follow instruction from ALLAH. Farming rituals among Magindanawn farmers are desired for prosperous production. The rituals before rice planting, like calling the name of stars such as *balatik, malala, mabu* and others, and calling the names of prominent people and different name of angels are not in conformity with Islamic teaching.

Keywords: Traditional Rice Farming, Farming Ritual Practices, Magindanawn, Southern Philippines

1. Introduction

Traditional Knowledge System has been recognized as effective vehicle in the process of Organic Agricultural Sustainable Development. Social Science Scientist agrees that is a repository of biologically loving technology that significantly contributed much to the sustainable development of Agriculture. Abas (1996) pointed out that Indigenous knowledge System is systemic process for the sustainable development of the World's Ecosystem.

Agriculture is the science or art of cultivating the soil for the production of crops. It is the basic industry of world, from which industries originate and upon which they depend. It deserves a high place in the world industries. It also deserves the attention of the best thinkers and the most skillful workers in the world (Phillips et.al., 1952). Along this, the holy Qur'an has this offer:

"Allah has spread out the earth for (His) Creatures: Therein are fruits and date palms, producing spathes (enclosing dates); fodder, and sweet-smelling plants. Then which of the favors of your Lord will ye deny?"(Quran 55:10-13).

Man was created by God as His vicegerent on earth. He is endowed with all the quality and power so that he would be able to work for his survival in order to serve God and live in harmony with nature and his fellowmen. All things in his environment were created to serve his needs. Thus, man is expected to exert his utmost effort to use whatever lays on his hands in accordance with the Divine will.

Kapangangawid (meaning Agriculture) is traditional farming system of the Magindanawn. It is the major source of their staple food such as Rice. Hence rituals are a part and inseparable to their farming system activities. Traditionally, the Magindanawn *tali-awid*(farmers) prefer to engage in kadtaligeba (slush-and burn farming) system of farming rather than the useof dadu *(plow)*, in any kind of landscape. Crop rotation was not within their knowledge; hence, they employed shifting cultivation of course with 3-5 years fallow period (Abas, 1997. And most importantly doing so, the Magindanawn farmers usually invoked rituals before clearing the new field. Sodusta (1993) stated that, Agricultures rituals are ceremonial forms wherein sacred symbols are involved that combine both emotion and metaphysical conceptions. She further explained that "symbol" means "any object, act, event, quality, or relation which serves as vehicle for conception,"[6] the farming rituals of the Magindanawn are observed to be heavily associated with superstitions and traditional beliefs. Thus this research work is purely a documentation of the traditional rice farming rituals and practices of the Magindanawn in Southern Philippines. Specifically aims: to identify the farming rituals and practices of the Magindanawn farmers; to descriptively analyze the rationale behind the practice of rituals in farming; and to determine the factors that made rituals persistence amidst the prevalence of the modern Agriculture.

2. Methodology

The research designed of this study is purely descriptive. It includes all the methods and procedure in allocating all necessary data. The study employed the ethnographic method in gathering information about farming rituals practices among Magindanawn rice farmers. Spradley (1979) define ethnographic research as a kind of research to describe a culture. The essential core of ethnographic activity is aimed to understand another way of life from the native's point of view. He further said that it also means learning from the people.

2.1. Research Location

This study was conducted in the Municipality of Datu, Paglas, province of Magindanaw, Cotabato in Region 12. The study was conducted from February to September, 2005. The Municipality of Datu Paglas has 23 registered barangays. Eleven of these barangays were chosen as the cites of the study namely: a) Brgy. Poblacion, the center of study; b) Brgy. Damalusay; c) Brgy. Damawato; d) Brgy. Datang (the farthest going to the Northeast); e) Brgy. Katil (the farthest going to the south); f) Brgy. Sepaka; g) Brgy. Bulud; h) Brgy. Lipaw); i) Brgy. Bunawan; and, j) Brgy. Lumuyon (the farthest barangay going to the East).

2.2. Respondents

From the 11 registered barangays of Datu Paglas, there are 3,553 Magindanawn farmers out of whom the researchers selected a total of 120 respondents.

2.3. Research Findings

Agriculture is the science or art of cultivating the soil for the production of crops. It is the basic industry of the world, from which industries originate and upon which they depend. It deserves a high place in the world industries. It also deserves the attention of the best thinkers and the most skillful workers in the world (Phillips et.al. 1952). Along this, the holy Qur'an has this offer:

"Allah has spread out the earth for (His) Creatures: Therein are fruits and date palms, producing spates (enclosing dates); fodder, and sweet-smelling plants. Then which of the favors of your Lord will ye deny?"(Quran 55:10-13).

Man was created by god as His vicegerent on earth. He is endowed with all the quality and power so that he would be able to work for his survival in order to serve God and live in harmony with nature and his fellowmen. All things in his environment were created to serve his needs. Thus, man is expected to exert his utmost effort to use whatever lies on his hands in accordance with the Divine will.

It is not vain when God placed Adam and Sittie Hawa (Eve) in the beautiful garden. The Holy Qur'an says:

"O Adam: dwell thou and thy life in the Garden, and enjoy (its good things as you wish: but approach not things tree, or ye run into harm and transgression." Quran 2:19).

Another verse in the Holy Qur'an says:

"God said: "Get ye down, with enmity Between yourselves. On earth will be your Dwelling-place and your means of livelihood for a time."

These verses imply that Adam and Sittie Hawa was the first person on earth who gives knowledge to mankind on how the crops were proven by sowing their seed. The holy Qur'an says:

"See ye seed that ye sow in the ground? It is ye that cause it to grow, or are we the Cause? Were it our Will, we could crumble it dry power, and ye would be left in the wonderment, (Saying, "We are ended left with debts (for Nothing)."(Quran 56:63-66).

Kapangangawid (in Magindanawn means agriculture) is traditionally the main source of income of the Magindanawn farmers. Traditionally, the Magindanawn *tali-awid* (farmers) prefer to engage in kadtaligeba (kaingin) system of farming rather than the use of dadu *(plow),* either in the palaw (mountain) or in basak (wet field). Crop rotation was not within their knowledge; hence, they transferred from one field to another field. But before doing so, the Magindanawn farmers usually invoked rituals before clearing the new rice field. Sodusta (1993) stated that, Agricultures rituals are ceremonial forms wherein sacred symbols are involved that combine both emotion and metaphysical conceptions. She further explained that "symbol" means "any object, act, event, quality, or relation which serves as vehicle for conception," the farming rituals of the Magindanawn are observed to be heavily associated with superstitions and traditional beliefs.

The existence of farming rituals among the Magindanawn was deeply motivated by the conception that every creature was treated equally by God and that this became a sacred customary practice of the Magindanawn farmers to perform rituals on the rice field.

Generally, the rationale behind farming rituals among Magindanawn farmers included:

1. Belief in the completeness of God. The one creator of all creatures, and all His creations were generally, governed by the law of God and the law of nature. To make his creation functional, the instructions of the maker had to be followed:

2. The palay as one of His creations or bounties was designed to follow the order of nature and the creature in-charge of the bounties had with them the instructions of God to be followed in order to make His creation functional and useful.

3. According to the Islamic point of view, angel Michael was one of the angels of God was instructed as in-charge of the bounties all over the world.

4. Human beings were instructed to follow the right instruction of God to seek all kinds of His bounties spread all over the world for their subsistence. This traditional right to be conducted in farming was given by God's instruction to the angel of bounty.

Accordingly, farming rituals among Magindanawn farmers are aimed to achieve maximum crop production. Without observing and practicing farming rituals, crop production would be lesser. It was also believed that rice, given proper conduct of rituals, would usually take a longer period to be consumed; on the other hand, rice without the proper rituals and care would be consumed in a relatively shorter time.

The researcher found out that the rationale behind the farming rituals was based on the beliefs that farming rituals among the Magindanawn was an act, or one way of communicating the soul of *uyag-uyag* (sustenance) which was the elder brother of the soul of human beings, according to the *tarsila* of the Magindanawn folk.

The rituals before clearing the new rice field such as *pagumba* (foretelling the future) and *pagapal* (food offering to evil spirits) were not in conformity with Islamic teachings. The Prophet Muhammad (peace be upon him) declared war on this deception which had no basis in knowledge, divine guidance, or a revealed scripture (Al-Qadarawi, 1984).[9] He recited to them what Allah had revealed to him:

Say: No one in the heavens and the earth

Known the unseen except Allah.

The rituals before planting, like calling the name of the stars such as *balatik* (morning), *malala* (crack), people and the different name of prominent people and the different names of angels like Gibril, Michael, Edzrael, and Edzrapel were invoked by the *apu-na*palay when started to plant rice in the field.

Figure 1. ApunaPalay setting the Ebpalayan.

The *apu-napalay* placed the ebpalayan (small house) at the center of the rice field believing that the spirit of the palay called *uyag-uyag* will stay there from the time of planting up to harvesting time to safeguard the rice plant figure 1. He also

performed rituals for greeting the soil before starting to plant it. As soon as the plants were already sprouting, the *kanduli* was a way of thanking and asking more mercy performed rituals in every stage of rice production figure 2. He visited the rice fields from time to time putting in mine and hearth that all destructive elements get away from the Palay perimeter, to ensure quality and prosperousproduction of rice.

Figure 2. The farmer offer Kaduli or thanks given to Almighty ALLAH for bounty harvest.

3. Conclusion

Based on the findings of this study, the following conclusions were drawn:

The researchers found out that the rationale behind farming rituals among Magindanawn farmers was based on the belief that farming rituals among Magindanawn farmers was an act or one way of communicating the soul of *uyag-uyag* (sustenance), which was the elder brother of human beings according to Magindanaw *tarsila* (history). Rituals were also used in communicating with evil spirits which may live in water, land, and in the forest.

The general rationale behind the farming rituals among Magindanawn farmers were as follows:

1. Belief in the completeness and oneness of God.

2. The palay (rice) as one of God's creations or bounties was design to follow the order of nature and the creature.

3. According to the Islamic point of view, angel Michael was instructed as in-charge of the bounties all over the world.

4. All human being were instructed to follow the instructions of God. Farming rituals among Magindanawn farmers were desired for greater production.

The researcher found out that the rituals involved in farming were as follows:

1. The calling of the names of stars, prominent names of people, angels and the mentioning of the names of four rightly guided caliphs.

2. Rituals for greeting the soil.

3. Rituals for sowing the rice/seed.

4. Rituals after sowing the rice.

5. Rituals for permission with peace.
6. Rituals when the plants are sprouting.
7. Rituals for *kanduli* (food offering).
8. Rituals during the vegetative stage of palay.
9. Rituals for *sabandiya* (appreciation).
10. Rituals for *penggemawan* (boating stage).
11. Rituals for curing the diseases of palay.
12. Rituals for the complete development of rice grain.
13. Rituals to cure underdeveloped rice grains of sickness.
14. Rituals for ripening the rice grain.
15. Rituals for harvesting of palay.
16. Rituals after cutting the first rice panicle.
17. Rituals for hauling the harvested rice.
18. Rituals for *tubali* (piling up) of rice.
19. Rituals for *pagapal* (food offering to evil spirits).
20. Rituals for clearing new rice farm with *pagumba* (foretelling the future).

The Magindanawn farmers adhered to traditional farm practices as well as to framing rituals due to the following reasons:

1. They were traditionally oriented.
2. They followed orthodox principles of farming.
3. These became *adat* or customary practices of the Magindanawn farmers, and
4. They lack knowledge on the new technology in rice farming. They believed that using artificial fertilizers without knowing the proper procedures of applying the fertilizer might even destroy the plant.

Furthermore, Magindanawn farmers were economically unstable. Majority of the Magindanawn farmers adhered to rituals in farming and other traditional farm practices because they could not afford and maintain to buy artificial fertilizers and chemicals and other pesticides due to their exorbitant prices. They were too expensive for the Magindanawn farmers. Therefore, they preferred to fertility of the soil by natural, organic means.

The researcher found out that the farming rituals which were in conformity with Islamic teachings were as follows:

1. Invoking the name of God before beginning to work was a declaration of divine permission.
2. Rituals for greeting the soil before planting rice or palay in the farm;
3. Rituals for curing diseases on rice plant;
4. Observing the position and characteristics of stars.
5. The giving of *zakat* after the threshing of rice;
6. Rituals for *kanduli*, permissible by Shari'ah values.

Farming rituals not in conformity with Islamic teachings were as follows:

1. *Pagumba* (foretelling the future).
2. *Pagapal* (food offerings to evil spirits).
3. Calling the names of stars, names of prominent people, angels of God and the names of rightfully guided caliphs;
4. All farming rituals not mentioned above were to be considered as traditional practices or *adat*/customary practices because these were not given sanctions by any *hadith* of Prophet Muhammad (peace be upon him.)

The researcher found out that during the duration of this study. The Magindanawn farmers were very much happy and cooperative in facilitating the collection of data. The key informants helped a lot in the collection of all necessary data of the study.

Another observation about the information was that some of them were reluctant in answering questions being asked but because of some strategy and techniques used, the researcher was able to get their sympathy. Mostly, the Magindanawn farmers depended on the *apu-napalay* in performing the rituals on the rice field. The *apu-napalay* managed to care the rice farm of every rice farmer through rituals. Almost all Magindanawn farmers who adhered to traditional practices were under the care of the *apu-napalay*.

A further observation during the duration of this study was that the Magindanawn farmers, especially orthodox ones, were really advocates of farming rituals with the exception of a few who were influenced by new technology through imitation, learning from their neighboring Christian farmers, and those who had the means to buy fertilizers and pesticides.

Finally, the farming rituals were mixed with Magindanawn indigenous terms, Arabic words and Malay. The researcher found difficulty in translating the different words which he was not familiar with.

References

[1] Abas, E. L. 1997. Indigenouse Agroforestry and Biodiversity Conservation Among the Magindanawn of Central Mindanao, Philippines. Unpublished Master Thesis. UPLB College Laguna, Philippines.

[2] Adams, Dorothy Ines. 1950. The role of rice rituals in Southeast Asia. Ann Arbor, University microfilm.

[3] Al-Qaradawi, Yusuf. 1984. The Lawful and the Prohibited in Islam. Beirut, Lebanon: the Holy Qur'an Publishing House,

[4] Chanco, Martha. 1980. Feast and rituals among the Karao of Eastern Benguet. University of the Philippines, Diliman Quezon City.

[5] Kamid, Magaluyan. 1982. The kabpagapal a maguindanaon folk rituals and its relation to Islam. University of the Philippines, Diliman, Quezon City.

[6] Lambrecht, Francis. The Mayawyaw rituals, Rice culture and rice rituals. Washington, D.C. USA. Catholic anthropological conference, 1932.

[7] Merino, Gonzalo. 1952. A Half-Century of the Philippine Agriculture. Graphic House, Inc., and the Bureau of Agriculture, Golden Jubilee Committee.

[8] Phillip, et al. Agriculture and Farm Life. New York: Second Revised Edition, the Macmillan Company, 1952.

[9] Sodusta, Jesucita L. 1983. Jamoyawon Ritual: A territorial Concept. Manila: A UP Diamond Jubilee Publication, UP Press, Quezon City.

[10] Westmark, Edward. 1968. Ritual and belief in Morocco, New York, University Book Inc. New Hyde Park.

Permissions

LIST OF CONTRIBUTORS

Ruzanna Robert Sadoyan
Scientific Center of Agriculture, Ministry of Agriculture, Echmiadzin, Republic of Armenia

Koech Oscar Kipchirchir, Kinuthia Robinson Ngugi, Mureithi Stephen Mwangi and Karuku George Njomo
Department of Land Resource Management and Agricultural Technology, University of Nairobi, Nairobi, Kenya

Wanjogu Raphael
National Irrigation Board, Mwea Irrigation Agricultural Development (MIAD) Centre, WANGURU, Kenya

Olubukola Omolara Babalola
Department of Biological Sciences, Landmark University, Omu-Aran, Kwara State, Nigeria
Department of Science Technology, Federal Polytechnic, Ado-Ekiti, Ekiti State, Nigeria

Afui Mathias Mih
Department of Botany and Plant Physiology, Faculty of Science, University of Buea, South West Region, Cameroon

Tonjock Rosemary Kinge
Department of Biological Sciences, Faculty of Science, University of Bamenda, North West Region, Cameroon

Hapu Arachchige Ruwani Kalpana Jayawardana
Department of Botany, Open University of Sri Lanka, Nawala, Sri Lanka

Mohamed Cassim Mohamed Zakeel
Department of Plant Sciences, Faculty of Agriculture, Rajarata University of Sri Lanka, Anuradhapura, Sri Lanka

Channa De Zoysa
Serendib Horticulture Technologies (Pvt) Ltd., Kalagedihena, Sri Lanka

Ismail Ali Abu-Zinada
Department of Plant Production & Protection, Faculty of Agriculture & Environment, Al-Azhar University-Gaza, Gaza Strip, Palestinian Territories

Martin Kagiki Njogu and Geoffrey Kingori Gathungu
Department of Plant Science, Chuka University, Chuka, Kenya

Peter Muchiri Daniel
Department of Plant Science and Crop Protection, University of Nairobi, Kangemi, Kenya
MoA, Wambugu Agriculture Training Centre, Nyeri, Kenya

Md. Farid Hossain
School of Agriculture and Rural Development, Bangladesh Open University, Gazipur, Bangladesh

Md. Anwarul Islam
School of Education, Bangladesh Open University, Gazipur, Bangladesh

Ram Kumar Shrestha
Tribhuvan University, Institute of Agriculture and Animal Science, Lamjung Campus, Lamjung, Nepal

José Antonio Valles Romero
Industrial Engineering Department, National Technological Institute, ITESHU, Mexico

Emilio Raymundo Morales Maldonado
Sustainable Agronomy Innovation Department, National Technological Institute, ITESHU, Mexico

Tarquinio Mateus Magalhães
Department of Forest Engineering, Eduardo Mondlane University, Main Campus, Maputo, Mozambique
Department of Forest and Wood Science, University of Stellenbosch, Stellenbosch, South Africa

Thomas Seifert
Department of Forest Engineering, Eduardo Mondlane University, Main Campus, Maputo, Mozambique
Department of Forest and Wood Science, University of Stellenbosch, Stellenbosch, South Africa

Ufele Angela Nwogor
Zoology Department, Nnamdi Azikiwe University, Awka, Nigeria

Mahendra Kumar Trivedi, Alice Branton, Dahryn Trivedi and Gopal Nayak
Trivedi Global Inc., Henderson, NV, USA

Mayank Gangwar and Snehasis Jana
Trivedi Science Research Laboratory Pvt. Ltd., Bhopal, Madhya Pradesh, India

Abba Halima Mohammed
Biology Unit, School of Basic and Remedial Studies, Gombe State University, Gombe, Nigeria

Sawa Fatima Binta Jahun, Gani Alhassan Mohammed and Abdul Suleiman Dangana
Department of Biological Sceinces, Abubakar Tafawa Balewa University, Bauchi, Nigeria

Esther Mwende Muindi
Department of Crop Science, Pwani University, Kilifi, Kenya

Jerome Mrema, Ernest Semu and Peter Mtakwa
Department of Soil Science, Sokoine University of Agriculture, Morogoro, Tanzania

Charles Gachene
Department of Land Resource Management, University of Nairobi, Nairobi, Kenya

Temesgen Bedassa Gudeta
Department of Biology, School of Natural Sciences, Madda Walabu University, Bale-Robe, Ethiopia

Mesfin Kassa
Soil science, Wolaita Sodo University, College of Agriculture, Wolaita Sodo, Ethiopia

Zemach Sorsa
Plant Breeding, Wolaita Sodo University, College of Agriculture, Wolaita Sodo, Ethiopia

Julian Witjaksono, Dahya and Asmin
The Assessment Institute for Agricultural Technology, Southeast Sulawesi, Indonesia

Benjamin E. Uchola
Faculty of Agriculture, Federal University, Dutsin-ma, Nigeria

Asa Ubong Andem, Daniel Enwongo Aniedi and Ebong Effiong Okon
Department of Agricultural Economics and Extension, University of Uyo, Uyo, Akwa Ibom State, Nigeria

Melese Bekele Hemade
Ethiopian Biodiversity Institute, Forest and Rangeland plants Biodiversity Directorate, Addis Ababa, Ethiopia

Wendawek Abebe
Hawassa University, Hawassa, Ethiopia

Peiqin Li, Zhou Wu and Tao Liu
Department of Forestry Pathology, College of Forestry, Northwest A&F University, Yangling, Shaanxi, China

Yanan Wang
Department of Landscape Architecture, College of Landscape Architecture and Arts, Northwest A&F University, Yangling, Shaanxi, China

John Mthandi, Fredrick C. Kahimba, Andrew K. P. R. Tarimo and Baandah. A. Salim
Department of Agri. Eng. and Land Planning, Sokoine University of Agriculture, Morogoro, Tanzania

Max W. Lowole
Department of Crop Science, Bunda College of Agriculture, Lilongwe, Malawi

R. G. Machhar, S. K. Patel, H. L. Kacha, U. M. Patel, G. D. Patel and R. Radha Rani
Krishi Vigyan Kendra, Anand Agricultural University, Campus Dahod, Gujarat, India

Syed Naeem Abbas, Mujeeb Sardar and Mamoona Wali Muhammad
Forest Education Division Pakistan, Forest Institute, Peshawar, Pakistan

Yawar Abbas and Syed Ali Haider
Department of Earth and Environmental Sciences, Bahria University, Islamabad, Pakistan

Rizwan Karim
Department of Forestry and Range Management, PMAS-Arid Agriculture University, Rawalpindi, Pakistan

Nawazish Ali
Department of Agriculture and Food Technology, Karakoram International University, Gilgit, Pakistan

Saeed Abbas
World Wide Fund for Nature-Pakistan, GCIC, NLI Colony Gilgit, Pakistan

Mosib B. Tagotong
College of Agriculture, Cotabato Foundation College of Science and Technology, Doroluman Arakan, Cotabato, Philippines

Onofre S. Corpuz
Research and Development Office, Cotabato Foundation College of Science and Technology, Doroluman Arakan, Cotabato, Philippines

Md. Halim Mahmud Bhuyan and Md. Toufiq Iqbal
Department of Agronomy and Agricultural Extension, University of Rajshahi, Rajshahi, Bangladesh

Most. Razina Ferdousi
Islamia Academy High School & Agriculture College, Bagha, Rajshahi, Bangladesh

Ahamed Khairul Hasan
Department of Agronomy, Bangladesh Agricultural University, Mymensingh, Bangladesh

Mahendra Kumar Trivedi, Alice Branton, Dahryn Trivedi and Gopal Nayak
Trivedi Global Inc., Henderson, USA

Mayank Gangwar and Snehasis Jana
Trivedi Science Research Laboratory Pvt. Ltd., Bhopal, Madhya Pradesh, India

Saavedra M. Mantikayan and Esmael L. Abas
College of Agriculture, Cotabato Foundation College of Science and Technology, Doroluman, Arakan, Cotabato Philippines

Index

A

Acid Soils, 101-103, 105, 107-109

Acidity, 30, 36-39, 62, 101-102, 107-108, 125

Agriculture, 1-2, 11, 17, 23, 25, 31, 33-37, 39-40, 48-50, 53-55, 60, 63, 68, 70, 77-78, 82, 99-101, 108-109, 121, 124, 128, 134-138, 140, 143, 160, 170, 175, 182, 186, 193-194, 198, 200, 202-203, 205

Agronomic Efficiency, 186, 188, 191-193

Androstachys Johnsonii Prain, 69-70, 78

Apsim, 160-163, 165-167, 169-170

Archachatina Marginata Snails, 13

B

Basal Area, 76, 145-146, 148-149

Bed Planting, 186-188, 190-194

Belowground Biomass, 69, 78

Benzylaminopurine, 40-47

Bio-organic Fertilizer, 182

Biodiversity, 10, 12, 90, 99, 134-138, 145, 205

Biofield Energy, 84-88, 195-200

Biomass Expansion Factor, 69

Bitter Kola, 141-144

C

Carcass Analysis, 13-15

Chemical Composition, 110, 113, 120

Cluster Properties, 37

Conventional Cotton, 129-133

Corn Production, 64

Corn Supply Chain, 64

Cropping System, 60-62, 169, 186

Cucumber (cucumis Sativus), 79-81

D

Dna Fingerprinting, 84-85, 87-88, 195-196, 200

Domestication, 80, 110, 119, 134-140

Drought Tolerance, 5, 7-11, 87

E

Economics, 122, 129, 133, 141, 144, 161, 193

Ecotourism, 175-181

Elaeis Guineensis Jacq, 25, 138

Environment, 6, 10, 12-14, 17-18, 21-22, 36, 50, 53-54, 57-60, 63, 68, 80, 88, 90-91, 99, 111, 113, 116, 135, 152, 161, 169, 175-176, 179-181, 194, 202-203

F

Fermented Fruit Juice, 182-184

Fermented Plant Juice, 182-183

Floriculture, 33, 35

Floristic Composition, 145-146, 151

Flower Induction, 33-35

Flowering Hormone, 33, 35

Fruit Properties, 37

G

Ganoderma, 25-26, 30-31

Generated Plants, 41, 46

Genetic Diversity, 84, 195, 198-201

Giberrelic Acid, 40

H

Herbaceous Species Diversity, 90, 99

Heterorhabditis Bacteriophora, 18-22, 24

Humic Acid, 50, 52-54

Hybrid Depression, 1-3

I

Irrigated Water Measurement, 188

Ixora Coccinea, 33-35

J

Jatropha Curcas, 110-123

K

Kanawa Forest Reserve (kfr) and Inventory, 90

L

Lime-aluminium-phosphate Interactions, 101

M

Maize Yields, 101

Mangrove Forest Plant, 55

Marketing, 55, 82, 141-144, 181

Mature Excreted (compost), 65

Mineral Fertilizers, 1-3, 53

Molasses, 55, 57

Momordica Charantia, 195-196, 200-201

N

Nipa Palm, 55-59

Nitrogen Concentration, 160, 164-165, 167-168

O

Oligochitosan, 152-155, 157-158

P

Paddy Field, 60, 62

Pasture Irrigation, 5

Pathogenic Fungi, 152-153, 155-158

Paundi Watershed, 60, 63

Pawpaw Leaves, 13-14

Pechay (brassica Pekinensis L.), 182

Phenology, 124

Phosphorus Fertilizer Rates, 53, 124, 126

Photorhabdus Luminescens, 18-24

Polymorphism, 1, 84, 86-88, 195, 197-200

Potato Yield, 50

R

Resultant Plants, 40

Revenue, 129-132, 142-143

Root System, 55, 70-71, 74-76, 88, 93

Rural Areas, 118, 141-144

S

Seed Sprouting, 40-41

Soil Sampling, 27, 63, 125

Soil Water Content, 160, 164, 166-167, 169

Soilwat Module, 160-162, 164, 167

Soybean, 25, 61, 82, 89, 108, 114, 133, 139, 159, 171-174, 200

Spread Pattern, 25-26, 28-29

Spread Pattern, 25-26, 28-29

Sprouted Tubers, 41

Symbiosis, Bioluminescence, 18

T

Technological Gap, 171-174

Tree Biomass, 69, 74, 76, 78

Tree Species, 77-78, 146, 148

Tuber Quality, 50

U

Urea Super Granule, 186, 188, 190, 192-194

V

Vegetative Growth, 3, 47, 50, 52-53, 61, 126, 191, 196

Vegetative Mass, 1-3

W

Water Stress Tolerance, 5-6, 8-9

Watermelon (citrillus Lanatus), 79-82

Wildlife, 17, 21, 82, 99, 134-136, 138-139, 176, 179-180

Y

Young Cotton Seedlings, 86

Z

Zanthoxylum Bungeanum, 152, 158